LONDON MATHEMATICAL SOCIETY LECTURE NOTE SERIES

Managing Editor: Professor M. Reid, Mathematics Institute,
University of Warwick, Coventry CV4 7AL, United Kingdom

The titles below are available from booksellers, or from Cambridge University Press at
www.cambridge.org/mathematics

287 Topics on Riemann surfaces and Fuchsian groups, E. BUJALANCE, A.F. COSTA & E. MARTÍNEZ (eds)
288 Surveys in combinatorics, 2001, J.W.P. HIRSCHFELD (ed)
289 Aspects of Sobolev-type inequalities, L. SALOFF-COSTE
290 Quantum groups and Lie theory, A. PRESSLEY (ed)
291 Tits buildings and the model theory of groups, K. TENT (ed)
292 A quantum groups primer, S. MAJID
293 Second order partial differential equations in Hilbert spaces, G. DA PRATO & J. ZABCZYK
294 Introduction to operator space theory, G. PISIER
295 Geometry and integrability, L. MASON & Y. NUTKU (eds)
296 Lectures on invariant theory, I. DOLGACHEV
297 The homotopy category of simply connected 4-manifolds, H.-J. BAUES
298 Higher operads, higher categories, T. LEINSTER (ed)
299 Kleinian groups and hyperbolic 3-manifolds, Y. KOMORI, V. MARKOVIC & C. SERIES (eds)
300 Introduction to Möbius differential geometry, U. HERTRICH-JEROMIN
301 Stable modules and the D(2)-problem, F.E.A. JOHNSON
302 Discrete and continuous nonlinear Schrödinger systems, M.J. ABLOWITZ, B. PRINARI & A.D. TRUBATCH
303 Number theory and algebraic geometry, M. REID & A. SKOROBOGATOV (eds)
304 Groups St Andrews 2001 in Oxford I, C.M. CAMPBELL, E.F. ROBERTSON & G.C. SMITH (eds)
305 Groups St Andrews 2001 in Oxford II, C.M. CAMPBELL, E.F. ROBERTSON & G.C. SMITH (eds)
306 Geometric mechanics and symmetry, J. MONTALDI & T. RATIU (eds)
307 Surveys in combinatorics 2003, C.D. WENSLEY (ed.)
308 Topology, geometry and quantum field theory, U.L. TILLMANN (ed)
309 Corings and comodules, T. BRZEZINSKI & R. WISBAUER
310 Topics in dynamics and ergodic theory, S. BEZUGLYI & S. KOLYADA (eds)
311 Groups: topological, combinatorial and arithmetic aspects, T.W. MÜLLER (ed)
312 Foundations of computational mathematics, Minneapolis 2002, F. CUCKER *et al* (eds)
313 Transcendental aspects of algebraic cycles, S. MÜLLER-STACH & C. PETERS (eds)
314 Spectral generalizations of line graphs, D. CVETKOVIĆ, P. ROWLINSON & S. SIMIĆ
315 Structured ring spectra, A. BAKER & B. RICHTER (eds)
316 Linear logic in computer science, T. EHRHARD, P. RUET, J.-Y. GIRARD & P. SCOTT (eds)
317 Advances in elliptic curve cryptography, I.F. BLAKE, G. SEROUSSI & N.P. SMART (eds)
318 Perturbation of the boundary in boundary-value problems of partial differential equations, D. HENRY
319 Double affine Hecke algebras, I. CHEREDNIK
320 L-functions and Galois representations, D. BURNS, K. BUZZARD & J. NEKOVÁŘ (eds)
321 Surveys in modern mathematics, V. PRASOLOV & Y. ILYASHENKO (eds)
322 Recent perspectives in random matrix theory and number theory, F. MEZZADRI & N.C. SNAITH (eds)
323 Poisson geometry, deformation quantisation and group representations, S. GUTT *et al* (eds)
324 Singularities and computer algebra, C. LOSSEN & G. PFISTER (eds)
325 Lectures on the Ricci flow, P. TOPPING
326 Modular representations of finite groups of Lie type, J.E. HUMPHREYS
327 Surveys in combinatorics 2005, B.S. WEBB (ed)
328 Fundamentals of hyperbolic manifolds, R. CANARY, D. EPSTEIN & A. MARDEN (eds)
329 Spaces of Kleinian groups, Y. MINSKY, M. SAKUMA & C. SERIES (eds)
330 Noncommutative localization in algebra and topology, A. RANICKI (ed)
331 Foundations of computational mathematics, Santander 2005, L.M PARDO, A. PINKUS, E. SÜLI & M.J. TODD (eds)
332 Handbook of tilting theory, L. ANGELERI HÜGEL, D. HAPPEL & H. KRAUSE (eds)
333 Synthetic differential geometry (2nd Edition), A. KOCK
334 The Navier–Stokes equations, N. RILEY & P. DRAZIN
335 Lectures on the combinatorics of free probability, A. NICA & R. SPEICHER
336 Integral closure of ideals, rings, and modules, I. SWANSON & C. HUNEKE
337 Methods in Banach space theory, J.M.F. CASTILLO & W.B. JOHNSON (eds)
338 Surveys in geometry and number theory, N. YOUNG (ed)
339 Groups St Andrews 2005 I, C.M. CAMPBELL, M.R. QUICK, E.F. ROBERTSON & G.C. SMITH (eds)
340 Groups St Andrews 2005 II, C.M. CAMPBELL, M.R. QUICK, E.F. ROBERTSON & G.C. SMITH (eds)
341 Ranks of elliptic curves and random matrix theory, J.B. CONREY, D.W. FARMER, F. MEZZADRI & N.C. SNAITH (eds)
342 Elliptic cohomology, H.R. MILLER & D.C. RAVENEL (eds)
343 Algebraic cycles and motives I, J. NAGEL & C. PETERS (eds)
344 Algebraic cycles and motives II, J. NAGEL & C. PETERS (eds)

345 Algebraic and analytic geometry, A. NEEMAN
346 Surveys in combinatorics 2007, A. HILTON & J. TALBOT (eds)
347 Surveys in contemporary mathematics, N. YOUNG & Y. CHOI (eds)
348 Transcendental dynamics and complex analysis, P.J. RIPPON & G.M. STALLARD (eds)
349 Model theory with applications to algebra and analysis I, Z. CHATZIDAKIS, D. MACPHERSON, A. PILLAY & A. WILKIE (eds)
350 Model theory with applications to algebra and analysis II, Z. CHATZIDAKIS, D. MACPHERSON, A. PILLAY & A. WILKIE (eds)
351 Finite von Neumann algebras and masas, A.M. SINCLAIR & R.R. SMITH
352 Number theory and polynomials, J. MCKEE & C. SMYTH (eds)
353 Trends in stochastic analysis, J. BLATH, P. MÖRTERS & M. SCHEUTZOW (eds)
354 Groups and analysis, K. TENT (ed)
355 Non-equilibrium statistical mechanics and turbulence, J. CARDY, G. FALKOVICH & K. GAWEDZKI
356 Elliptic curves and big Galois representations, D. DELBOURGO
357 Algebraic theory of differential equations, M.A.H. MACCALLUM & A.V. MIKHAILOV (eds)
358 Geometric and cohomological methods in group theory, M.R. BRIDSON, P.H. KROPHOLLER & I.J. LEARY (eds)
359 Moduli spaces and vector bundles, L. BRAMBILA-PAZ, S.B. BRADLOW, O. GARCÍA-PRADA & S. RAMANAN (eds)
360 Zariski geometries, B. ZILBER
361 Words: Notes on verbal width in groups, D. SEGAL
362 Differential tensor algebras and their module categories, R. BAUTISTA, L. SALMERÓN & R. ZUAZUA
363 Foundations of computational mathematics, Hong Kong 2008, F. CUCKER, A. PINKUS & M.J. TODD (eds)
364 Partial differential equations and fluid mechanics, J.C. ROBINSON & J.L. RODRIGO (eds)
365 Surveys in combinatorics 2009, S. HUCZYNSKA, J.D. MITCHELL & C.M. RONEY-DOUGAL (eds)
366 Highly oscillatory problems, B. ENGQUIST, A. FOKAS, E. HAIRER & A. ISERLES (eds)
367 Random matrices: High dimensional phenomena, G. BLOWER
368 Geometry of Riemann surfaces, F.P. GARDINER, G. GONZÁLEZ-DIEZ & C. KOUROUNIOTIS (eds)
369 Epidemics and rumours in complex networks, M. DRAIEF & L. MASSOULIÉ
370 Theory of p-adic distributions, S. ALBEVERIO, A.YU. KHRENNIKOV & V.M. SHELKOVICH
371 Conformal fractals, F. PRZYTYCKI & M. URBAŃSKI
372 Moonshine: The first quarter century and beyond, J. LEPOWSKY, J. MCKAY & M.P. TUITE (eds)
373 Smoothness, regularity and complete intersection, J. MAJADAS & A. G. RODICIO
374 Geometric analysis of hyperbolic differential equations: An introduction, S. ALINHAC
375 Triangulated categories, T. HOLM, P. JØRGENSEN & R. ROUQUIER (eds)
376 Permutation patterns, S. LINTON, N. RUŠKUC & V. VATTER (eds)
377 An introduction to Galois cohomology and its applications, G. BERHUY
378 Probability and mathematical genetics, N. H. BINGHAM & C. M. GOLDIE (eds)
379 Finite and algorithmic model theory, J. ESPARZA, C. MICHAUX & C. STEINHORN (eds)
380 Real and complex singularities, M. MANOEL, M.C. ROMERO FUSTER & C.T.C WALL (eds)
381 Symmetries and integrability of difference equations, D. LEVI, P. OLVER, Z. THOMOVA & P. WINTERNITZ (eds)
382 Forcing with random variables and proof complexity, J. KRAJÍČEK
383 Motivic integration and its interactions with model theory and non-Archimedean geometry I, R. CLUCKERS, J. NICAISE & J. SEBAG (eds)
384 Motivic integration and its interactions with model theory and non-Archimedean geometry II, R. CLUCKERS, J. NICAISE & J. SEBAG (eds)
385 Entropy of hidden Markov processes and connections to dynamical systems, B. MARCUS, K. PETERSEN & T. WEISSMAN (eds)
386 Independence-friendly logic, A.L. MANN, G. SANDU & M. SEVENSTER
387 Groups St Andrews 2009 in Bath I, C.M. CAMPBELL *et al* (eds)
388 Groups St Andrews 2009 in Bath II, C.M. CAMPBELL *et al* (eds)
389 Random fields on the sphere, D. MARINUCCI & G. PECCATI
390 Localization in periodic potentials, D.E. PELINOVSKY
391 Fusion systems in algebra and topology M. ASCHBACHER, R. KESSAR & B. OLIVER
392 Surveys in combinatorics 2011, R. CHAPMAN (ed)
393 Non-abelian fundamental groups and Iwasawa theory, J. COATES *et al* (eds)
394 Variational problems in differential geometry, R. BIELAWSKI, K. HOUSTON & M. SPEIGHT (eds)
395 How groups grow, A. MANN
396 Arithmetic differential operators over the p-adic integers, C.C. RALPH & S.R. SIMANCA
397 Hyperbolic geometry and applications in quantum chaos and cosmology, J. BOLTE & F. STEINER (eds)
398 Mathematical models in contact mechanics, M. SOFONEA & A. MATEI
399 Circuit double cover of graphs, C.-Q. ZHANG
400 Dense sphere packings: a blueprint for formal proofs, T. HALES

London Mathematical Society Lecture Note Series: 400

Dense Sphere Packings

A blueprint for formal proofs

THOMAS C. HALES
University of Pittsburgh

CAMBRIDGE
UNIVERSITY PRESS

CAMBRIDGE
UNIVERSITY PRESS

University Printing House, Cambridge CB2 8BS, United Kingdom

One Liberty Plaza, 20th Floor, New York, NY 10006, USA

477 Williamstown Road, Port Melbourne, VIC 3207, Australia

314-321, 3rd Floor, Plot 3, Splendor Forum, Jasola District Centre, New Delhi - 110025, India

103 Penang Road, #05-06/07, Visioncrest Commercial, Singapore 238467

Cambridge University Press is part of the University of Cambridge.

It furthers the University's mission by disseminating knowledge in the pursuit of education, learning and research at the highest international levels of excellence.

www.cambridge.org
Information on this title: www.cambridge.org/9780521617703

First published 2012

A catalogue record for this publication is available from the British Library

ISBN 978-0-521-61770-3 Paperback

Contents

	Preface	*page* vii
	PART ONE OVERVIEW	1
1	**Close Packing**	3
	1.1 History	3
	1.2 Face-Centered Cubic	6
	1.3 Hexagonal-Close Packing	10
	1.4 Gauss	13
	1.5 Thue	14
	1.6 Dense Packings in a Nutshell	17
	PART TWO FOUNDATIONS	23
2	**Trigonometry**	25
	2.1 Background Knowledge	25
	2.2 Trig Identities	27
	2.3 Vector Geometry	33
	2.4 Angle	39
	2.5 Coordinates	51
	2.6 Cycle	55
	2.7 Chapter Summary	60
3	**Volume**	61
	3.1 Background in Measure	61
	3.2 Primitive Volume	63
	3.3 Finiteness and Volume	70

4 Hypermap 72
4.1 Background on Permutations 73
4.2 Definitions 74
4.3 Walkup 78
4.4 Planarity 83
4.5 Path 86
4.6 Subquotient 91
4.7 Generation 95

5 Fan 112
5.1 Definitions 112
5.2 Topology 117
5.3 Planarity 121
5.4 Polyhedron 132

PART THREE THE KEPLER CONJECTURE 143

6 Packing 145
6.1 The Primitive State of Our Subject Revealed 145
6.2 Rogers Simplex 150
6.3 Cells 167
6.4 Clusters 180
6.5 Counting Spheres 185

7 Local Fan 194
7.1 Localization 194
7.2 Modification 203
7.3 Polarity 209
7.4 Main Estimate 212

8 Tame Hypermap 235
8.1 Definition 236
8.2 Contravening Hypermap 238
8.3 Contravention is Tame 243
8.4 Admissibility 248
8.5 Linear Programs 250
8.6 Strong Dodecahedral Theorem 252

***Appendix A* Credits** 257
References 261
General index 264
Notation index 269

Preface

"I think there's a revolution in mathematics around the corner. I think that . . . people will look back on the fin-de-siècle of the twentieth century and say 'Then is when it happened' (just like we look back at the Greeks for inventing the concept of proof and at the nineteenth century for making analysis rigorous). I really believe that. And it amazes me that no one seems to notice.

"Never before have the Platonic mathematical world and the physical world been this similar, this close. Is it strange that I expect leakage between these two worlds? That I think the proof strings will find their way to the computer memories?. . .

"What I expect is that some kind of computer system will be created, a proof checker, that all mathematicians will start using to check their work, their proofs, their mathematics. I have no idea what shape such a system will take. But I expect some system to come into being that is past some threshold so that it is practical enough for real work, and then quite suddenly some kind of 'phase transition' will occur and everyone will be using that system."

–Freek Wiedijk [49]

Alecos: Christos has a problem with the 'foundational quest'!

Christos: Wrong! I have two problems with your version *of it! One, it didn't fail and, two, it wasn't a tragedy! Granted, there are some tragic parts! But the ending is happy, as in the 'Oresteia'!*

Apostolos: Happy for whom? Cantor, going insane? Gödel starving himself to death out of paranoia? Hilbert or Russell and their psychotic sons? Or Frege with–

Christos: 'The meaning is in the ending!' you said so yourself! So, follow the quest for ten more years and you get a brand-new triumphant finale with the creation of the computer, which is the quest's real hero! Your problem is, simply, that you see it as a story of people!

Apostolos: Well, stories do tend to be about people!

Christos: So, choose the right people! And show what they really did! All we we learn of the great von Neumann is he said 'It's over' when he heard Gödel!

Alecos: But it was over in a sense, wasn't it? Pop went Hilbert's 'no ignorabimus'!

Christos: But then came the quest's jeune premier, its parsifal ... Alan Turing! He said 'Ok, we can't prove everything! So, let's see what we can prove!' and to define proof, he invented, in 1936, a theoretical machine which contains all the ideas of the computer!... which, after the war, he and von Neumann, the quest's proudest sons, brought to full life!

–Doxiadis and Papadimitriou, Logicomix [10]

"Despite the unusual nature of the proof, the editors of the Annals of Mathematics agreed to publish it, provided it was accepted by a panel of twelve referees. In 2003, after four years of work, the head of the referee's panel Gábor Fejes Tóth (son of László Fejes Tóth) reported that the panel were '99% certain' of the correctness of the proof."

– Wikipedia entry on the Kepler conjecture

"Sometimes fixing a 1 percent defect takes 500 percent effort."

– Joel Spolsky, Joel on Software [42]

"Every one fully persuaded is a fool."

– Barthasar Gracián, the Art of Worldly Wisdom [17]

The Kepler Conjecture

In 1611, Johannes Kepler wrote a booklet in which he asserted that the familiar cannonball arrangement of congruent balls in space achieves the highest possible density. This assertion has become known as the Kepler conjecture. This book presents a proof.

As early as 1831, Gauss established a special case of the conjecture, by proving that the cannonball arrangement is optimal among all *lattices* [14]. Later in the nineteenth century, Thue solved the corresponding problem in two dimensions, showing that the hexagonal arrangement of disks in the plane achieves optimal density [45] and [46]. Hilbert, in his famous list of mathematical problems, made the Kepler conjecture part of his eighteenth problem. In 1953, Fejes Tóth formulated a general strategy to confirm the Kepler conjecture, but lacked the computational resources to carry it out [12]. The conjecture was finally resolved in 1998, even though the full proof was not published until 2006 [22]. Section 1.1 gives additional historical background.

The Kepler conjecture has become a test of the capability of computers to deliver a reliable mathematical proof. The original proof involved many long computer calculations that led a team of referees to exhaustion. This book has redesigned the proof in a way that makes the correctness of the computer proof as transparent as possible.

Formal Proofs

After all is said and done, a proof is only as reliable as the processes that are used to verify its correctness. The ultimate standard of proof is a formal proof, which is nothing other than an unbroken chain of logical inferences from an explicit set of axioms. While this may be the mathematical ideal of proof, actual mathematical practice generally deviates significantly from the ideal.

In recent years, as part of this project, I have been increasingly preoccupied by the processes that mathematicians rely on to ensure the correctness of complex proofs. A century ago, Russell's paradox and other antinomies threatened set theory with fires of destruction. Researchers from Frege to Gödel solved the problem of rigor in mathematics and found a theoretical solution but did not extinguish the fire at the foundations of mathematics because they omitted the practical implementation. Some, such as Bourbaki, have even gone so far as to claim that "formalized mathematics cannot in practice be written down in full" and call such a project "absolutely unrealizable" [7, pp. 10–11].

While it is true that formal proofs may be too long to print, computers –

which do not have the same limitations as paper – have become the natural host of formal mathematics. In recent decades, logicians and computer scientists have reworked the foundations of mathematics, putting them in an efficient form designed for real use on real computers.

For the first time in history, it is possible to generate and verify every single logical inference of major mathematical theorems. This has now been done for many theorems, including the four-color theorem, the prime number theorem, the Jordan curve theorem, the Brouwer fixed point theorem, and the fundamental theorem of calculus. Freek Wiedijk reports that 87% of a list of one hundred famous theorems have now been checked formally [50]. The list of remaining theorems contains two particular challenges: the independence of the Continuum hypothesis and Fermat's Last theorem.

Some mathematicians remain skeptical of the process because computers have been used to generate and verify the logical inferences. Computers are notoriously imperfect, with flaws ranging from software bugs to defective chips. Even if a computer verifies the inferences, who verifies the verifier, or then verifies the verifier of the verifier? Indeed, it would be unscientific of us to place an unmerited trust in computers.

The choice comes down to two competing verification processes. The first is the traditional process of referees, which depends largely on the luck of the draw – some referees are meticulous, while others are careless. The second process is formal computer verification, which is less dependent on the whims of a particular referee. In my view, the choice between the conventional process by human referee and computer verification is as evident as the choice between a sundial and an atomic clock in science.

The standard of proof I have adopted is the highest scientific standard available by current technology. The introduction of steel in architecture is not a mere reinforcement of wood and stone; it changes the world of structural possibilities. There is no longer any reason to limit proofs to ten thousand pages when our technology supports a million pages.

The style of formal proofs is different from that of conventional ones. It is easier to formalize several short snappy proofs than a few intricate ones. Humans enjoy surprising new perspectives, but computers benefit from repetition and standardization. Despite these differences, I have sought proofs that might bring pleasure to the human reader while providing precise instructions for the implementation in silicon.

Conventions

To make formalization proceed more smoothly, long proofs have been broken into a sequence of smaller claims. Each claim starts a new paragraph and is set in italics. The second sentence of the paragraph begins with the word *indeed* when the proof of the claim is direct and with the word *otherwise* when the proof is indirect by contradiction.

Lemmas and theorems that are marked with an asterisk appear out of the natural logical sequence. Care should be taken to avoid logical gaps when they are cited.

The pronoun *we* is used inclusively for the author and reader as we work our way through the proofs in this book. The pronoun *I* refers to the author alone.

The asterisk $*$ is used as a wildcard symbol in patterns. It replaces a term in contexts where the name of the term is not relevant. It can also denote a bound variable. For example, the function $f(*, y)$ of a single variable is obtained from f by evaluating the second argument at a fixed value y.

The union of the family X of sets is written as $\bigcup X$ or as $\bigcup_{x \in X} x$ without any difference in meaning. The first form is preferred because of its economy. We also use both expressions $\bigcap X$ and $\bigcap_{x \in X} x$ for the intersection of a family of sets.

The documentation of the computer calculations for the Kepler conjecture has evolved over time. The 1998 preprint version of the proof of the Kepler conjecture contains long appendices that list hundreds of calculations that enter into the proof. These appendices were cut from the published version of the proof because it is more useful to store the computer part of the proof at a computer code repository that is permanent, versioned, and freely available. The computer code and documentation are housed at *Google Code project hosting*. Separate documentation, which is available at the project site, describes the computer calculations that appear in this book. When this book uses an external calculation, it is marked in italic font as a *computer calculation*[1] [21].

A Blueprint

The book is a blueprint for formal proofs because it gives the design of the formal proof to be constructed. The parts of this book that cover the text portions of the proof of the Kepler conjecture are being formally verified in the proof assistant HOL Light. I dream of a fully formally verified solution to the

[1] `[notation]` This explains notation.

proof that includes the computer portions of the proof as well. Details about and credits for this large team effort appear in Appendix A.

Decisions about what to include in this book have been shaped by the list of theorems already available in the library of the proof assistant *HOL Light*. For example, this book accepts basic point-set topology and measure theory because they have been formalized by Harrison [26].

The book is divided into three parts, the first of which describes the major ideas, methods, and organization of the proof.

The part on foundations provides background material about constructions in discrete geometry. The first of these chapters covers trigonometric identities and basic vector geometry. The second treats volume from an elementary point of view. The third chapter covers planar graph theory from a purely combinatorial perspective. The fourth chapter continues with planar graphs, now from a geometric perspective.

The final part of the book gives the solution to the packing problem. The first of these chapters gives a top-level overview of the major steps of the proof, describing how the problem can be reduced from a problem with infinitely many variables to one in finitely many variables. The remaining chapters in this part flesh out the proof.

The final section of the book views dense sphere packings from a larger perspective. It resolves another longstanding conjecture in discrete geometry: Bezdek's strong dodecahedral conjecture.

Simplifications

Many simplifications of the original proof have been found over the past several years. These simplifications are published here for the first time. Gonthier has reworked the proof of the four-color theorem to avoid the use of the Jordan curve theorem, using instead the much simpler notion of Möbius contour from the theory of hypermaps. I have followed Gonthier's lead.

The optimality of the face-centered cubic packing is an assertion about infinite space-filling packings. For computational purposes, it is useful to reduce the sphere packing problem to finite packings. A *correction term* is associated with each different reduction from infinite packings to finite packings. Ferguson and I worked together to produce the original proof of the Kepler conjecture. The two of us considered a large number of different correction terms, seeking one that would simplify the computations as much as possible. In a discussion of the solution of the packing problem, I wrote that "correction terms are extremely flexible and easy to construct, and soon Samuel Ferguson and I realized that every time we encountered difficulties in solving the minimization problem, we could adjust f [the correction term] to skirt the difficulty.... If I were to revise the proof to produce a simpler one, the first thing I would do would be to change the correction term once again. It is the key to a simpler proof" [19]. Marchal has recently found a simple correction term, giving a new way to reduce from infinite packings to finite packings [31]. This book implements his reduction step.

There are many other improvements of the proof that are not visible in the book because they are implemented in computer code, including a reduction of the number of lines of computer code from over 187,000 to about 10,000. Needless to say, the quickest way to be sure that a block of computer code will not execute a bug is to delete the code altogether.

Thomas C. Hales
Pittsburgh, PA

PART ONE

OVERVIEW

1

Close Packing

1.1 History

This section gives a brief history of the study of dense sphere packings. Further details appear at [43] and [20]. The early history of sphere packings is concerned with the face-centered cubic (FCC) packing, a familiar pyramid arrangement of congruent balls used to stack cannonballs at war memorials and oranges at fruit stands (Figure 1.1).

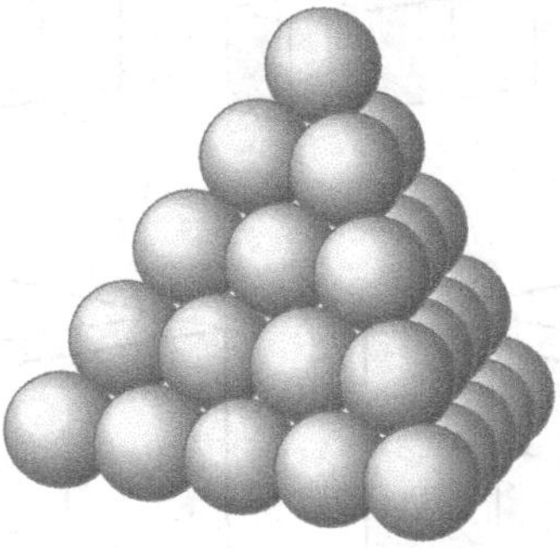

Figure 1.1 The face-centered cubic (FCC) packing.

1.1.1 Sanskrit sources

The study of the mathematical properties of the FCC packing can be traced[1] to a Sanskrit work (the Āryabhaṭīya of Āryabhaṭa) composed around 499 CE. The following passage gives the formula for the number of balls in a pyramid

[1] I am obliged to Plofker [35].

pile with triangular base as a function of the number of balls along an edge of the pyramid [40].

> *For a series [lit. "heap"] with a common difference and first term of 1, the product of three [terms successively] increased by 1 from the total, or else the cube of [the total] plus 1 diminished by [its] root, divided by 6, is the total of the pile [lit. "solid heap"].*

In modern notation, the passage gives two formulas for the number of balls in a pyramid with n balls along an edge (Figure 1.2):

$$\frac{n(n+1)(n+2)}{6} = \frac{(n+1)^3 - (n+1)}{6}. \qquad (1.1)$$

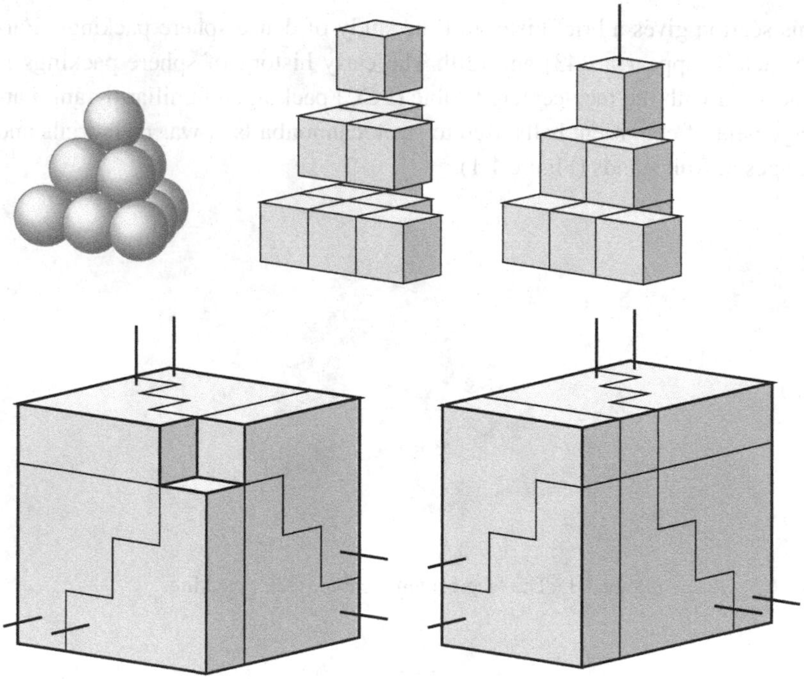

Figure 1.2 Derivation of Sanskrit formula (1.1). A cannonball packing can be converted to unit cubes in a staircase aligned along the rear column. Six staircase shapes fill an $(n+1)^3$ cube without its diagonal of $n+1$ unit cubes, or a rectangle of dimensions n by $n+1$ by $n+2$.

1.1.2 Harriot and Kepler

The modern mathematical study of spheres and their close packings can be traced to Harriot. His work – unpublished, unedited, and largely undated – shows a preoccupation with sphere packings. He seems to have first taken an interest in packings at the prompting of Sir Walter Raleigh. At the time, Harriot was Raleigh's mathematical assistant, and Raleigh gave him the problem of determining formulas for the number of cannonballs in regularly stacked piles. Harriot interpreted the number of balls in a pyramid as an entry in Pascal's triangle[2] (Figure 1.3). Through his study of triangular and pyramidal numbers, Harriot later discovered finite difference interpolation [3]. Shirley, Harriot's biographer, writes that it was his study of cannonball arrangements in the late sixteenth century that "led him inevitably to the corpuscular or atomic theory of matter originally deriving from Lucretius and Epicurus" [39, p. 242].

Kepler became involved in sphere packings through his correspondence with Harriot around 1606–1607 on the topic of optics. Harriot, the atomist, attempted to understand reflection and refraction of light in atomic terms. Kepler favored a more classical explanation of reflection and refraction in terms of what Kargon describes as "the union of two opposing qualities – transparence and opacity" [27, p.26]. Harriot was stunned that Kepler would be satisfied by such reasons.

Despite Kepler's initial reluctance to adopt an atomic theory, he was eventually swayed and published an essay in 1611 that explores the consequences of a theory of matter composed of small spherical particles. Kepler's essay describes the FCC packing and asserts that "the packing will be the tightest possible, so that in no other arrangement could more pellets be stuffed into the same container" [28]. This assertion has come to be known as the Kepler conjecture. This book gives a proof of this conjecture.

1.1.3 Newton and Gregory

The next episode in the history of this problem, a debate between Isaac Newton and David Gregory, centered on the question of how many congruent balls can be arranged to touch a given ball. The analogous question in two dimensions is readily answered; six pennies, but no more, can be arranged to touch a central penny. In three dimensions, Newton said that the maximum was twelve balls, but Gregory claimed that thirteen might be possible.

The Newton–Gregory problem was not solved until centuries later (Figure 1.4). The first proper proof was obtained by van der Waerden and Schütte in

[2] Harriot was well-versed in Pascal's triangle long before Pascal.

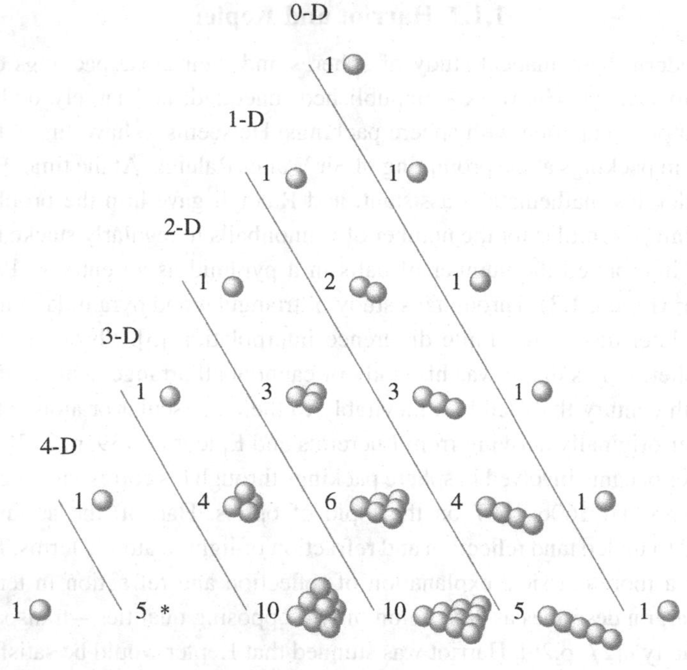

Figure 1.3 The binomial coefficient $\binom{d+n}{n}$ gives the general formula for the number of balls in a d-dimensional pyramid of side $n+1$. As Harriot observed, the recursion of Pascal's triangle $\binom{d+n}{n} = \binom{d+(n-1)}{n-1} + \binom{(d-1)+n}{n}$ can be interpreted as a partition of a pyramid of side $n+1$ into a pyramid of side n resting on a pyramidal base of side $n+1$ in dimension $d-1$.

1953 [38]. An elementary proof appears in Leech [30]. Although a connection between the Newton–Gregory problem and Kepler's problem is not obvious, Fejes Tóth successfully linked the problems in 1953 [12].

1.2 Face-Centered Cubic

The FCC packing is the familiar pyramid arrangement of balls on a square base as well as a pyramid arrangement on a triangular base. The two packings differ only in their orientation in space. Figure 1.5 shows how the triangular base packing fits between the peaks of two adjacent square based pyramids.

Density, defined as a ratio of volumes, is insensitive to changes of scale. For convenience, it is sufficient to consider balls of unit radius. This means that the distance between centers of balls in a packing is always at least 2. We identify

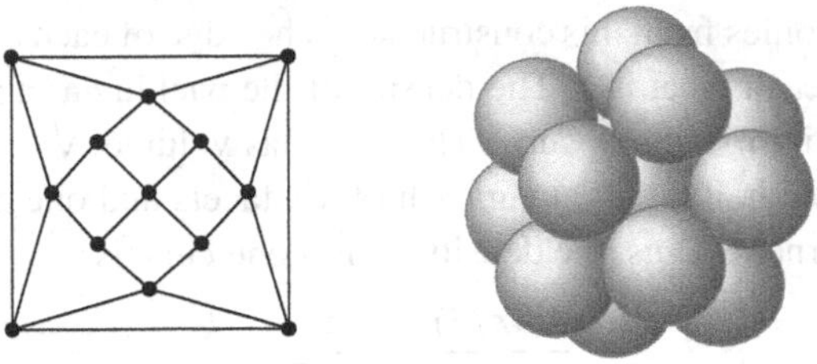

Figure 1.4 Newton's claim – twelve is the maximum number of congruent balls that can be tangent to a given congruent ball– was confirmed in the 1953. Musin and Tarasov only recently proved that the arrangement shown here is the unique arrangement of thirteen congruent balls that shrinks the thirteen by the least possible amount to permit tangency [32]. Each node of the graph represents one of the thirteen balls and each edge represents a pair of touching balls. The node at the center of the graph corresponds to the uppermost ball in the second frame. The other twelve balls are perturbations of the FCC tangent arrangement.

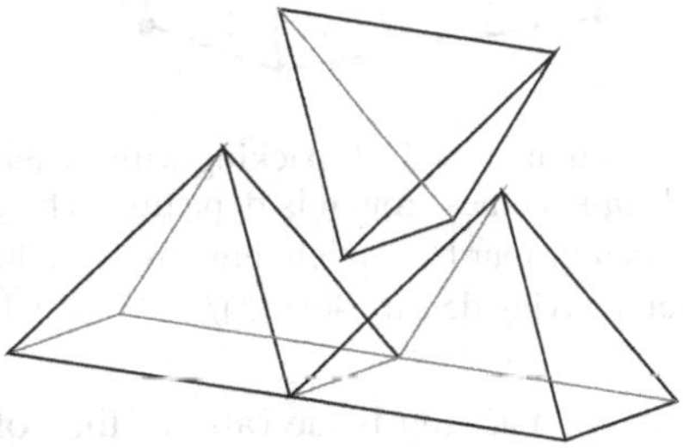

Figure 1.5 The pyramid on a square base is the same lattice packing as the pyramid on a triangular base. The only differences are the orientation of the lattice in space and the exposed facets of the lattice. Their orientation and exposed facets are matched as shown.

a packing with its set V of centers. For our purposes, a packing is just a set of points in $\mathbb{R}^3$ in which the elements are separated by distances of at least 2.

The density of a packing is the ratio of the volume occupied by the balls to the volume of a large container. The purpose of a finite container is to prevent the volumes from becoming infinite. To eliminate the distortion of the packing caused by the shape of the its boundary, we take the limit of the densities within an increasing sequence of spherically shaped containers, as the diameter tends to infinity.

The FCC packing is obtained from a cubic lattice, by inserting a ball at each of the eight extreme points of each cube and then inserting a another ball at the center of each of the six facets of each cube (Figure 1.6). The name *face-*

centered cubic comes from this construction. The edge of each cube is $\sqrt{8}$, and the diagonal of each facet is 4. The density of the packing as a whole is equal to the density within a single cube. The cube has volume $\sqrt{8}^3$ and contains a total of four balls: half a ball along each of six facets and one eighth a ball at each of eight corners. Thus, the density within one cube is

$$\frac{4(4\pi/3)}{\sqrt{8}^3} = \frac{\pi}{\sqrt{18}}.$$

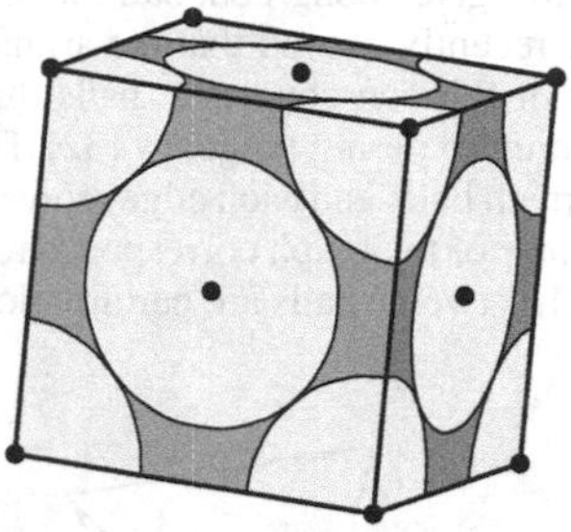

Figure 1.6 The intersection of the FCC packing with a cube of side $\sqrt{8}$. The name *face-centered cubic* comes from this depiction. The cube has volume $\sqrt{8}^3$ and contains a total of four balls (eight eighths from the corners and six halves from the facets), giving density $4(4\pi/3)/\sqrt{8}^3 = \pi/\sqrt{18}$.

The density $\pi/\sqrt{18}$ of the packing is the ratio of the volume $4\pi/3$ of a ball to the volume of a fundamental domain of the FCC lattice. The volume of the fundamental domain is therefore $4\sqrt{2}$. A fundamental domain of the FCC lattice is a parallelepiped that can be dissected into two regular tetrahedra and one regular octahedron (Figure 1.7). The FCC packing is then an alternating tiling by tetrahedra and octahedra in 2:1 ratio. A tetrahedron scaled by a factor of two consists of one tetrahedron at each extreme point and one octahedron in the center (Figure 1.8). By similarity, the total volume is $8 = 2^3$ times the volume of each smaller tetrahedron. This dissection exhibits the volume of a regular octahedron as exactly four times the volume of a regular tetrahedron of the same edge length. As a result, the volume of a regular tetrahedron of side 2 is $1/6$ the volume of the fundamental domain, or $2\sqrt{2}/3$.

The density of the FCC packing is the weighted density of the densities of the tetrahedron and octahedron. Write δ_{tet} and δ_{oct} for these densities. Explicitly, δ_{tet} is the ratio of the volume of the part within the tetrahedron of the unit balls (at the four extreme points) to the full volume of the tetrahedron. As tetrahedra fill $1/3$ of volume of the fundamental domain and an octahedron fills the

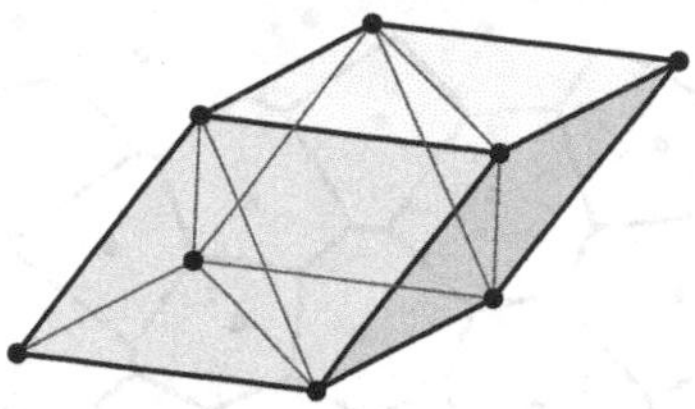

Figure 1.7 The fundamental domain of the FCC lattice can be partitioned into two regular tetrahedra and a regular octahedron. The fundamental domain tiles space. Tetrahedra and octahedra tile space in the ratio 2:1.

Figure 1.8 A regular tetrahedron whose edge is two units can be partitioned into four unit-edge tetrahedra and one unit-edge octahedron at its center. Similarly, a regular octahedron whose edge is two units can be partitioned into six unit-edge octahedra and eight unit-edge tetrahedra.

other 2/3,

$$\frac{\pi}{\sqrt{18}} = \frac{1}{3}\delta_{tet} + \frac{2}{3}\delta_{oct}.$$

As above, we identify a packing with the set V of centers of the balls. The *Voronoi cell* of a point v in a packing V is defined as the set of all points in $\mathbb{R}^3$ (or more generally in $\mathbb{R}^n$) that are at least as close to v as to any other point of V (Figure 1.9). Each Voronoi cell of the FCC packing is a rhombic dodecahedron (Figure 1.10), which is constructed from an inscribed cube by placing a square based pyramid (with height half as great as an edge of its square base) on each of the six facets.

Rhombic dodecahedra, being the Voronoi cells of the FCC packing, tile space. In each rhombic dodecahedron, we may color the inscribed cube black and the six square-based pyramids white. In the tiling, the black cubes fill the black spaces of an infinite three-dimensional checkerboard, and the white pyramids fill the white spaces.

A Voronoi cell contains an inscribed black cube of side $\sqrt{2}$ and a total of one white cube, for a total volume of $4\sqrt{2}$, which is again the volume of the

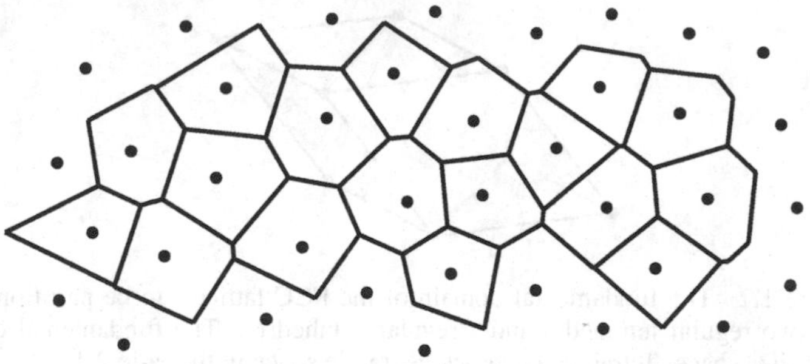

Figure 1.9 Voronoi cells of a two-dimensional sphere packing.

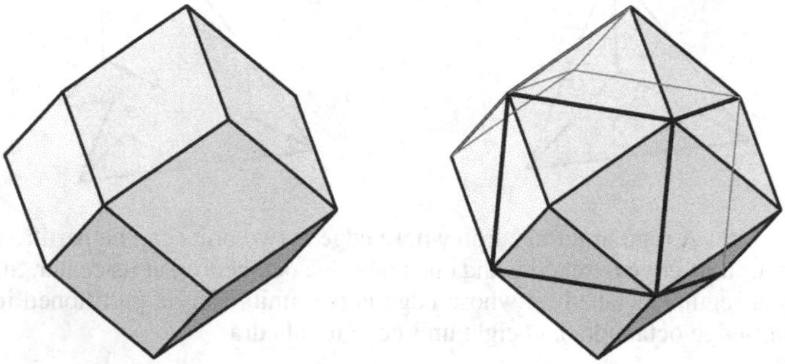

Figure 1.10 The Voronoi cell of the FCC packing is a rhombic dodecahedron. It can be constructed by placing a square-based pyramid along each facet of an inscribed cube.

fundamental domain. The density of the FCC packing is the ratio of the volume of a ball to the volume of its Voronoi cell, which gives $\pi/\sqrt{18}$ yet again.

1.3 Hexagonal-Close Packing

There is a popular and persistent misconception that the FCC packing is the only packing with density $\pi/\sqrt{18}$. The hexagonal-closed packing (HCP) has the same density.

In the FCC packing, each ball is tangent to twelve others in the same fixed arrangement. We call it the *FCC pattern*. Likewise, in the HCP, each ball is tangent to twelve others in the same arrangement (Figure 1.11). We call it the *HCP pattern*. The FCC pattern and HCP patterns are different from each other. In the FCC pattern, four different planes through the center give a regular hexagonal cross section, while the HCP pattern has only one such plane.

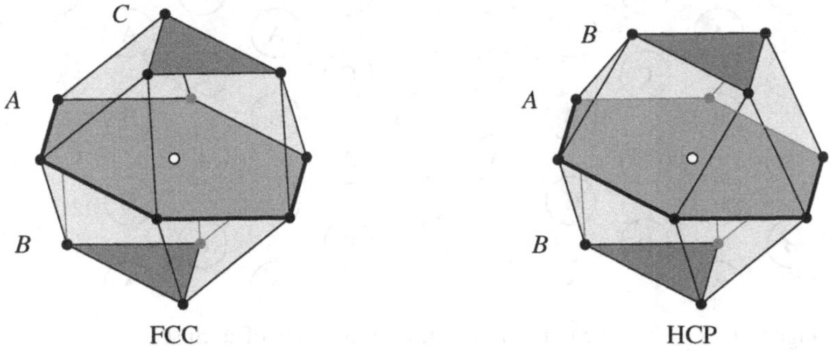

Figure 1.11 The patterns of twelve neighboring points in the FCC and HCP packings. In both cases, the convex hull of the twelve points is a polyhedron with six squares and eight triangles, but the top layer of the HCP pattern is rotated 60 degrees with respect to the FCC pattern. The FCC pattern is a cuboctahedron. In the HCP pattern, there is a uniquely determined plane of reflectional symmetry, containing six of the twelve points.

There are, in fact, uncountably many packings of density $\pi/\sqrt{18}$ in which the tangent arrangement around each ball is either the FCC pattern or the HCP pattern.

A *hexagonal layer* (Figure 1.12) is a translate of the two-dimensional hexagonal lattice (also known as the triangular lattice). That is, it is a translate of the planar lattice generated by two vectors of length 2 and angle $2\pi/3$. The FCC packing is an example of a packing built from hexagonal layers.

If L is a hexagonal layer, then a second hexagonal layer L' can be placed parallel to the first so that each lattice point of L' has distance 2 from three different points of L, which is the smallest possible distance from first layer. A choice of a unit normal vector $\mathbf{e}$ to the plane of L determines an upward direction. There are two different positions in which L' can be closely placed above L (Figure 1.12). Each successive layer (L, L', L'', and so forth) offers two further choices for the placement of that layer. Running through different sequences of choices gives uncountably many packings. In each of these packings the tangent arrangement around each ball is the FCC or HCP arrangement.

As a packing is constructed, each layer may be labeled A, B, or C depending on three possible orthogonal projections to a fixed plane with normal vector $\mathbf{e}$. Each layer carries a different label from the layers immediately above and below it. In the FCC packing, the successive layers are A, B, C, A, B, C, and so forth. In the HCP packing, the successive layers are A, B, A, B, and so forth. If the vertices of a triangle are labeled A, B, and C, then the succession of labels

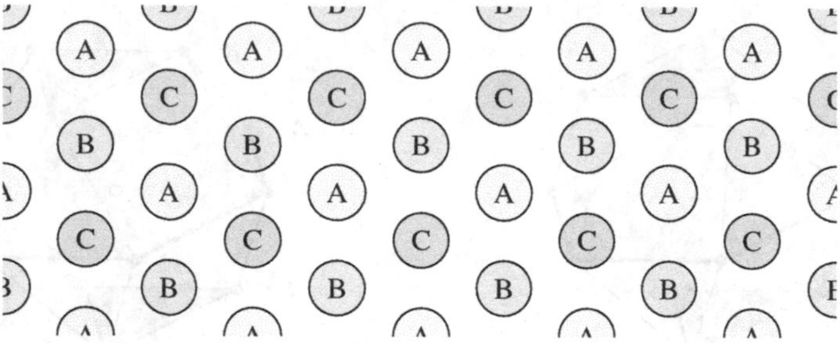

Figure 1.12 If the centers in one hexagonal layer of a close packing are placed at the sites marked A, then the hexagonal layer above it will occupy all the sites marked B, or all of the sites marked C. In general, each hexagon layer of a close packing is determined by its label A, B, or C, which must always differ from the label of the layer below. The HCP packing is $\ldots ABABAB\ldots$. The FCC packing is $\ldots ABCABCABC\ldots$. There are infinitely many other close packings, consisting of hexagonal layers, corresponding to sequences of A, B, C.

is a walk along the vertices of the triangle, and inequivalent walks through the triangle describe different packings.

The different walks through a triangle give all possible packings of infinitely many congruent balls in which each tangent arrangement is either the FCC pattern or the HCP pattern [9]. To see that there are no other possibilities, we first assume that every ball of V is surrounded by the FCC pattern. Adjacent FCC patterns interlock in a unique way that forces V itself to crystallize into the FCC packing. This completes the proof in this case.

Now we assume that a packing V contains some ball (centered at $\mathbf{u}$) in the HCP pattern. Its uniquely determined plane of reflectional symmetry contains $\mathbf{u}$ and the centers of six others arranged in a regular hexagon. If $\mathbf{v}$ is the center of one of the six other balls in the plane of symmetry, its tangent arrangement of twelve balls must include $\mathbf{u}$ and an additional four of the twelve balls around $\mathbf{u}$. These five centers around $\mathbf{v}$ are not a subset of the FCC pattern, but extend uniquely to a HCP pattern. Around $\mathbf{u}$ and $\mathbf{v}$, the HCP patterns have the same plane of symmetry. In this way, as soon as some center has the HCP pattern, the pattern propagates along the plane of symmetry to create a hexagonal layer L.

Once a packing V contains a single hexagonal layer, the condition that each ball be tangent to twelve others forces a hexagonal layer L' above L and another hexagonal layer below L. Thus, a single hexagonal layer forces an infinite sequence of close-packed hexagonal layers. The position of each layer over the previous layer is described by the labels A, B, and C of the triangle. This completes the proof that the different walks through a triangle give all possibilities.

1.4 Gauss

Gauss proved that the FCC packing has the greatest density of any lattice packing in three-dimensional Euclidean space. There is a short proof that does not require any calculations.

Proof Start with an arbitrary lattice V in which every point has distance at least 2 from every other. Center a unit ball at each point in the lattice. In a lattice of greatest density, some pair of balls touch. The lattice property then forces the balls into parallel infinite linear strings like beads on a string. Two of these infinite parallel strings touch if the lattice is optimal. The lattice property then constrains the strings in parallel sheets. On each sheet the touching parallel strings form a rhombic tiling. Each parallel sheet sits as snugly as possible on the sheet below in an optimal lattice. In such an arrangement, a ball (centered at $\mathbf{v}_0$) of one sheet touches three balls (centered at $\mathbf{v}_1, \mathbf{v}_2, \mathbf{v}_3$) on the next layer down (Figure 1.13).

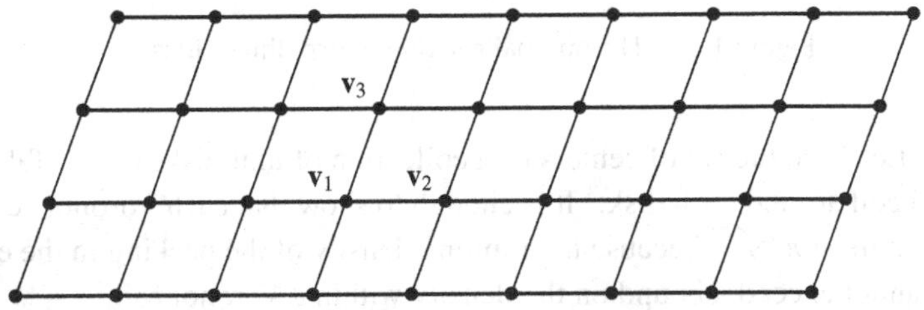

Figure 1.13 At one stage of the proof that the FCC lattice is the optimal packing among lattices, we show that an optimal lattice consists of parallel sheets of a rhombic tiling, and that a ball from one sheet rests on three balls centered at $\mathbf{v}_1, \mathbf{v}_2, \mathbf{v}_3$ in the layer below.

As the balls on each sheet form a rhombic tile, two of the distances between $\mathbf{v}_1, \mathbf{v}_2, \mathbf{v}_3$, corresponding to two edges of the rhombus, are equal to 2. This means that $\mathbf{v}_0$ together with two of $\mathbf{v}_1, \mathbf{v}_2, \mathbf{v}_3$ form an equilateral triangle.

From the perspective of the plane containing this equilateral triangle, the lattice property forces this entire plane, as well as parallel planes, to be tiled with equilateral triangles. From the earlier argument, each of these planes sits as snugly as possible on the sheet below. A ball of one sheet touches the three balls in an equilateral triangle on the layer below. These four balls form a regular tetrahedron, which uniquely identifies the lattice as the FCC. □

1.5 Thue

As mentioned in the preface, Thue solved the packing problem for congruent disks in the plane. The optimal packing is the hexagonal packing (Figure 1.14). The density of this packing is $\pi/\sqrt{12}$, that is, the ratio of the area of a unit disk to the area of a hexagon of inradius one. Thue's theorem admits an elementary proof that we sketch. Casselman has an interactive demo of this solution [8].

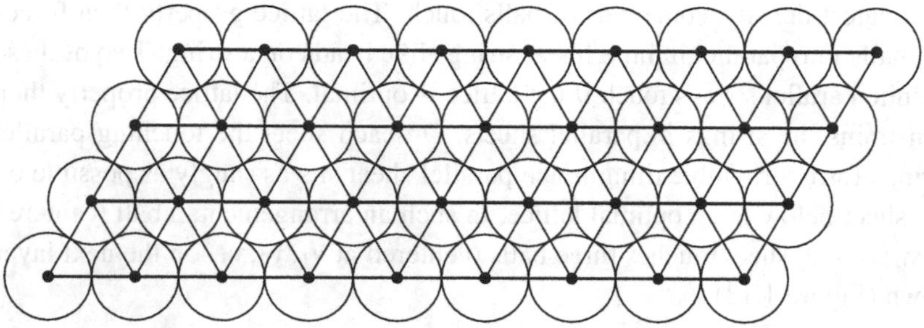

Figure 1.14 The optimal packing in two dimensions.

Proof Let V be the set of centers of a collection of unit disks in $\mathbb{R}^2$. Take the Voronoi cell around each disk.[3] It is enough to show that each Voronoi cell has density at most $\pi/\sqrt{12}$ because the limiting density of the packing in the entire plane cannot exceed a bound on the density within a Voronoi cell.

Truncate the Voronoi cell by intersecting it with a disk of radius $r = 2/\sqrt{3}$. The density increases as the volume of the cell is made smaller, so if the truncated Voronoi cell has density at most $\pi/\sqrt{12}$, then so does the untruncated Voronoi cell.

There is not a point $\mathbf{w}$ in the plane that has distance less than r from three disk centers $\mathbf{v}_1, \mathbf{v}_2, \mathbf{v}_3$. Otherwise, one of the three angles γ at $\mathbf{w}$ formed by pairs $(\mathbf{v}_i, \mathbf{v}_j)$ of points is at most $2\pi/3$, and $\cos\gamma \geq -0.5$. The *law of cosines* applied to the triangle $\mathbf{w}, \mathbf{v}_i, \mathbf{v}_j$ with angle γ and sides a, b, and c gives the contradiction

$$4 \leq c^2 = a^2 + b^2 - 2ab\cos\gamma \leq a^2 + b^2 + ab < 3r^2 = 4.$$

[3] Voronoi cells of packings in any dimension $\mathbb{R}^n$ are defined by the same rule as we gave above for $\mathbb{R}^3$.

Thus, the boundary of the truncated Voronoi cell consists of circular arcs and chords of the circle of radius r, as shown in Figure 1.15.

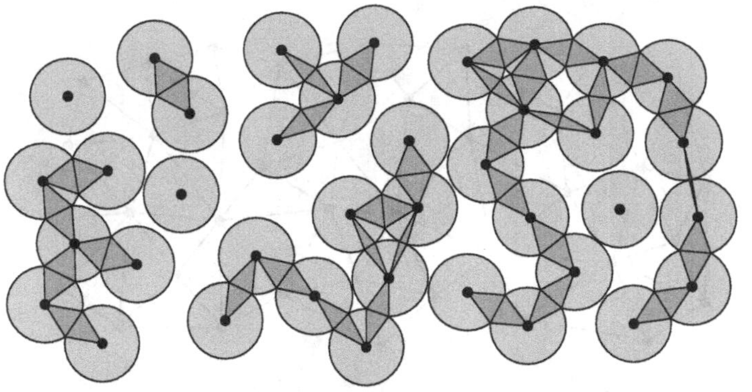

Figure 1.15 This partition of the plane gives a proof of Thue's theorem. The disks have radius $2/\sqrt{3}$. Each shaded sector and each triangle in an arbitrary packing has density at most $\pi/\sqrt{12}$.

The parts of the Voronoi cell that lie within a circular sector have density $1/r^2 = 3/4 < \pi/\sqrt{12}$. A simple calculation shows that the part of a Voronoi cell that lies within a triangle has density

$$\frac{\theta}{r^2 \cos\theta \sin\theta} \tag{1.2}$$

for some $0 \le \theta \le \pi/6$. An easy optimization gives the maximum at $\theta = \pi/6$ with value $\pi/\sqrt{12}$. This completes the proof of Thue's theorem. □

In some ways it it unfortunate that the problem in two dimensions is so elementary. It gives only meager hints about how to solve the problem in three dimensions such as the value of Voronoi cells and the usefulness of truncation. The optimization problem on triangles in Equation 1.2 generalizes to n-dimensions. But beyond these simple observations, little from the proof of Thue's theorem prepares us for higher dimensions.

There are other proofs of Thue's theorem, including one by Fejes Tóth that uses the *Delaunay triangulation* of a packing V in the plane (or in n-dimensions). A Delaunay triangulation of V is a triangulation of Euclidean space into simplices with extreme points in V such that no point of V lies in the interior of any circumscribing circle of any of the simplices (Figure 1.16). If V is *saturated*,[4] then a Delaunay triangulation of V exists. Each Delaunay

[4] A packing V is saturated if it is not a proper subset of any other packing V'. To maximize density, it is useful to increase the density by saturating the packing with additional points.

triangle in a saturated packing V has circumradius at most 2 because otherwise an additional point can be placed at the center of the circumscribing circle, contrary to saturation.

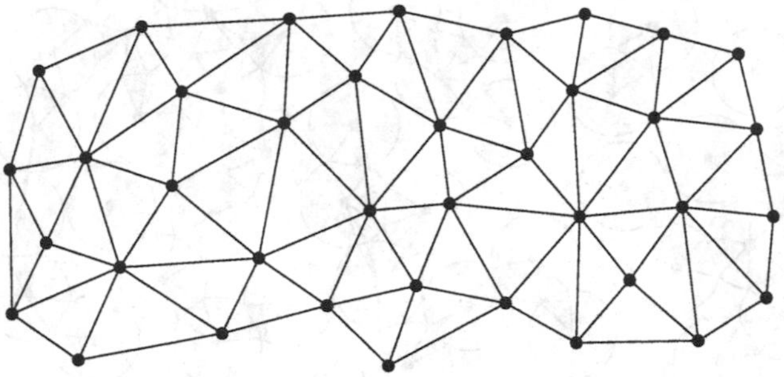

Figure 1.16 Delaunay triangles of a two-dimensional sphere packing.

Proof By admitting the existence of a Delaunay triangulation, the proof of the packing problem for saturated packings V in two dimensions becomes elementary. Each Delaunay triangle contains a portion of a disk at each of its three vertices. The three interior angles of a triangle sum to π, giving half a disk per triangle. If we show that each triangle has area at least $\sqrt{3}$, then it follows that the density of the packing is at most $(\pi/2)/\sqrt{3} = \pi/\sqrt{12}$. The problem thus reduces to an area minimization problem. To decrease the area of a triangle $\{\mathbf{v}_0, \mathbf{v}_1, \mathbf{v}_2\}$, we first replace it with a smaller similar triangle with shortest edge (say $\mathbf{v}_1\mathbf{v}_2$) of length 2. The third vertex $\mathbf{v}_0$ is constrained to have distance at least 2 from $\mathbf{v}_1$ and $\mathbf{v}_2$, and to have circumradius at most 2. The constraints on $\mathbf{v}_0$ form three circular arcs as shown in Figure 1.17.

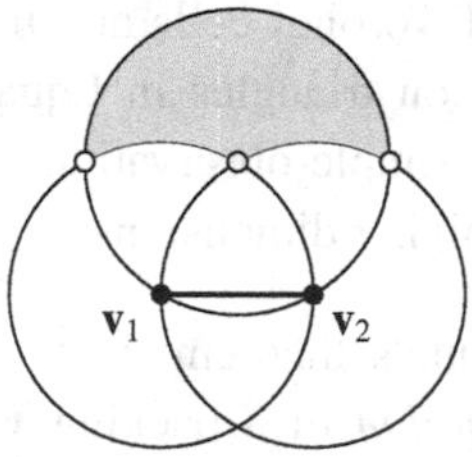

Figure 1.17 The horizontal segment is a fixed edge of length 2 of a Delaunay triangle. The shaded region constrains the position of the third vertex of the Delaunay triangle. The white dots indicate the three positions of the third vertex that minimize the area of the Delaunay triangle.

The minimizing triangle is determined by the point $\mathbf{v}_0$ closest to the line through $\mathbf{v}_1$ and $\mathbf{v}_2$. There are three such triangles, each with area exactly $\sqrt{3}$. This completes the proof. □

1.6 Dense Packings in a Nutshell

This section describes the proof of the Kepler conjecture in general, without getting embroiled in detail. The entire book is a blueprint with all the electrical schematics, plumbing, and ventilation systems. This section is the tourist brochure.

The Kepler conjecture asserts that no packing of congruent balls in three-dimensional Euclidean space has density greater than the density $\pi/\sqrt{18} \approx 0.74048$ of the FCC packing. For a contradiction, we suppose that an explicit counterexample exists to the Kepler conjecture in the form of a packing of balls of unit radius with density greater than $\pi/\sqrt{18}$. Additional balls may be added to this packing until saturation is reached. The saturation of a counterexample may push its density even higher.

We present the proof in four stages. Undefined terms are clarified in the discussion that follows.

1. A geometric partition of space, adapted to a saturated counterexample V, reduces the problem to finite packing W that gives a counterexample to a particular inequality. In notation established below, the particular inequality is $\mathcal{L}(W, \mathbf{0}) \le 12$ for every finite packing $W \subset B(\mathbf{0}, 2.52)$, where $B(\mathbf{p}, r)$ denotes the open ball of radius r centered at $\mathbf{p}$. The counterexample satisfies $\mathcal{L}(W, \mathbf{0}) > 12$.
2. The finite packing W is transformed into another finite packing that violates the same inequality and that has a few additional properties that make it a *contravening* packing.
3. The combinatorial structure of W is encoded as a hypermap. A list is made of the purely combinatorial properties of W. A hypermap with these properties is said to be *tame*.
4. A computer generates an explicit list, enumerating tame hypermaps up to isomorphism. Linear programs, which are adapted to each tame hypermap in the enumeration, certify that none of the combinatorial possibilities can be realized geometrically as a finite packing $W \subset \mathbb{R}^3$.

From the nonexistence of a counterexample W, it follows that there is no saturated counterexample V to the Kepler conjecture.

1.6.1 geometric partition

The first stage of the proof defines a geometric partition of space and uses it to reduce the Kepler conjecture to an optimization problem in a finite number of variables.

We recall that a saturated packing is identified with the discrete set V of centers of the congruent balls. Also, as above, the Voronoi cell $\Omega(V, \mathbf{v})$ associated with $\mathbf{v} \in V$ is the polyhedron formed by all points of $\mathbb{R}^3$ that are at least as close to $\mathbf{v}$ as to any other $\mathbf{w} \in V$.

The Voronoi cell at $\mathbf{v}$ can be further partitioned into Rogers simplices, each of which is determined by a facet of the Voronoi cell, an edge of the facet, and an extreme point of the edge. The Rogers simplex is defined to be the convex hull of four points: $\mathbf{v} \in V$, the closest point $\mathbf{v}_1$ to $\mathbf{v}$ on the given facet, the closest point $\mathbf{v}_2$ to $\mathbf{v}_1$ on the edge, and the extreme point $\mathbf{v}_3$ of the edge (Figure 1.18).

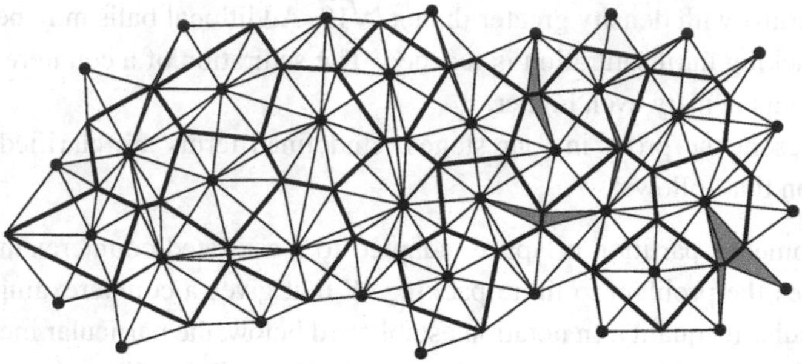

Figure 1.18 Rogers simplices of a two-dimensional sphere packing. Heavy edges are facets of Voronoi cells. The Rogers simplices that are not right triangles are shaded.

We dissect and combine the Rogers simplices somewhat further to make them into *Marchal cells* (Figure 1.19). The exact rules for the construction of Marchal cells do not concern us here. The rules depend on which of the points $\mathbf{v}_1, \ldots, \mathbf{v}_3$ have distance less than $\sqrt{2}$ from $\mathbf{v}$.

The function $\mathcal{L}(V, \mathbf{v})$ is defined as

$$\mathcal{L}(V, \mathbf{v}) = \sum_{\mathbf{w} \in V \setminus \{v\}} L(\|\mathbf{w} - \mathbf{v}\|/2), \tag{1.3}$$

where L is the piecewise linear function that has a linear graph from $(x, y) = (1, 1)$ to $(0, 1.26)$ and is equal to zero for $x \geq 1.26$. (The constants 1.26 and $2.52 = 2(1.26)$ appear throughout the proof as parameters used in truncation.) The sum in the definition of $\mathcal{L}$ is actually finite for every packing V because only finitely many terms lie in the support of L.

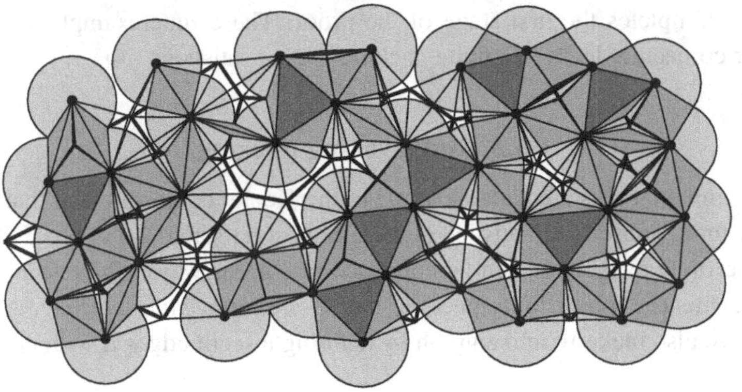

Figure 1.19 Marchal cells of a two-dimensional sphere packing.

Next, a function $G : V \to \mathbb{R}$ is defined geometrically in terms of the volumes, solid angles, and dihedral angles of Marchal cells. We do not give the definition here because it is rather complex. The function G has the following two fundamental properties.

1. If $\mathcal{L}(V, \mathbf{v}) \leq 12$, then

$$4\sqrt{2} \leq \mathrm{vol}(\Omega(V, \mathbf{v})) + G(\mathbf{v}).$$

2. There exists $C > 0$ such that the points of V in a ball $B(\mathbf{0}, r)$ of radius $r \geq 1$ satisfy

$$\sum_{\mathbf{v} \in V \cap B(\mathbf{0}, r)} G(\mathbf{v}) < Cr^2.$$

The constant $4\sqrt{2}$ is the volume of the Voronoi cell of the FCC packing.

From these fundamental properties and from the assumption that V is a saturated counterexample, it follows that $\mathcal{L}(V, \mathbf{v}) > 12$ for some $\mathbf{v} \in V$. Indeed, if $\mathcal{L}(V, \mathbf{v}) \leq 12$ for all $\mathbf{v} \in V$, then the fundamental properties imply that on average the Voronoi cells of V have volume at least that of the FCC packing, up to a negligible error term Cr^2. From this, it follows that the density of the packing V is at most that of the FCC packing.

Returning to the counterexample V, we select $\mathbf{v} \in V$ such that $\mathcal{L}(V, \mathbf{v}) > 12$. By the translational invariance of the problem, we may assume that $\mathbf{v} = \mathbf{0}$. Then

$$\mathcal{L}(W, \mathbf{0}) = \sum_{\mathbf{w} \in W} L(\|\mathbf{w}\|/2) > 12, \tag{1.4}$$

where W is the finite set $\{\mathbf{w} \in V \ : \ 0 < \|\mathbf{w}\| \leq 2.52\}$.

This completes the first stage of the proof. The counterexample V to the Kepler conjecture leads to a finite packing W that satisfies (1.4).

1.6.2 contravening packing

We assume that V is a counterexample to the Kepler conjecture and that $W \subset V$ is a finite subset that satisfies (1.4). The second stage of the proof shows that the finite packing W can be enhanced in various ways. The result of the enhancement is a new finite packing that is a *contravening packing*. At this stage, we also make W into a graph by defining a set of edges E with nodes in W.

For example, the value of $\mathcal{L}$ depends only on the norms $\|\mathbf{w}\|$, and L is a decreasing function, so that any rearrangement of the points of W that does not increase the norms strengthens the inequality (1.4).

The finite packing W determines a graph (W, E) with node set W. The set of edges is defined by $\{\mathbf{v}, \mathbf{w}\} \in E$ if

$$2 \leq \|\mathbf{v} - \mathbf{w}\| \leq 2.52.$$

This graph is called the *standard fan* of W.

We can get a crude idea about what W must look like by studying the set of normalized points $\mathbf{w}/\|\mathbf{w}\|$ in the unit sphere. These points are extreme points of spherical polygons that partition the unit sphere. As we know that the sum of the areas of the polygons equals the area 4π of the sphere, we can extract bits of information about W from estimates of the areas of the polygons. Analysis along these lines leads to the conclusion that some finite packing W has the following properties.

1. $W \subset B(\mathbf{0}, 2.52)$.
2. $\mathcal{L}(W) > 12$.
3. The cardinality of W is thirteen, fourteen, or fifteen.
4. W maximizes the function $\mathcal{L}$.
5. Join points $\mathbf{v}/\|\mathbf{v}\|$ and $\mathbf{w}/\|\mathbf{w}\|$ with a geodesic arc on the unit sphere if $\{\mathbf{v}, \mathbf{w}\} \in E$. Then the arcs do not meet except at the endpoints and give a planar graph. Moreover, the angle between each pair of consecutive arcs at a vertex is less that π. In particular, the spherical polygons cut out by the arcs are geodesically convex.

A finite packing W with these properties is called a *contravening* packing.

1.6.3 tame hypermap

The starting point of the third stage of the proof is a contravening packing W and the corresponding planar graph (W, E). The result of this stage is a *tame hypermap* (described below).

By definition, a *planar graph* is a graph that admits a *planar* embedding. By contrast, a graph, endowed with a fixed embedding into the plane, is a *plane graph*. A planar graph has too little structure for our purposes because it does not single out a particular embedding and the plane graph has too much structure because it gives a topological object where combinatorics alone should suffice. A hypermap gives just the right amount of structure. It is a purely combinatorial notion, yet encodes the relations among nodes, edges, and faces determined by the embedding. An entire chapter of this book is about hypermaps.

The graph (W, E) of a contravening packing W determines a planar hypermap $\mathrm{hyp}(W, E)$. We study the following question: what purely combinatorial properties of the hypermap $\mathrm{hyp}(W, E)$ can be derived from the assumption that W is a contravening packing? For example, the cardinality of a contravening packing W is thirteen, fourteen, or fifteen. Hence, the hypermap has thirteen, fourteen, or fifteen nodes. Later chapters of the book revolve around the combinatorial properties of the hypermap.

The final chapter of the proof compiles all of these combinatorial properties into a long list. Although the exact details of the list are not significant, the list of combinatorial properties severely constrains the set of possible hypermaps.

Any hypermap satisfying all of these properties is said to be *tame*. This list of properties appears in Definition 8.7.

1.6.4 linear programming

The fourth and final stage completes the proof the nonexistence of the contravening packing W. At the beginning of this stage, $\mathrm{hyp}(W, E)$ is a tame hypermap. The list of defining properties of a tame hypermap are sufficiently restrictive that an explicit finite list can be generated of every tame hypermap, up to isomorphism. This list is generated by computer. The details of the algorithm are described in the chapter on hypermaps.

Equipped with an explicit list of possible combinatorial structures, we move to the proof's end game. At this stage, because of the computer generated list of tame hypermaps, the cardinality and combinatorial structure of W are explicit.

Certain properties of W (and its associated hypermap) can be encoded as a system of linear inequalities. For each tame hypermap, a computer solves one or more linear programs that test for feasible solutions to the system of linear inequalities. In each case, the computer produces a certificate that shows that no feasible solution exists. It follows that no tame hypermap can be realized in the form hyp(W, E). Each tame hypermap, which represents a combinatorially feasible arrangement, is geometrical infeasible. It follows that W, and hence also V, do not exist.

As no counterexample exists, the proof of the Kepler conjecture ensues.

PART TWO

FOUNDATIONS

2

Trigonometry

Summary. *This part of this book, which is the first of the four foundational chapters, presents a systematic development of trigonometry, volume, hypermap, and fan. There is a separate chapter on each of these topics. The purpose of the this material is to build a bridge between the foundations of mathematics, as presented in formal theorem proving systems such as HOL Light, and the solution to the packing problem.*

In this chapter, trigonometry is developed analytically. Basic trigonometric functions are defined by their power series representations, and calculus of a single real variable is used to develop the basic properties of these functions. Basic vector geometry is presented.

2.1 Background Knowledge

2.1.1 formal proof

We repeat that our purpose is to give a blueprint of the formal proof of Kepler's conjecture that no packing of congruent balls in three-dimensional Euclidean space has density greater than the familiar cannonball packing. The blueprint of a formal proof is not the same as a formal proof, which is a fleeting pattern of bits in a computer. The book describes to the reader[1] how to construct the computer code that produces and then reliably reproduces that pattern of bits.

[1] "The words will be minced into atomized search-engine keywords ... copied millions of times by algorithms ... scanned, rehashed, and misrepresented by crowds.... And yet it is you, the person, the rarity among my readers, I hope to reach" –Jaron Lanier [29].

A more traditional book might take as its starting point the imagined mathematical background of a typical reader. The blueprint of a formal proof starts instead with the current mathematical background of a formal proof assistant. I surveyed the knowledge base of my formal proof assistant and compared it with what is needed in the construction of our formal proof. It turns out that the proof assistant already has an adequate background in real analysis, basic topology, and plane trigonometry, including the trigonometric addition laws, and formulas for derivatives. Since the proof assistant already has a significant library of theorems in real analysis and point-set topology, we use background facts in these areas wherever they help.

However, when this project began, the proof assistant lacked the background in some of the less frequently used trigonometric identities and has had nothing at all about spherical trigonometry. While it had adequate command of general concepts of vector geometry in n-dimensional Euclidean space, its library on three-dimensional analytic geometry was spotty. For example, dihedral angles and cylindrical and spherical coordinates were missing from the system.

I imagine the typical reader to have a much stronger background in trigonometry and analytic geometry than the proof assistant, which, after all, is still in its youth. The mathematician might want to jump directly to Definition 2.35, which specifies subsets $\mathrm{aff}_\pm$ of affine space. This definition gives a compact notation that encompasses many of the standard polyhedra (points, lines, planes, rays, half-planes, half-spaces, convex hulls, affine hulls) that appear throughout the book. From there, the reader can consult the definition of two important polynomials Δ and υ, make a note of the unorthodox notation $\mathrm{arc}(a, b, c)$ for the angle opposite c of a triangle with sides of lengths a, b, c, stop a moment to admire Euler's formula for the solid angle of a spherical triangle; and then jump directly to the final section, which introduces polar cycle.

Polar cycle is a familiar concept, wrapped in an unfamiliar way for the sake of the proof assistant: take a finite set of points in the plane, order them by increasing angle, and then take the cyclic permutation on the points induced by this order. The azimuth cycle is the corresponding permutation in three dimensions, ordering points by increasing azimuth angle (longitude) in spherical coordinates. Although intuitively clear, our proof assistant demands extra assistance at this point.

2.1.2 real analysis

This chapter assumes general facts about real analysis at the level of a typical undergraduate textbook. In particular, it assumes a general working knowledge of set theory and basic properties of the set of *natural numbers*, $\mathbb{N} = \{0, 1, \ldots\}$, and the field $\mathbb{R}$ of real numbers. By convention, $0 \in \mathbb{N}$. In real analysis, the chapter assumes basic properties of convergence, absolute convergence, limits, and differentiation. The term *real analysis* is to be interpreted broadly to include even the most elementary facts of real arithmetic, including results that do not involve limits.

2.1.3 Tarski arithmetic

Certain sentences in real arithmetic can be expressed with nothing more than the usual logical operations (the connectives *and*, *or*, *implies*, *logical negation*); the ring operations (addition, subtraction, and multiplication) for the real numbers; comparison ((=) and (>)) of real numbers; the constants 0 and 1; real-valued variables; and quantifiers (universal and existential) over the real numbers. Such sentences are said to belong to the *Tarski arithmetic*. For example, the sentence

$$\exists x.\ x^7 - 4x - 3 = 0 \ \wedge\ x > 0 \tag{2.1}$$

falls within the Tarski arithmetic (after expanding the exponent x^7 as $x \cdot x \cdot x \cdot x \cdot x \cdot x \cdot x$ and the constants $4 = 1+1+1+1$ and $3 = 1+1+1$). Starting with Tarski, researchers have developed algorithms to decide the truth of any sentence in the Tarski arithmetic [44], [6]. Although these algorithms are generally too slow to be of practical use, it is useful to identify such sentences. To follow the details of proofs, reader should have the skill to solve particularly simple problems in the Tarski arithmetic such as determining that the sentence (2.1) is true.

2.2 Trig Identities

2.2.1 sine and cosine

The cosine and sine functions are defined[2] by their infinite series:

$$\cos(x) = 1 - x^2/2! + x^4/4! \cdots, \qquad \sin(x) = x - x^3/3! + x^5/5! \cdots. \tag{2.2}$$

[2] This is how the trigonometric functions were originally defined in the proof assistant HOL Light. More recently, complex analysis has been developed in HOL Light sufficient for the analytic proof of the prime number theorem [25]. The cosine and sine are now defined in the system as the real and complex parts of the exponential function e^{ix}. To simplify the exposition, this section presents the original definitions.

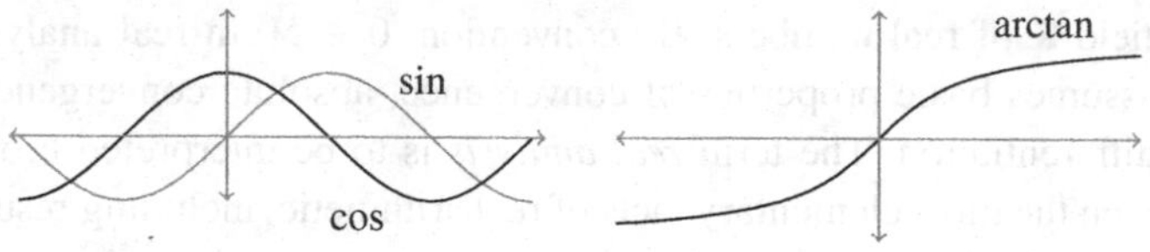

Figure 2.1 Trigonometric and inverse trigonometric functions.

By real analysis, convergence is absolute for every real number x. Each series can be evaluated at 0:

$$\cos(0) = 1, \qquad \sin(0) = 0. \tag{2.3}$$

These series may be differentiated term by term to establish the identities:

$$\frac{d}{dx}\cos(x) = -\sin(x), \qquad \frac{d}{dx}\sin(x) = \cos(x). \tag{2.4}$$

Powers $(\cos(x))^n$ and $(\sin(x))^n$ are conventionally written $\cos^n(x)$ and $\sin^n(x)$.

If two functions are the *unique* solution of an ordinary differential equation with given initial conditions, then the two functions are equal. This observation gives a method to prove many functional identities, including trigonometric identities. The next two lemmas take this approach, by certifying a trigonometric identity with a function f that satisfies the ordinary differential equation $f' = 0$ with initial condition $f(0) = 0$.

Lemma 2.5

$$\sin^2(x) + \cos^2(x) = 1.$$

Proof By real analysis and (2.4), the derivative of $f(x) = \cos^2(x) + \sin^2(x)$ is identically zero, so the function itself is constant. From (2.3), it follows that $f(x) = f(0) = 1$. □

Lemma 2.6

$$\sin(x + y) = \sin(x)\cos(y) + \cos(x)\sin(y)$$
$$\cos(x + y) = \cos(x)\cos(y) - \sin(x)\sin(y).$$

Proof The proof is an exercise in real analysis. Fix y. Let

$$f(x) = (\cos(x + y) - \cos(x)\cos(y) + \sin(x)\sin(y))^2 + (\sin(x + y) - \sin(x)\cos(y) - \cos(x)\sin(y))^2.$$

The derivative of f is identically zero. The function is therefore constant. Also,

$f(0) = 0$. Thus, f is identically zero. If a sum of real squares is zero, the individual terms are zero. The identities follow. □

Lemma 2.7 *The cosine is an even function. The sine is an odd function. That is,*

$$\cos(-x) = \cos(x), \quad \sin(-x) = -\sin(x).$$

Proof The result can be checked directly from the definition of the trigonometric functions as power series. A second proof can be given by differentiation, as follows. By real analysis, the derivative of

$$(\cos(-x) - \cos(x))^2 + (\sin(-x) + \sin(x))^2$$

is identically zero. Complete the proof as in the proof of Lemma 2.6. □

2.2.2 periodicity

It is known that the cosine has a unique root between 0 and 2. The constant π is defined to be twice that root. Thus, by definition

$$\cos(\pi/2) = 0,$$
$$\cos(x) > 0, \quad \text{when } 0 < x < \pi/2. \tag{2.8}$$

The cosine is in fact nonnegative on the interval $[0, \pi/2]$:

$$\cos(x) \geq 0, \quad 0 \leq x \leq \pi/2. \tag{2.9}$$

Lemma 2.10 *The sine function is nonnegative on* $[0, \pi/2]$ *and* $\sin(\pi/2) = 1$.

Proof The proof is an exercise in real analysis. The derivative of the sine is nonnegative between 0 and $\pi/2$. The value of the sine at 0 is 0. It follows that sin is nonnegative on $[0, \pi/2]$. It is enough to check that $\sin^2(\pi/2)$ equals 1. Then $\sin^2(\pi/2) = 1 - \cos^2(\pi/2) = 1$. □

Lemma 2.11

$$\sin(\pi/2 - x) = \cos(x),$$
$$\cos(\pi/2 - x) = \sin(x).$$

Proof Apply the addition law for the sine function (Lemma 2.6),

$$\sin(\pi/2 - x) = \sin(\pi/2)\cos(-x) + \cos(\pi/2)\sin(-x)$$

and use $\sin(\pi/2) = 1$ and $\cos(\pi/2) = 0$. Then use that cos is an even function. The second identity is similar. □

Similarly, $\cos(\pi/2 + x) = -\sin(x)$, $\sin(\pi/2 + x) = \cos(x)$. Further,

$$\begin{aligned}
\sin(\pi + x) &= \cos(\pi/2 + x) = -\sin(x),\\
\cos(\pi + x) &= -\sin(\pi/2 + x) = -\cos(x),\\
\sin(2\pi + x) &= -\sin(\pi + x) = \sin(x),\\
\cos(2\pi + x) &= -\cos(\pi + x) = \cos(x).
\end{aligned} \tag{2.12}$$

Lemma 2.13 *The sine function is nonnegative on* $[0, \pi]$.

Proof By Lemma 2.10, sin is nonnegative on $[0, \pi/2]$. Furthermore, for $x \in [\pi/2, \pi]$,

$$\sin(x) = -\sin(-x) = \sin(\pi - x) \geq 0.$$

□

2.2.3 tangent

Definition 2.14 (tangent) Let $\tan(x) = \sin(x)/\cos(x)$, when $\cos(x) \neq 0$.

Lemma 2.15 *If* $\cos(x) \neq 0$, $\cos(y) \neq 0$, *and* $\cos(x + y) \neq 0$ *then*

$$\tan(x + y) = \frac{\tan(x) + \tan(y)}{1 - \tan(x)\tan(y)}.$$

Proof Divide the first line of Lemma 2.6 by the second line of the same lemma. Then use the definition of the tangent. □

Lemma 2.16

$$\tan(\pi/4) = 1.$$

Proof

$$\tan(\pi/4) = \sin(\pi/2 - \pi/4)/\cos(\pi/4) = \cos(\pi/4)/\cos(\pi/4) = 1.$$

□

Lemma 2.17 *The function* tan *is strictly increasing and one-to-one on the domain* $(-\pi/2, \pi/2)$.

Proof By a derivative test, the tangent is strictly increasing on $(-\pi/2, \pi/2)$. By real arithmetic, a strictly increasing function is one-to-one. □

2.2.4 arctangent

This section reviews the properties of the arctangent function.

Definition 2.18 (arctangent) By the inverse function theorem of real analysis and properties of tan, there is a unique function arctan : $\mathbb{R} \to \mathbb{R}$ with image $(-\pi/2, \pi/2)$ such that

$$\tan(\arctan x) = x. \tag{2.19}$$

(See Figure 2.1.)

Additional properties of the arctangent function are exercises in real analysis. If $-\pi/2 < x < \pi/2$, then also $\arctan(\tan(x)) = x$. In particular,

$$\arctan(1) = \arctan(\tan(\pi/4)) = \pi/4. \tag{2.20}$$

The function arctan is differentiable with derivative

$$\frac{d}{dx}\arctan(x) = \frac{1}{1+x^2}. \tag{2.21}$$

The derivative is positive, and the function arctan is strictly increasing. Proofs in this book often need to use $\arctan(y/x)$ as x approaches 0. For this, the following variant of arctan is preferable because it clears the denominator.

Definition 2.22 ($\arctan_2$)

$$\arctan_2 : \mathbb{R}^2 \to (-\pi, \pi].$$

$$\arctan_2(x,y) = \begin{cases} \arctan(y/x), & x > 0 \\ \pi/2 - \arctan(x/y), & y > 0 \\ \pi + \arctan(y/x), & x < 0,\ y \geq 0 \\ -\pi/2 - \arctan(x/y), & y < 0 \\ \pi, & x = y = 0. \end{cases}$$

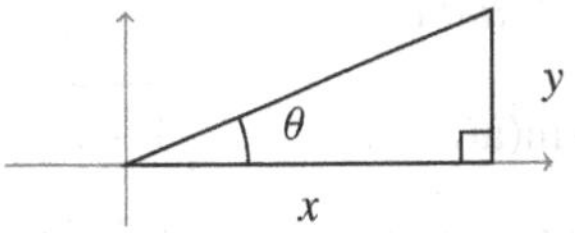

Figure 2.2 The function $\arctan_2$ gives the polar angle θ of (x, y).

There is some overlap between cases. Nevertheless, trig identities similar to those already established show that this function is well-defined. For example, to check the equality of the first two cases ($x > 0$ and $y > 0$), we compute the tangent of both sides, which is sufficient, since both sides lie between $(-\pi/2, \pi/2)$ and tan is one-to-one:

$$\tan(\arctan(y/x)) = y/x = 1/\tan(\arctan(x/y)) = \tan(\pi/2 - \arctan(x/y)).$$

We can give a more intuitive description of the function $\arctan_2$: the polar angle of (x, y) with the branch cut along the negative axis. That is, $x = r\cos\theta$ and $y = r\sin\theta$ for some $r \geq 0$, where $\theta = \arctan_2(x, y)$. This definition avoids all the case distinctions of Definition 2.22.

The ANSI C programming language implements this function as *arctan2*. Note that some programming languages implement this function with the two arguments in reverse: (y, x).

2.2.5 inverse trig

We prefer the arctangent over other inverse trigonometric functions because its domain is the entire field of real numbers, its range is bounded, and its derivative is a rational function. Wherever angles appear in this book, the arctangent is apt to appear as well. Other inverse trigonometric functions are generally reduced to the arctangent. This section defines the arccos function and shows how it can be expressed in terms of $\arctan_2$.

Definition 2.23 (arccos) By the inverse function theorem of real analysis, there exists a unique function $\arccos y$ on the interval $[-1, 1]$ that takes values in $[0, \pi]$ and that is the inverse function of cos:

$$y \in [-1, 1] \Rightarrow \cos(\arccos y) = y$$
$$x \in [0, \pi] \Rightarrow \arccos(\cos x) = x.$$

Lemma 2.24 *If* $y \in [-1, 1]$, *then*

$$\sin(\arccos(y)) = \sqrt{1 - y^2}.$$

Proof The range of $\arccos(y)$ is $[0, \pi]$. On this interval, sin is nonnegative. By real analysis, it is enough to check that the squares of the two nonnegative numbers are equal. It then an arithmetic consequence of the circle identity (Lemma 2.5) and Definition 2.23. □

The following lemma shows how to rewrite arccos in terms of $\arctan_2$.

Lemma 2.25 *If* $y \in [-1, 1]$, *then*

$$\arccos(y) + \arctan_2\left(\sqrt{1 - y^2}, y\right) = \pi/2.$$

Proof The brief justification is simply that $\arccos(y/z)$ gives one acute angle of a right triangle with hypotenuse z and sides x and y, and $\arctan_2(x, y)$ gives the other acute angle. The two acute angles of a right triangle have sum $\pi/2$.

A bit more detail is needed for an argument that can be turned into a formal proof. The endpoints $y = \pm 1$ can be checked directly from definitions. If $y \in (-1, 1)$, $\beta = \arccos(y)$, and

$$\alpha = \arctan\left(y/\sqrt{1 - y^2}\right) = \arctan_2\left(\sqrt{1 - y^2}, y\right),$$

then arithmetic gives $-\pi/2 < \pi/2 - \beta < \pi/2$, and $-\pi/2 < \alpha < \pi/2$. By the injectivity of the function tan, it is therefore enough to check that $\tan(\pi/2-\beta) = \tan(\alpha)$. But

$$\tan(\pi/2 - \beta) = \frac{\cos(\beta)}{\sin(\beta)} = \frac{y}{\sin(\arccos(y))} = \frac{y}{\sqrt{1 - y^2}} = \tan(\alpha).$$

□

2.3 Vector Geometry

This section reviews vector geometry in $\mathbb{R}^N$, including products (scalar and dot), inequalities (triangle and Cauchy–Schwarz), and hulls (convex and affine).

2.3.1 Euclidean space

Definition 2.26 ($\mathbb{R}^N$, vector) For any finite set N, define $\mathbb{R}^N$ as the set of functions $\mathbf{v} : N \to \mathbb{R}$. Write v_i for the value of the function $\mathbf{v}$ at $i \in N$. A function in $\mathbb{R}^N$ is called a *vector*. The zero vector $\mathbf{0}$ is the function that is identically zero.

Vectors are written in a bold face: $\mathbf{u}$, $\mathbf{v}$, $\mathbf{w}$, $\mathbf{p}$, $\mathbf{q}$, and so forth. As a general notational practice, there is a general tendency to use $\mathbf{u}$, $\mathbf{v}$, and $\mathbf{w}$ to denote vectors that are constrained to lie in some previously determined subset $V \subset \mathbb{R}^N$ and to use $\mathbf{p}$ and $\mathbf{q}$ to denote vectors that run without restriction over all of $\mathbb{R}^N$.

No distinction is made between vectors and points in $\mathbb{R}^N$, and none is made between $\mathbb{R}^N$ and Euclidean space. Write $\mathbb{R}^n$ as an alias of $\mathbb{R}^N$ when $n \in \mathbb{N}$ and $N = \{0, \ldots, n - 1\}$.

Definition 2.27 (vector addition, scalar multiplication) Two standard arithmetic operations, addition and scalar multiplication, are defined on the set $\mathbb{R}^N$. These operations are the pointwise addition and scalar multiplication of functions:

$$(\mathbf{u}+\mathbf{v})_i = u_i + v_i.$$
$$(t\mathbf{u})_i = tu_i, \quad t \in \mathbb{R}. \tag{2.28}$$

Define the difference of two vectors to be $\mathbf{u} - \mathbf{v} = \mathbf{u} + (-1)\mathbf{v}$.

The operations on $\mathbb{R}^N$ satisfy the axioms of a *vector space*. In particular, addition is commutative and associative.

Definition 2.29 (dot product) The *dot product* $(\cdot)$ is the bilinear binary operation on $\mathbb{R}^N$ defined by

$$\mathbf{u} \cdot \mathbf{v} = \sum_{i \in N} u_i v_i.$$

The dot product satisfies the following properties:

$$\begin{aligned} \mathbf{u} \cdot (\mathbf{v}+\mathbf{w}) &= \mathbf{u}\cdot\mathbf{v} + \mathbf{u}\cdot\mathbf{w} \\ (\mathbf{u}+\mathbf{v})\cdot\mathbf{w} &= \mathbf{u}\cdot\mathbf{w} + \mathbf{v}\cdot\mathbf{w} \\ (t\mathbf{u})\cdot\mathbf{w} &= t(\mathbf{u}\cdot\mathbf{w}) = \mathbf{u}\cdot(t\mathbf{w}) \\ 0 &\le \mathbf{u}\cdot\mathbf{u}. \end{aligned} \tag{2.30}$$

Definition 2.31 (norm) The *norm* of a vector $\mathbf{u} \in \mathbb{R}^N$ is

$$\|\mathbf{u}\| = \sqrt{\mathbf{u}\cdot\mathbf{u}}.$$

By real arithmetic, $\|\mathbf{u}\| = 0$ if and only if $\mathbf{u} = \mathbf{0}$. Moreover, $\|t\mathbf{u}\| = |t|\,\|\mathbf{u}\|$.

Lemma 2.32 (Cauchy–Schwarz inequality)

$$|\mathbf{u}\cdot\mathbf{v}| \le \|\mathbf{u}\|\,\|\mathbf{v}\|.$$

Furthermore, the case $\pm\mathbf{u}\cdot\mathbf{v} = \|\mathbf{u}\|\,\|\mathbf{v}\|$ *of equality holds exactly when* $\|\mathbf{v}\|\mathbf{u} = \pm\|\mathbf{u}\|\mathbf{v}$ *(with matching signs).*

Proof This is an exercise in real arithmetic. Let $\mathbf{w} = \|\mathbf{v}\|\mathbf{u} \pm \|\mathbf{u}\|\mathbf{v}$. The expansion of $\mathbf{w}\cdot\mathbf{w}$ gives

$$0 \le \mathbf{w}\cdot\mathbf{w} = 2\|\mathbf{u}\|^2\|\mathbf{v}\|^2 \pm 2\|\mathbf{u}\|\,\|\mathbf{v}\|(\mathbf{u}\cdot\mathbf{v}) = 2\|\mathbf{u}\|\,\|\mathbf{v}\|(\|\mathbf{u}\|\,\|\mathbf{v}\| \pm (\mathbf{u}\cdot\mathbf{v})).$$

If $2\|\mathbf{u}\|\,\|\mathbf{v}\| = 0$, then $\mathbf{u}$ or $\mathbf{v}$ is zero, and the result easily ensues. Otherwise divide both sides of the inequality by the positive number $2\|\mathbf{u}\|\,\|\mathbf{v}\|$ to get the result. □

Lemma 2.33 (triangle inequality)

$$\|\mathbf{u}+\mathbf{v}\| \le \|\mathbf{u}\| + \|\mathbf{v}\|.$$

Equality holds exactly when $\|\mathbf{v}\|\mathbf{u} = \|\mathbf{u}\|\mathbf{v}$.

Proof This is an exercise in real arithmetic. Both sides are nonnegative; it is enough to compare the squares of both sides. By the Cauchy–Schwarz inequality,

$$\|\mathbf{u}+\mathbf{v}\|^2 = \mathbf{u}\cdot\mathbf{u} + 2\mathbf{u}\cdot\mathbf{v} + \mathbf{v}\cdot\mathbf{v} \le \mathbf{u}\cdot\mathbf{u} + 2\|\mathbf{u}\|\,\|\mathbf{v}\| + \mathbf{v}\cdot\mathbf{v} = (\|\mathbf{u}\| + \|\mathbf{v}\|)^2.$$

The case of equality follows from the case of equality in the Cauchy–Schwarz inequality. □

2.3.2 affine geometry

Most of the following definitions apply to n-dimensional Euclidean space; however, this book uses them only in two and three dimensions. The first definition gives the affine span of a finite set. For example, the affine span of two distinct points is a line; the affine span of three independent points is a plane. By placing additional positivity constraints on the linear combinations, the definitions extend to a large assortment of other geometric objects such as rays, half-planes, convex hulls, and cones. Each of these comes in two versions: an open version defined by strict inequality and a closed version defined by weak inequality. For example, the closed half-plane includes a bounding line and the open half-plane does not. In this chapter, open and closed are not topological notions; rather, they indicate the semialgebraic conditions of strict and weak inequality.

Definition 2.34 (affine hull) A set $A \subset \mathbb{R}^N$ is *affine*, if for every finite nonempty subset $S \subset A$ and every function $t : S \to \mathbb{R}$ such that $\sum_{\mathbf{v}\in S} t(\mathbf{v}) = 1$, we have

$$\sum_{\mathbf{v}\in S} t(\mathbf{v})\mathbf{v} \in A.$$

The *affine hull*, $\mathrm{aff}(S)$, of a set $S \subset \mathbb{R}^N$ is the smallest affine set containing S. That is, the affine hull of S is the intersection of all affine sets containing S.

Definition 2.35 (affine) If $V = \{\mathbf{v}_1, \mathbf{v}_2, \ldots, \mathbf{v}_k\}$ and $V' = \{\mathbf{v}_{k+1}, \ldots, \mathbf{v}_n\}$ are finite subsets of $\mathbb{R}^N$, then set

$$\operatorname{aff}_{\pm}(V, V') = \{t_1\mathbf{v}_1 + \cdots t_n\mathbf{v}_n \;:\; t_1 + \cdots + t_n = 1, \pm t_j \geq 0, \text{ for } j > k\},$$
$$\operatorname{aff}^0_{\pm}(V, V') = \{t_1\mathbf{v}_1 + \cdots t_n\mathbf{v}_n \;:\; t_1 + \cdots + t_n = 1, \pm t_j > 0, \text{ for } j > k\}.$$

To lighten the notation for singleton sets, abbreviate $\operatorname{aff}_{\pm}(\{\mathbf{v}\}, V')$ to $\operatorname{aff}_{\pm}(\mathbf{v}, V')$.

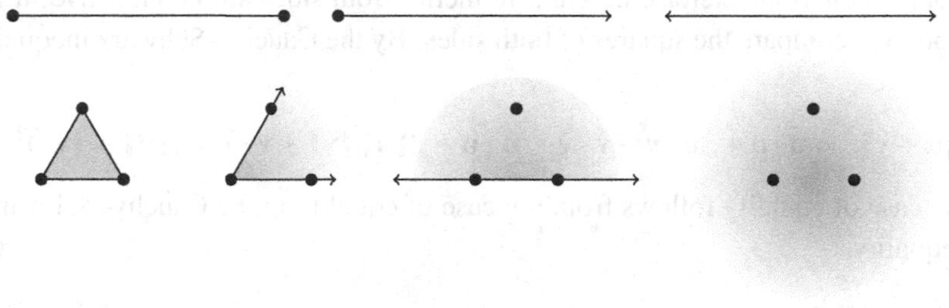

Figure 2.3 When $\operatorname{card}(V) + \operatorname{card}(V') - 1 \in \{1, 2\}$, the set $\operatorname{aff}_{+}(V, V')$ is a segment, ray, or line; simplex, blade, half-plane, or plane.

Remark 2.36 When $n + 1 = \operatorname{card}(V) + \operatorname{card}(V')$, the generic set $\operatorname{aff}_{+}(V, V')$ is an n-dimensional polyhedron bounded by $\operatorname{card}(V')$ hyperplanes. For example, $n = 1$, gives a segment, a ray, or a line (Figure 2.3). When $n = 2$, the set is a 2-simplex, a planar wedge bounded by two lines, a half-plane, or a plane. When $n = 3$, the set is a 3-simplex; an unbounded connected region in space bounded by one, two, or three intersecting planes; or all of $\mathbb{R}^3$.

Definition 2.37 (convex hull) A subset $C \subset \mathbb{R}^N$ is *convex*, if for every $\mathbf{v}, \mathbf{w} \in C$ and every $t \in [0, 1]$,

$$t\mathbf{v} + (1 - t)\mathbf{w} \in C.$$

If $S \subset \mathbb{R}^N$, then let $\operatorname{conv}(S)$ be the smallest convex set (or equivalently, the intersection of all convex sets) containing S. It is called the *convex hull.*

When the set is finite, the convex hull takes the following form.

Lemma 2.38 *If* $V = \{\mathbf{v}_1, \mathbf{v}_2, \ldots, \mathbf{v}_n\} \subset \mathbb{R}^N$, *then*

$$\operatorname{conv} V = \operatorname{aff}_{+}(\emptyset, V).$$

Lemma 2.39 *If* $V \subset \mathbb{R}^N$ *is finite, then* $\operatorname{aff}_{\pm}(V, \emptyset) = \operatorname{aff}^0_{\pm}(V, \emptyset)$ *is the affine hull of* V.

Proof Both proofs are left as exercises for the reader. □

In the following definition of a cone, the point $\mathbf{v}$ is the apex, and V is a generating set for the positive directions. In the special case that V is a singleton $\{\mathbf{w}\}$, the cone gives a ray originating at $\mathbf{v}$ and passing through $\mathbf{w}$. Later chapters call a set of the form $\mathrm{aff}_+(\mathbf{v}, \{\mathbf{u}_1, \mathbf{u}_2\})$ a *blade*. Blades are planar sets bounded by two rays originating at $\mathbf{v}$.

Definition 2.40 (line, collinear, parallel) Any set of the form $\mathrm{aff}\{\mathbf{v}, \mathbf{w}\}$ is a *line* when $\mathbf{v} \neq \mathbf{w}$. A set that is contained in some $\mathrm{aff}\{\mathbf{v}, \mathbf{w}\}$ is *collinear*. If $\{\mathbf{0}, \mathbf{v}, \mathbf{w}\}$ is collinear, then $\mathbf{v}$ and $\mathbf{w}$ are said to be *parallel*. Also, $\{\mathbf{v}, \mathbf{w}\}$ is said to be a parallel set.

Definition 2.41 (plane, half plane, coplanar) An affine hull $A = \mathrm{aff}\{\mathbf{u}, \mathbf{v}, \mathbf{w}\}$ is a *plane* when $\{\mathbf{u}, \mathbf{v}, \mathbf{w}\}$ is not collinear. A set $\mathrm{aff}_\pm(\{\mathbf{u}, \mathbf{v}\}, \{\mathbf{w}\})$ is a *half-plane* when $\{\mathbf{u}, \mathbf{v}, \mathbf{w}\}$ is not collinear. A set that is contained in some $\mathrm{aff}\{\mathbf{u}, \mathbf{v}, \mathbf{w}\}$ is *coplanar*.

Definition 2.42 (half space) A set $\mathrm{aff}_\pm(\{\mathbf{u}, \mathbf{v}, \mathbf{w}\}, \{\mathbf{v}'\})$ is a *half-space*, when $\{\mathbf{u}, \mathbf{v}, \mathbf{w}, \mathbf{v}'\}$ is not coplanar. Under the substitution of $\mathrm{aff}_\pm$ for $\mathrm{aff}^0_\pm$, it is called an *open half-space*.

2.3.3 parallelepiped

The following polynomial, Δ, appears in many different functions related to the geometry of three dimensions. The formula following the definition shows that it is closely related to the square of the volume of a parallelepiped. The interpretation as volume is not relevant until the next chapter, but its nonnegativity is immediately relevant.

Definition 2.43 (Δ) Let

$$\begin{aligned}\Delta(x_1, \ldots, x_6) = {} & x_1x_4(-x_1 + x_2 + x_3 - x_4 + x_5 + x_6) \\ & + x_2x_5(x_1 - x_2 + x_3 + x_4 - x_5 + x_6) \\ & + x_3x_6(x_1 + x_2 - x_3 + x_4 + x_5 - x_6) \\ & - x_2x_3x_4 - x_1x_3x_5 - x_1x_2x_6 - x_4x_5x_6.\end{aligned}$$

Remark 2.44 (Cayley–Menger determinant) The polynomial Δ appears in the following context. Cayley and Menger found a formula for the square of the determinant D of the matrix with rows $\mathbf{v}_1-\mathbf{v}_0,\ldots,\mathbf{v}_n-\mathbf{v}_0$ for arbitrary vectors $\mathbf{v}_i \in \mathbb{R}^n$. Set

$$x_{ij} = \|\mathbf{v}_i - \mathbf{v}_j\|^2, \tag{2.45}$$

arranged as entries of a matrix $[x_{ij}]$. Write $\underline{1}$ for a row vector of length n with entries that are all equal to $1 \in \mathbb{R}$. They found that elementary matrix manipulations give an identity of determinants:

$$D^2 = \frac{(-1)^{n-1}}{2^n} \begin{vmatrix} [x_{ij}] & {}^t\underline{1} \\ \underline{1} & 0 \end{vmatrix}. \tag{2.46}$$

The right-hand side is a polynomial in the squares of the edge lengths.

A calculation of the determinant on the right when $n = 3$ yields the polynomial Δ.

$$4D^2 = \Delta(x_{01}, x_{02}, x_{03}, x_{23}, x_{13}, x_{12}).$$

The left-hand side is evidently a square and the polynomial on the right is nonnegative, whenever the variables x_{ij} satisfy (2.45) for some vectors $\mathbf{v}_0,\ldots,\mathbf{v}_3 \in \mathbb{R}^3$. Moreover, D and hence also Δ is positive when the set of four vectors is not coplanar.

Background 2.47 (matrix theory) *Very little matrix theory is required in this book. The next lemma is a rare exception. Its proof requires various very basic facts about* 3×3 *matrices and determinants. The determinant of a product of two matrices is the product of determinants. The transpose of a matrix A has the same determinant as A. The determinant of a matrix A is zero if and only if there exists a (row) vector* $\mathbf{u}$ *such that* $\mathbf{u}A = \mathbf{0}$.

Lemma 2.48 *Let* $V = \{\mathbf{v}_0, \mathbf{v}_1, \mathbf{v}_2, \mathbf{v}_3\} \subset \mathbb{R}^3$. *Let* $x_{ij} = \|\mathbf{v}_i - \mathbf{v}_j\|^2$. *Then* $\Delta(x_{ij}) \ge 0$. *Moreover, the set V is coplanar if and only if* $\Delta(x_{ij}) = 0$.

Proof The proof is an exercise in matrix theory and real arithmetic. (The statement also falls within the scope of Tarski arithmetic.) This lemma can be proved directly as follows, without recourse to the general Cayley–Menger theorem.

Let A be the 3×3 matrix with rows $\mathbf{v}_i - \mathbf{v}_0$. Then $D^2 = \det(A)^2 = \det(A\,{}^tA)$. Each entry of the product $A\,{}^tA$ is a dot product $(\mathbf{v}_i - \mathbf{v}_0)\cdot(\mathbf{v}_j - \mathbf{v}_0)$, which can be expressed in terms of the constants x_{ij} by the following identity:

$$\begin{aligned} 2(\mathbf{v}_i - \mathbf{v}_0)\cdot(\mathbf{v}_j - \mathbf{v}_0) &= (\mathbf{v}_i - \mathbf{v}_0)\cdot(\mathbf{v}_i - \mathbf{v}_0) + (\mathbf{v}_j - \mathbf{v}_0)\cdot(\mathbf{v}_j - \mathbf{v}_0) \\ &\quad - (\mathbf{v}_i - \mathbf{v}_j)\cdot(\mathbf{v}_i - \mathbf{v}_j) \\ &= x_{i0} + x_{j0} - x_{ij}. \end{aligned} \tag{2.49}$$

A computation of the determinant then gives $4D^2 = \Delta$. Thus, $D^2 \geq 0$ implies $\Delta \geq 0$.

Also $\Delta = 0$ if and only if $D = 0$, which holds if and only if $\mathbf{u} A = \mathbf{0}$ for some vector $\mathbf{u}$. By the definition of coplanar, this holds if and only if V is coplanar. □

Remark 2.50 The calculation of the general Cayley–Menger formula (2.46) for $n + 1$ points in $\mathbb{R}^n$ is based on the same method as the 3×3 case; the identity (2.49) gives a rewrite rule for each matrix entry of $\det(A\,{}^tA)$ as a linear combination of the variables x_{ij}. Row and column operations then put the matrix in a form in which each matrix entry is a single variable x_{ij}.

Remark 2.51 The volume of a 4-simplex in $\mathbb{R}^3$ is zero. This implies that Cayley–Menger determinant for $\mathbf{v}_0, \ldots, \mathbf{v}_4 \in \mathbb{R}^3$ is zero. This gives a polynomial relation between the $10 = \binom{5}{2}$ squared edge lengths x_{ij}. The relation is quadratic in the tenth edge, say $x = x_{04}$, expressing it in terms of the other nine. The leading coefficient of the quadratic polynomial is nonzero if $\{\mathbf{v}_1, \mathbf{v}_2, \mathbf{v}_3\}$ is not collinear.

2.4 Angle

Until now, the discussion of trigonometric functions has been purely analytic. This section interprets them geometrically. It covers fundamental identities in both Euclidean and spherical trigonometry, including the law of cosines, the law of sines, the spherical law of cosines, and a beautiful formula that Euler and Lagrange gave for the area of a spherical triangle.

If $\mathbf{v}, \mathbf{w}$ are nonzero vectors, then by the Cauchy–Schwarz inequality,

$$-1 \leq \frac{\mathbf{v} \cdot \mathbf{w}}{\|\mathbf{v}\|\,\|\mathbf{w}\|} \leq 1.$$

The middle term lies in the domain of the function arccos. The value of this function is the angle in the following definition.

Definition 2.52 (angle, arclength) Let $\mathbf{u}, \mathbf{v}, \mathbf{w}$ be vectors with $\mathbf{u} \neq \mathbf{v}, \mathbf{w}$. Define

$$\mathrm{arc}_V(\mathbf{u}, \{\mathbf{v}, \mathbf{w}\}) = \arccos\left(\frac{(\mathbf{v} - \mathbf{u}) \cdot (\mathbf{w} - \mathbf{u})}{\|\mathbf{v} - \mathbf{u}\|\,\|\mathbf{w} - \mathbf{u}\|}\right).$$

The value of this function is the *angle* at $\mathbf{u}$ formed by $\mathbf{v}$ and $\mathbf{w}$.

By the relation between arccos and $\arctan_2$ (Lemma 2.25),

$$\mathrm{arc}_V(\mathbf{0}, \{\mathbf{v}, \mathbf{w}\}) = \frac{\pi}{2} - \arctan_2\left(\sqrt{(\|\mathbf{v}\|^2\|\mathbf{w}\|^2 - (\mathbf{v} \cdot \mathbf{w})^2)}, \mathbf{v} \cdot \mathbf{w}\right). \tag{2.53}$$

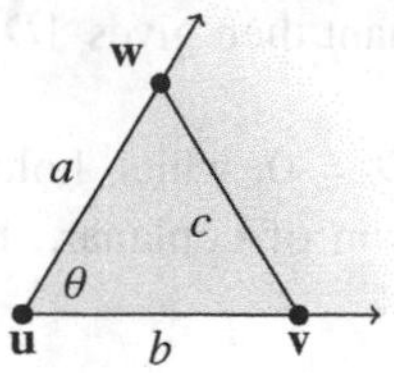

Figure 2.4 The angle $\theta = \mathrm{arc}_V(\mathbf{u}, \{\mathbf{v}, \mathbf{w}\}) = \mathrm{arc}(a, b, c)$.

The notation arc_V for angle comes from its interpretation as the length of a geodesic arc on a unit sphere centered at **u** from point **v** to **w**. The subscript V is a reminder that the function arguments are vectors. The function arc, without the subscript, gives the angle as a function of the three edge lengths of a triangle.

Definition 2.54 (arc) Define

$$\mathrm{arc}(a, b, c) = \arccos\left(\frac{a^2 + b^2 - c^2}{2ab}\right).$$

If the triangle inequalities hold:

$$a + b \geq c, \quad b + c \geq a, \quad c + a \geq b$$

and if $a, b > 0$, then

$$2ab = (\mp a + b + c)(a \mp b \pm c) \pm (a^2 + b^2 - c^2) \geq \pm(a^2 + b^2 - c^2)$$

and the argument of arccos in the definition of arc falls within its domain.

Lemma 2.55 (law of cosines) *Let* **u**, **v**, **w** *be vectors with* $\mathbf{v} \neq \mathbf{u}$, $\mathbf{w} \neq \mathbf{u}$. *Let* $a = \|\mathbf{w} - \mathbf{u}\|$, $b = \|\mathbf{v} - \mathbf{u}\|$, *and* $c = \|\mathbf{v} - \mathbf{w}\|$. *Let* $\gamma = \mathrm{arc}_V(\mathbf{u}, \{\mathbf{v}, \mathbf{w}\})$. *Then*

$$c^2 = a^2 + b^2 - 2ab\cos\gamma.$$

Also,

$$\mathrm{arc}_V(\mathbf{u}, \{\mathbf{v}, \mathbf{w}\}) = \mathrm{arc}(a, b, c).$$

Proof By the definition of arc_V, the definition of arccos, and (2.49),

$$2ab\cos\gamma = 2(\mathbf{w} - \mathbf{u}) \cdot (\mathbf{v} - \mathbf{u}) = a^2 + b^2 - c^2.$$

This identity can be solved for γ and gives the final statement of the lemma. □

Definition 2.56 (υ) Let υ (the symbol is a Greek upsilon, which is written with a wider stroke than a roman vee) be the polynomial

$$\upsilon(x, y, z) = -x^2 - y^2 - z^2 + 2xy + 2yz + 2zx.$$

This polynomial is nonnegative under conditions described by the following lemma.

Lemma 2.57 *Let $V = \{\mathbf{v}_0, \mathbf{v}_1, \mathbf{v}_2\} \subset \mathbb{R}^3$. Let $x_{ij} = \|\mathbf{v}_i - \mathbf{v}_j\|^2$. Then*

$$\upsilon(x_{01}, x_{12}, x_{02}) \geq 0.$$

Moreover, the set V is collinear if and only if $\upsilon(x_{ij}) = 0$.

Proof The polynomial factors

$$\upsilon(a^2, b^2, c^2) = 16s(s-a)(s-b)(s-c), \tag{2.58}$$

where $s = (a+b+c)/2$. If a, b, c are the lengths of the sides of a triangle, then $a, b, c \geq 0$ and the triangle inequality (Lemma 2.33) holds for all orderings of sides: $(b + c - a) \geq 0$ and so forth. Non-negativity $0 \leq \upsilon(a^2, b^2, c^2)$ follows from the triangle inequality applied to each factor in the factorization of υ: $2(s - a) = (b + c - a) \geq 0$ and so forth. The case of equality in the lemma is the case of equality in the triangle inequality. □

An alternative way to view nonnegativity is that υ, like Δ, is the square of a Cayley–Menger determinant (2.46).

$$0 \leq (2D)^2 = -\begin{vmatrix} 0 & a^2 & b^2 & 1 \\ a^2 & 0 & c^2 & 1 \\ b^2 & c^2 & 0 & 1 \\ 1 & 1 & 1 & 1 \end{vmatrix} = \upsilon(a^2, b^2, c^2).$$

Section 2.4.1 further identifies the determinant D as the norm of a cross product. Volume and area are the topics of the next chapter, but it is appropriate at this point to consider a formula for the area of a triangle. By means of formula (2.58) for υ, Heron's classical formula for the area of a triangle with sides a, b, c can be put in the form

$$\sqrt{\upsilon(a^2, b^2, c^2)}/4.$$

Lemma 2.59 (law of sines) *Assume that $a, b > 0$ and $a + b \geq c$, $b + c \geq a$, and $c + a \geq b$. Let $\gamma = \mathrm{arc}(a, b, c)$. Then*

$$2ab \sin \gamma = \sqrt{\upsilon(a^2, b^2, c^2)}.$$

Proof Both sides are nonnegative, so it is enough to check that their squares are equal. By the definition of arc, we have

$$4a^2b^2\sin^2\gamma = 4a^2b^2(1-\cos^2\gamma) = (4a^2b^2 - (a^2+b^2-c^2)^2) = \upsilon(a^2,b^2,c^2).$$

□

Another useful relation writes arc in terms of $\arctan_2$.

$$\mathrm{arc}(a,b,c) = \pi/2 - \arctan_2\Big(\sqrt{\upsilon(a^2,b^2,c^2)}, a^2+b^2-c^2\Big). \tag{2.60}$$

This follows directly from Lemma 2.25 and the definitions of arc and υ.

2.4.1 cross product

This book makes infrequent use of the cross product. A definition and the most basic properties suffice.

Definition 2.61 (cross product) Let $\mathbf{v} = (x,y,z)$ and $\mathbf{w} = (x',y',z')$. Let the cross product be defined by

$$\mathbf{v}\times\mathbf{w} = (yz'-y'z, zx'-xz', xy'-yx').$$

Lemma 2.62 *Any two vectors* $\mathbf{v},\mathbf{w}\in\mathbb{R}^3$ *satisfy*

$$\|\mathbf{v}\times\mathbf{w}\| = \|\mathbf{v}\|\,\|\mathbf{w}\|\sin\gamma,$$

where $\gamma = \mathrm{arc}_V(\mathbf{0},\{\mathbf{v},\mathbf{w}\})$. *Also,* $\mathbf{v}\cdot(\mathbf{v}\times\mathbf{w}) = \mathbf{w}\cdot(\mathbf{v}\times\mathbf{w}) = 0$.

Proof This proof is an exercise in real arithmetic and basic trigonometry. Both the left and right sides are nonnegative, so it is enough to compare the squares of both sides. The square of the left-hand side is

$$\begin{aligned}\|\mathbf{v}\times\mathbf{w}\|^2 &= (yz'-y'z)^2 + (zx'-xz')^2 + (xy'-yx')^2\\ &= (x^2+y^2+z^2)(x'^2+y'^2+z'^2) - (xx'+yy'+zz')^2\\ &= \|\mathbf{v}\|^2\|\mathbf{w}\|^2 - (\mathbf{v}\cdot\mathbf{w})^2 \qquad (2.63)\\ &= \|\mathbf{v}\|^2\|\mathbf{w}\|^2(1-\cos^2\gamma)\\ &= \|\mathbf{v}\|^2\|\mathbf{w}\|^2\sin^2\gamma.\end{aligned}$$

The second assertion of the lemma follows by arithmetic directly from the definitions of the dot and cross products. □

Lemma 2.64 *For any* $\mathbf{v},\mathbf{w}\in\mathbb{R}^3$, *the set* $\{\mathbf{0},\mathbf{v},\mathbf{w}\}$ *is collinear if and only if* $\mathbf{v}\times\mathbf{w} = \mathbf{0}$.

Proof By Equation (2.63),

$$\mathbf{v} \times \mathbf{w} = \mathbf{0} \quad \text{if and only if} \quad \|\mathbf{v}\| \, \|\mathbf{w}\| = |(\mathbf{v} \cdot \mathbf{w})|.$$

This is the case of equality in the Cauchy–Schwarz inequality, which is given as

$$\|\mathbf{v}\| \, \mathbf{w} = \pm \|\mathbf{w}\| \, \mathbf{v}.$$

This is equivalent to the collinearity of $\{\mathbf{0}, \mathbf{v}, \mathbf{w}\}$. □

Lemma 2.65

$$\mathbf{u} \times \mathbf{v} = -\mathbf{v} \times \mathbf{u}, \quad (\mathbf{u} \times \mathbf{v}) \cdot \mathbf{w} = (\mathbf{v} \times \mathbf{w}) \cdot \mathbf{u}, \quad (\mathbf{u} \times \mathbf{v}) \times \mathbf{w} = (\mathbf{u} \cdot \mathbf{w})\,\mathbf{v} - (\mathbf{v} \cdot \mathbf{w})\,\mathbf{u}.$$

Proof These are arithmetic consequences of the definition of cross product. □

2.4.2 dihedral angle

A dihedral angle of a tetrahedron is the angle formed between two of its facets. In general, the dihedral angle refers to the angle formed by two half-planes delimited by a common line. The dihedral angle is determined by a pair $\{\mathbf{v}_0, \mathbf{v}_1\}$ of points on the delimiting line and another pair $\{\mathbf{v}_2, \mathbf{v}_3\}$ of two points on the respective half-planes.

Definition 2.66 (dihedral angle) When $\mathbf{v}_0 \neq \mathbf{v}_1$, write $\mathrm{dih}_V(\{\mathbf{v}_0, \mathbf{v}_1\}, \{\mathbf{v}_2, \mathbf{v}_3\})$ for the angle $\gamma \in [0, \pi]$ formed by

$$\bar{\mathbf{w}}_2 = (\mathbf{w}_1 \cdot \mathbf{w}_1)\mathbf{w}_2 - (\mathbf{w}_1 \cdot \mathbf{w}_2)\mathbf{w}_1 \text{ and } \bar{\mathbf{w}}_3 = (\mathbf{w}_1 \cdot \mathbf{w}_1)\mathbf{w}_3 - (\mathbf{w}_1 \cdot \mathbf{w}_3)\mathbf{w}_1,$$

where $\mathbf{w}_i = \mathbf{v}_i - \mathbf{v}_0$. We call it the dihedral angle formed by $\mathbf{v}_2$ and $\mathbf{v}_3$ along $\{\mathbf{v}_0, \mathbf{v}_1\}$.

The subscript V is a reminder that the dihedral angle takes vector arguments. Later, a second version, without the subscript, computes the angle as a function of the lengths of edges of a tetrahedron. As the notation suggests, the dihedral angle depends only on the unordered pairs $\{\mathbf{v}_0, \mathbf{v}_1\}$, $\{\mathbf{v}_2, \mathbf{v}_3\}$.

The dihedral angle can be interpreted as the planar angle between two rays, obtained by projection of the two half-planes to a plane orthogonal to both of them. Up to positive scalars, $\bar{\mathbf{w}}_2$ and $\bar{\mathbf{w}}_3$ are the projections of $\mathbf{w}_2$ and $\mathbf{w}_3$ to the plane through the origin orthogonal to the vector $\mathbf{w}_1$. The dihedral angle is the angle between the projections $\bar{\mathbf{w}}_2$ and $\bar{\mathbf{w}}_3$ at $\mathbf{0}$.

Remark 2.67 The dihedral angle is unchanged if $\mathbf{w}_1$ is replaced with $t\mathbf{w}_1$ with $t \neq 0$. The dihedral angle is unchanged if $\mathbf{w}_2$ is replaced with $t_2\mathbf{w}_2 + t_1\mathbf{w}_1$

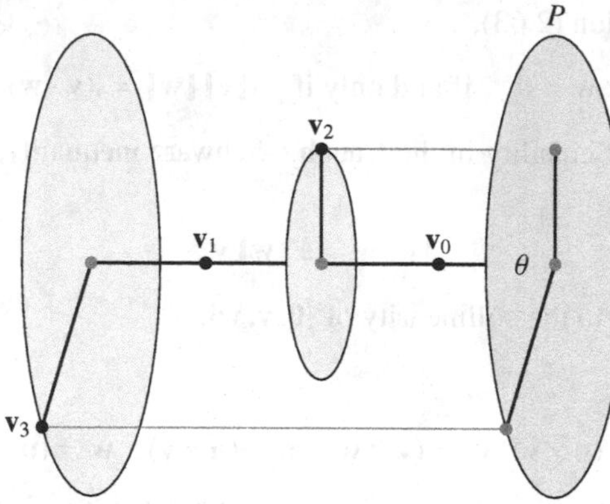

Figure 2.5 The dihedral angle $\theta = \mathrm{dih}_V(\{\mathbf{v}_0, \mathbf{v}_1\}, \{\mathbf{v}_2, \mathbf{v}_3\})$ is calculated by projection of $\mathbf{v}_2$ and $\mathbf{v}_3$ to a plane P with normal $\mathbf{v}_1 - \mathbf{v}_0$. The azimuth angle (Definition 2.79) is closely related to the dihedral angle, but depends on the ordering $(\mathbf{v}_0, \mathbf{v}_1, \mathbf{v}_2, \mathbf{v}_3)$ and takes values between 0 and 2π, unlike the dihedral angle, which takes values between 0 and π.

with $0 < t_2$ and t_1 arbitrary because such points project along the same ray. It is unchanged if $\mathbf{w}_3$ is replaced with $t_3\mathbf{w}_3 + t_1\mathbf{w}_1$ with $0 < t_3$ and t_1 arbitrary, because such points project along the same ray. In particular, the dihedral angle formed by $\mathbf{w}_2$ and $\mathbf{w}_3$ along $\{\mathbf{0}, \mathbf{w}_1\}$ is the same as that formed by $\mathbf{w}_2/\|\mathbf{w}_2\|$ and $\mathbf{w}_3/\|\mathbf{w}_3\|$ along $\mathbf{w}_1/\|\mathbf{w}_1\|$.

The dihedral angle is degenerate and is not be used when $\mathbf{w}_1 = \mathbf{0}$, $\bar{\mathbf{w}}_2 = \mathbf{0}$, or $\bar{\mathbf{w}}_3 = \mathbf{0}$. Equivalently, degeneracy occurs when $\{\mathbf{v}_0, \mathbf{v}_1, \mathbf{v}_2\}$ or $\{\mathbf{v}_0, \mathbf{v}_1, \mathbf{v}_3\}$ is a collinear set.

Lemma 2.68 *Let $\mathbf{v}_0, \ldots, \mathbf{v}_3 \in \mathbb{R}^3$ be given with $\mathbf{v}_0 \neq \mathbf{v}_1$. Then*

$$\mathrm{dih}_V(\{\mathbf{v}_0, \mathbf{v}_1\}, \{\mathbf{v}_2, \mathbf{v}_3\})$$

is the angle at $\mathbf{0}$ formed by $\mathbf{w}_1 \times \mathbf{w}_2$ and $\mathbf{w}_1 \times \mathbf{w}_3$, where $\mathbf{w}_i = \mathbf{v}_i - \mathbf{v}_0$.

Proof For any $\mathbf{u}, \mathbf{v}, \mathbf{w} \in \mathbb{R}^3$ with $\mathbf{u} \cdot \mathbf{w} = \mathbf{v} \cdot \mathbf{w} = 0$, we use Lemma 2.65 to compute

$$\begin{aligned}(\mathbf{u} \times \mathbf{w}) \cdot (\mathbf{v} \times \mathbf{w}) &= -(\mathbf{u} \times \mathbf{w}) \cdot (\mathbf{w} \times \mathbf{v}) \\ &= -\mathbf{v} \cdot ((\mathbf{u} \times \mathbf{w}) \times \mathbf{w}) \\ &= -\mathbf{v} \cdot (-(\mathbf{w} \cdot \mathbf{w})\mathbf{u}) \\ &= (\mathbf{u} \cdot \mathbf{v})(\mathbf{w} \cdot \mathbf{w}).\end{aligned}$$

That is, $* \times \mathbf{w}$ preserves dot products, up to a scalar $(\mathbf{w} \cdot \mathbf{w})$. Thus, if $\mathbf{w} \neq \mathbf{0}$, the angle formed by $\mathbf{u}$ and $\mathbf{v}$ is equal to the angle formed by $\mathbf{u} \times \mathbf{w}$ and $\mathbf{v} \times \mathbf{w}$.

The dihedral angle is the angle formed by

$$\bar{\mathbf{w}}_2 = (\mathbf{w}_1 \cdot \mathbf{w}_1)\mathbf{w}_2 - (\mathbf{w}_1 \cdot \mathbf{w}_2)\mathbf{w}_1 = (\mathbf{w}_1 \times \mathbf{w}_2) \times \mathbf{w}_1$$
$$\bar{\mathbf{w}}_3 = (\mathbf{w}_1 \cdot \mathbf{w}_1)\mathbf{w}_3 - (\mathbf{w}_1 \cdot \mathbf{w}_3)\mathbf{w}_1 = (\mathbf{w}_1 \times \mathbf{w}_3) \times \mathbf{w}_1.$$

Let $\mathbf{u} = \mathbf{w}_1 \times \mathbf{w}_2$, $\mathbf{v} = \mathbf{w}_1 \times \mathbf{w}_3$, and $\mathbf{w} = \mathbf{w}_1$. The preceding calculation shows that the angle formed by $\bar{\mathbf{w}}_2 = \mathbf{u} \times \mathbf{w}$ and $\bar{\mathbf{w}}_3 = \mathbf{v} \times \mathbf{w}$ is equal to the angle formed by $\mathbf{u}$ and $\mathbf{v}$. The lemma ensues. □

Lemma 2.69 (spherical law of cosines) *Let γ be the dihedral angle formed by $\mathbf{v}_2$ and $\mathbf{v}_3$ along $\{\mathbf{v}_0, \mathbf{v}_1\}$. Let a, b, and c be the angle at $\mathbf{v}_0$ between $\mathbf{v}_3$ and $\mathbf{v}_1$, $\mathbf{v}_2$ and $\mathbf{v}_1$, and $\mathbf{v}_2$ and $\mathbf{v}_3$, respectively. Assume that $\{\mathbf{v}_0, \mathbf{v}_1, \mathbf{v}_2\}$ and $\{\mathbf{v}_0, \mathbf{v}_1, \mathbf{v}_3\}$ are not collinear. Then*

$$\cos\gamma = \frac{\cos c - \cos a \cos b}{\sin a \sin b}.$$

Remark 2.70 The spherical law of cosines is the most fundamental identity of spherical trigonometry. A *spherical triangle* is a figure formed by three points on a unit sphere, together with three minimal geodesic arcs on the sphere that connect each pair of points. In the lemma, a, b, and c are the arclengths of the sides of a spherical triangle with vertices $\mathbf{v}_2/\|\mathbf{v}_2\|$, $\mathbf{v}_3/\|\mathbf{v}_3\|$, and $\mathbf{v}_1/\|\mathbf{v}_1\|$, when $\mathbf{v}_0 = \mathbf{0}$. See Figure 2.6. Also, γ measures the angle of the spherical triangle opposite the side c.

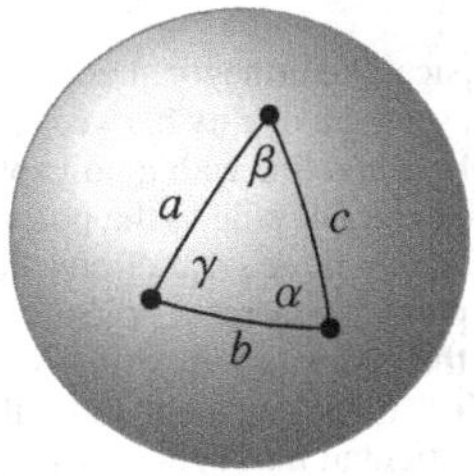

Figure 2.6 The spherical law of cosines gives the angle γ of a spherical triangle in terms of its edge lengths a, b, and c. The polar form of the spherical law of cosines gives the side c in terms of the angles α, β, and γ.

Proof The proof is an exercise based on previously established trigonometric identities. Let $\mathbf{w}_i = \mathbf{v}_i - \mathbf{v}_0$. An earlier remark states that the dihedral angle is unchanged if $\mathbf{w}_2$, $\mathbf{w}_3$, and $\mathbf{w}_1$ are replaced by $\mathbf{w}_2/\|\mathbf{w}_2\|$, $\mathbf{w}_3/\|\mathbf{w}_3\|$, $\mathbf{w}_1/\|\mathbf{w}_1\|$, respectively. Hence, we may assume without loss of generality that $\|\mathbf{w}_2\| = \|\mathbf{w}_3\| = \|\mathbf{w}_1\| = 1$.

Let $\bar{\mathbf{w}}_2$ and $\bar{\mathbf{w}}_3$ be the vectors in Definition 2.66. The law of cosines gives

$$\cos\gamma = \frac{\bar{\mathbf{w}}_2 \cdot \bar{\mathbf{w}}_3}{\|\bar{\mathbf{w}}_2\|\,\|\bar{\mathbf{w}}_3\|}.$$

The unit normalizations of $\mathbf{w}_3, \mathbf{w}_2, \mathbf{w}_1$ give

$$\|\bar{\mathbf{w}}_2\|^2 = \bar{\mathbf{w}}_2\cdot\bar{\mathbf{w}}_2 = (\mathbf{w}_2-(\mathbf{w}_1\cdot\mathbf{w}_2)\mathbf{w}_1)\cdot(\mathbf{w}_2-(\mathbf{w}_1\cdot\mathbf{w}_2)\mathbf{w}_1) = 1-(\mathbf{w}_1\cdot\mathbf{w}_2)^2 = \sin^2 b.$$

So $\|\bar{\mathbf{w}}_2\| = \sin b$. Similarly, $\|\bar{\mathbf{w}}_3\| = \sin a$. These calculations give the denominator in the spherical law of cosines. An expansion of the dot product gives the numerator:

$$\begin{aligned}\bar{\mathbf{w}}_2 \cdot \bar{\mathbf{w}}_3 &= (\mathbf{w}_2 - (\mathbf{w}_1 \cdot \mathbf{w}_2)\mathbf{w}_1) \cdot (\mathbf{w}_3 - (\mathbf{w}_1 \cdot \mathbf{w}_3)\mathbf{w}_1)\\ &= (\mathbf{w}_2 \cdot \mathbf{w}_3) - (\mathbf{w}_1 \cdot \mathbf{w}_2)(\mathbf{w}_1 \cdot \mathbf{w}_3)\\ &= \cos c - \cos a \cos b.\end{aligned}$$

The identity ensues. □

The spherical law of cosines gives the angles of a spherical triangle as a function of its sides. In spherical geometry, a duality[3] exists between angles and sides of a triangle. As a result, formulas in spherical trigonometry tend to come in pairs. The spherical law of cosines gives the angle of a spherical triangle as a function of its edge lengths. The polar form of the formula gives the edge length of a spherical triangle as a function of its angles. Up to signs, the polar formula has the same form as the law of cosines.

[3] In three-dimensional Euclidean space, the orthogonal complement of a plane through the origin is a line through the origin, giving a duality between planes and lines through the origin. The intersection of each plane and line with a unit sphere at the origin yields a duality between great circles and antipodal pairs of points (the poles of the great circle). The three edges of a spherical triangle ABC lie on three great circles that determine three antipodal pairs of points. From each of the three pairs, a coherent choice can be made between the two poles (with the preferred pole closer to the opposite vertex of ABC). These three poles are the vertices of the polar triangle $A'B'C'$. Each statement about the triangle ABC can be dualized to a statement about $A'B'C'$. In particular, the edges a, b, c and angles α, β, γ of ABC are related to those $a', b', \ldots$ of $A'B'C'$ by

$$a + \alpha' = \pi, \quad a' + \alpha = \pi,$$

and so forth.

Lemma 2.71 (spherical law of cosines - polar form) *Let* $\{\mathbf{v}_0, \mathbf{v}_1, \mathbf{v}_2, \mathbf{v}_3, \} \subset \mathbb{R}^3$. *Let* α, β, γ *be the dihedral angles:*

$$\alpha = \mathrm{dih}_V(\{\mathbf{v}_0, \mathbf{v}_2\}, \{\mathbf{v}_3, \mathbf{v}_1\})$$
$$\beta = \mathrm{dih}_V(\{\mathbf{v}_0, \mathbf{v}_3\}, \{\mathbf{v}_2, \mathbf{v}_1\})$$
$$\gamma = \mathrm{dih}_V(\{\mathbf{v}_0, \mathbf{v}_1\}, \{\mathbf{v}_3, \mathbf{v}_2\}).$$

Let c *be the angle between* $\mathbf{v}_2$ *and* $\mathbf{v}_3$ *at* $\mathbf{v}_0$. *Assume that* $\{\mathbf{v}_0, \mathbf{v}_2, \mathbf{v}_1\}$, $\{\mathbf{v}_0, \mathbf{v}_2, \mathbf{v}_3\}$, *and* $\{\mathbf{v}_0, \mathbf{v}_3, \mathbf{v}_1\}$ *are not collinear. Then*

$$\cos c = \frac{\cos\gamma + \cos\alpha\cos\beta}{\sin\alpha\sin\beta}.$$

Proof What follows is a direct computational proof that avoids polarity and is an application of established trigonometric identities. Let a be the angle between $\mathbf{v}_1$ and $\mathbf{v}_3$, and let b be the angle between $\mathbf{v}_1$ and $\mathbf{v}_2$ at $\mathbf{v}_0$. Let $A = \cos a$, $B = \cos b$, $C = \cos c$, $A' = \sin a$, $B' = \sin b$, $C' = \sin c$. The spherical law of cosines gives

$$\sin^2\beta = 1 - \left(\frac{B - AC}{A'C'}\right)^2 = \frac{p}{A'^2C'^2},$$

where $p = 1 - A^2 - B^2 - C^2 + 2ABC$. In particular, $p \geq 0$. A computation of $\sin^2\alpha$ and the remaining terms in the same way gives

$$\sin\alpha\sin\beta = \frac{p}{A'B'C'^2}$$

$$\cos\gamma + \cos\alpha\cos\beta = \frac{C - AB}{A'B'} + \frac{A - BC}{B'C'}\frac{B - AC}{A'C'} = \frac{pC}{A'B'C'^2}.$$

The result follows by real arithmetic. □

The following lemma gives a formula for the dihedral angle of a tetrahedron along an edge in terms of its edge lengths. The familiar polynomials υ and Δ appear once again.

Lemma 2.72 *Let* $\mathbf{v}_0, \mathbf{v}_1, \mathbf{v}_2, \mathbf{v}_3$ *be vectors with* $\{\mathbf{v}_0, \mathbf{v}_1, \mathbf{v}_2\}$ *not collinear, and* $\{\mathbf{v}_0, \mathbf{v}_1, \mathbf{v}_3\}$ *not collinear. Let* γ *be the dihedral angle formed by* $\mathbf{v}_2$ *and* $\mathbf{v}_3$ *along* $\{\mathbf{v}_0, \mathbf{v}_1\}$. *Let*

$$(x_1, \ldots, x_6) = (x_{01}, x_{02}, x_{03}, x_{23}, x_{13}, x_{12}), \text{ where } x_{ij} = \|\mathbf{v}_i - \mathbf{v}_j\|^2.$$

Let Δ_4 *be the partial derivative of* $\Delta(x_1, \ldots, x_6)$ *with respect to* x_4. *The dihedral angle* $\gamma = \mathrm{dih}_V(\{\mathbf{v}_0, \mathbf{v}_1\}, \{\mathbf{v}_2, \mathbf{v}_3\})$ *is given by*

$$\gamma = \arccos\left(\frac{\Delta_4(x_1, \ldots, x_6)}{\sqrt{\upsilon(x_1, x_2, x_6)\upsilon(x_1, x_3, x_5)}}\right).$$

It is also given by

$$\gamma = \frac{\pi}{2} - \arctan_2\left(\sqrt{4x_1\Delta(x_1, \ldots, x_6)}, \Delta_4(x_1, \ldots, x_6)\right).$$

Proof We use the notation $\mathbf{w}_i, \bar{\mathbf{w}}_i$ established in Definition 2.66. Let $\beta = \mathrm{arc}_V(\mathbf{v}_0, \{\mathbf{v}_1, \mathbf{v}_2\})$. The assumptions give $\bar{\mathbf{w}}_2 \neq \mathbf{0}$ and $\bar{\mathbf{w}}_3 \neq \mathbf{0}$. By expanding definitions and dot products and by the law of sines,

$$\bar{\mathbf{w}}_2 \cdot \bar{\mathbf{w}}_2 = (\mathbf{w}_1 \cdot \mathbf{w}_1)((\mathbf{w}_1 \cdot \mathbf{w}_1)(\mathbf{w}_2 \cdot \mathbf{w}_2) - (\mathbf{w}_1 \cdot \mathbf{w}_2)^2) = x_1^2 x_2 \sin^2\beta = \frac{1}{4}x_1\upsilon(x_1, x_2, x_6).$$

Similarly,

$$\bar{\mathbf{w}}_3 \cdot \bar{\mathbf{w}}_3 = \frac{1}{4}x_1\upsilon(x_1, x_3, x_5)$$

and by dot product formula (2.49),

$$\begin{aligned}\bar{\mathbf{w}}_2 \cdot \bar{\mathbf{w}}_3 &= (\mathbf{w}_1 \cdot \mathbf{w}_1)((\mathbf{w}_1 \cdot \mathbf{w}_1)(\mathbf{w}_2 \cdot \mathbf{w}_3) - (\mathbf{w}_1 \cdot \mathbf{w}_2)(\mathbf{w}_1 \cdot \mathbf{w}_3)) \\ &= x_1\left(\frac{x_1(x_2 + x_3 - x_1)}{2} - \frac{(x_1 + x_2 - x_6)(x_1 + x_3 - x_5)}{4}\right) \\ &= x_1\Delta_4(x_1, \ldots, x_6)/4.\end{aligned}$$

The result follows in terms of arccos.

The translation to $\arctan_2$ uses the arccos-$\arctan_2$ identity (Lemma 2.25) and the following polynomial identity

$$\upsilon(x_1, x_2, x_6)\upsilon(x_1, x_3, x_5) - \Delta_4(x_1, \ldots, x_6)^2 = 4x_1\Delta(x_1, \ldots, x_6).$$

□

2.4.3 Euler triangle

The expression $\alpha_1 + \alpha_2 + \alpha_3 - \pi$ is *Girard's formula* (known first to Harriot) for the area of a spherical triangle with angles α_1, α_2, α_3. We return to this formula in the next chapter (3.24), when area and volume are treated. Although the statement and proof do not explicitly mention area, the following lemma can be interpreted as an alternative formula discovered by Euler and Lagrange for the area of a spherical triangle.

Lemma 2.73 (Euler triangle) *Let $\mathbf{v}_0, \mathbf{v}_1, \mathbf{v}_2, \mathbf{v}_3$ be points in $\mathbb{R}^3$. Let*

$$(y_1, \ldots, y_6) = (y_{01}, y_{02}, y_{03}, y_{23}, y_{13}, y_{12}), \text{ where } y_{ij} = \|\mathbf{v}_i - \mathbf{v}_j\|.$$

Set $x_i = y_i^2$. and

$$p = y_1 y_2 y_3 + y_1(\mathbf{w}_2 \cdot \mathbf{w}_3) + y_2(\mathbf{w}_1 \cdot \mathbf{w}_3) + y_3(\mathbf{w}_1 \cdot \mathbf{w}_2).$$

where $\mathbf{w}_i = \mathbf{v}_i - \mathbf{v}_0$. Let

$$\alpha_i = \mathrm{dih}_V(\{\mathbf{v}_0, \mathbf{v}_i\}, \{\mathbf{v}_j, \mathbf{v}_k\})$$

where $\{i, j, k\} = \{1, 2, 3\}$. Assume that $\Delta(x_1, \ldots, x_6) > 0$. Then

$$\alpha_1 + \alpha_2 + \alpha_3 - \pi = \pi - 2\arctan_2(\Delta(x_1, \ldots, x_6)^{1/2}, 2p).$$

Before we jump into the details of the proof, it helps to understand why a formula of this general form should exist. Each angle α_i equals a single arctangent (Lemma 2.72). The addition law for arctangent, which is obtained by inverting the additional law for the tangent (Lemma 2.15), rewrites the sum $\alpha_1+\alpha_2+\alpha_3$ of arctangents as a single arctangent, or as twice a single arctangent if the double angle formula is invoked. Euler's formula is a precise formula for the sum of arctangents in the form $2\arctan_2(\cdots)$.

In practice, it is easier to carry out the details of the proof by a slightly different strategy. We can check that the derivatives of the two sides of the identity are equal as rational functions. The domain is connected, and from this it follows that the two sides differ by at most a constant. By calculating a particular test value, we see that the two sides are precisely equal.

Proof This proof is an exercise in real analysis and established trigonometric identities. According to an earlier remark, the dihedral angles are unchanged if the vectors $\mathbf{w}_i$ are rescaled so that $\|\mathbf{w}_i\| = 1$. By inspection, the given formula is also unchanged under rescalings: the factor p is homogeneous of degree three under a change $\mathbf{w}_i \mapsto t\mathbf{w}_i$ for $t > 0$, and so is $\sqrt{\Delta}$ by the formula for Δ. Thus, without loss of generality, $\|\mathbf{w}_i\| = 1$ for $i = 1, 2, 3$. Consequently, $y_1 = y_2 = y_3 = 1$. It is convenient to use different notation $a = x_4$, $b = x_5$, $c = x_6$ for the other variables. The expansion of the dot products in p by the dot product law gives

$$2p = 8 - (a + b + c).$$

Also, the definitions of Δ and υ give

$$\Delta(x_1, \ldots, x_6) = \Delta(1, 1, 1, a, b, c) = \upsilon(a, b, c) - abc.$$

Since $\Delta > 0$ by assumption, the arctangent formula in Lemma 2.72 applies for the dihedral angles α_i. After this substitution (and clearing a factor of three), the desired identity takes the form $f(a, b, c) = 0$, where

$$f(a, b, c) = -\pi/2 - \sum_{i=1}^{3} \arctan\left(u_i/\sqrt{\Delta}\right) + 2\arctan\left(2p/\sqrt{\Delta}\right),$$

for some rational functions u_i of a, b, c. The aim is to prove this trig identity holds whenever $\Delta > 0$.

To see that the function f does not depend on a, we fix (b, c) and differentiate f with respect to a. The partial derivative $\partial f/\partial a$ has the form $g(a, b, c)/\sqrt{\Delta}$ for some rational function g of a, b, c. The denominator of g has no real zero. Algebraic simplification of this rational function shows that the polynomial numerator of $g(a, b, c)$ is identically zero. (Euler himself did not shun brute force [11].)

By real analysis, the derivative of f is zero, and the function f is constant along any segment in $\mathbb{R}^3$ along which Δ is positive. The remaining part of the proof constructs two segments along which Δ is positive. The first connects $f(a, b, c)$ to $f(a, 2, 2)$, provided the variables are ordered appropriately. The second connects $(a, 2, 2)$ to $(2, 2, 2)$. From this construction it follows that $f(a, b, c) = f(2, 2, 2)$. The last step is to evaluate the constant $f(2, 2, 2)$. Arithmetic gives $\Delta = 4$, $2p = 2$, $u_1 = u_2 = u_3 = 0$, when $a = b = c = 2$. Finally,

$$f(a, b, c) = f(2, 2, 2) = -\pi/2 + 2\arctan(1) = 0.$$

Let us return to the construction of the two segments. By the triangle inequality, $a = \|\mathbf{v}_2 - \mathbf{v}_3\|^2 \leq (\|\mathbf{v}_2 - \mathbf{v}_0\| + \|\mathbf{v}_3 - \mathbf{v}_0\|)^2 = 4$. If equality holds, then $\{\mathbf{v}_0, \mathbf{v}_2, \mathbf{v}_3\}$ is collinear and $\{\mathbf{v}_0, \ldots, \mathbf{v}_4\}$ is coplanar. From this it follows that $\Delta = 0$, which is contrary to assumption. Similarly, $a = 0$ implies that $\Delta = 0$. Hence $0 < a < 4$. Similarly, $0 < b < 4$ and $0 < c < 4$. By the *pigeonhole* principle, two of the real numbers a, b, c must lie in the same subinterval $[0, 2]$ or $[2, 4]$. To fix notation, assume that b and c lie in the same subinterval.

The polynomial Δ is positive[4] *on the linear segment from* (a, b, c) *to* $(a, 2, 2)$. Indeed, for $0 \leq t \leq 1$, Tarski arithmetic gives

$$\begin{aligned}
&\Delta(1, 1, 1, a, b(1 - t) + 2t, c(1 - t) + 2t) \\
&\qquad = \Delta(1, 1, 1, a, b, c) + t(2 - t)(a(b - 2)(c - 2) + (b - c)^2) \\
&\qquad \geq \Delta(1, 1, 1, a, b, c) \\
&\qquad > 0.
\end{aligned}$$

The polynomial Δ is positive on the linear segment from $(a, 2, 2)$ *to* $(2, 2, 2)$.

[4] This paragraph follows the book's general convention of typesetting claims in italic.

Indeed,

$$\Delta(1, 1, 1, a, 2, 2) = a(4 - a) > 0.$$

The rest of the proof has been sketched above. □

2.5 Coordinates

This section establishes the existence and basic properties of the standard coordinate systems: polar coordinates, spherical coordinates, and cylindrical coordinates.

2.5.1 azimuth angle

For every pair of real numbers x and y, there are real numbers r and θ such that

$$x = r\cos\theta, \quad y = r\sin\theta. \tag{2.74}$$

If x and y are both zero, then take $r = 0$, and (2.74) holds for all choices of θ. If x and y are not both zero, then take $0 < r$, and θ is uniquely determined (up to multiples of 2π). By convention, we take $0 \le \theta < 2\pi$.

Definition 2.75 (frame, positive, adapted) A tuple $(\mathbf{e}_1, \mathbf{e}_2, \mathbf{e}_3)$ of vectors in $\mathbb{R}^3$ is a *frame* if $\mathbf{e}_i \cdot \mathbf{e}_j$ and $\|\mathbf{e}_i\| = 1$ for all i and j. A tuple $(\mathbf{e}_1, \mathbf{e}_2, \mathbf{e}_3)$ is *positive* if $(\mathbf{e}_1 \times \mathbf{e}_2) \cdot \mathbf{e}_3 = 1$. A tuple $(\mathbf{e}_1, \mathbf{e}_2, \mathbf{e}_3)$ is *adapted* to $(\mathbf{v}_0, \mathbf{v}_1, \mathbf{v}_2)$ if

$$\mathbf{e}_1 = (\mathbf{v}_1 - \mathbf{v}_0)/\|\mathbf{v}_0 - \mathbf{v}_1\| \quad \text{and} \quad \mathbf{e}_2 \in \mathrm{aff}_+^0(\{\mathbf{v}_0, \mathbf{v}_1\}, \mathbf{v}_2).$$

.

Lemma 2.76 (orthonormalization) *Assume that the set* $\{\mathbf{v}_0, \mathbf{v}_1, \mathbf{v}_2\} \subset \mathbb{R}^3$ *is not collinear. Then the unique positive frame adapted to* $\{\mathbf{v}_0, \mathbf{v}_1, \mathbf{v}_2\}$ *is* $(\mathbf{e}_1, \mathbf{e}_2, \mathbf{e}_3)$, *where*

$$\begin{aligned}
\mathbf{e}_1 &= \mathbf{w}_1/\|\mathbf{w}_1\|, \\
\mathbf{e}_2 &= \bar{\mathbf{w}}_2/\|\bar{\mathbf{w}}_2\|, \quad \bar{\mathbf{w}}_2 = \mathbf{w}_2 - (\mathbf{e}_1 \cdot \mathbf{w}_2)\mathbf{e}_1, \\
\mathbf{e}_3 &= \mathbf{e}_1 \times \mathbf{e}_2,
\end{aligned}$$

and where $\mathbf{w}_i = \mathbf{v}_i - \mathbf{v}_0$.

Proof It follows by basic vector arithmetic that $(\mathbf{e}_1, \mathbf{e}_2, \mathbf{e}_3)$ is a positive frame adapted to $\{\mathbf{v}_0, \mathbf{v}_1, \mathbf{v}_2\}$. The choices of vectors $\mathbf{e}_1$ and $\mathbf{e}_2$ are dictated by the definition of adapted frame. The choice of $\mathbf{e}_3$ is dictated by the definition of positive frame. □

Lemma 2.77 (cylindrical coordinates) *Let* $\mathbf{v}_0$ *and* $\mathbf{v}_1$ *be distinct points in* $\mathbb{R}^3$. *Let* $(\mathbf{e}_1, \mathbf{e}_2, \mathbf{e}_3)$ *be a positive frame where* $\mathbf{e}_1 = (\mathbf{v}_1 - \mathbf{v}_0)/\|\mathbf{v}_1 - \mathbf{v}_0\|$. *Then every* $\mathbf{p} \in \mathbb{R}^3$ *that is not in the line* $\mathrm{aff}(\mathbf{v}_0, \mathbf{v}_1)$ *can be uniquely expressed in the form*

$$\mathbf{p} = \mathbf{v}_0 + r\cos\psi\,\mathbf{e}_2 + r\sin\psi\,\mathbf{e}_3 + h(\mathbf{v}_1 - \mathbf{v}_0),$$

for some $0 < r$, $0 \le \psi < 2\pi$, $h \in \mathbb{R}$. *Furthermore, assume that* $\mathbf{p}_1$ *and* $\mathbf{p}_2$ *do not lie in the line* $\mathrm{aff}(\mathbf{v}_0, \mathbf{v}_1)$. *Then there exist unique* $\psi, \theta, r_1, r_2, h_1, h_2$ *such that* $0 \le \psi < 2\pi$, $0 \le \theta < 2\pi$, $0 < r_1$, $0 < r_2$, *and*

$$\begin{aligned}\mathbf{p}_1 &= \mathbf{v}_0 + r_1\cos\psi\,\mathbf{e}_2 + r_1\sin\psi\,\mathbf{e}_3 + h_1(\mathbf{v}_1 - \mathbf{v}_0),\\ \mathbf{p}_2 &= \mathbf{v}_0 + r_2\cos(\psi+\theta)\,\mathbf{e}_2 + r_2\sin(\psi+\theta)\,\mathbf{e}_3 + h_2(\mathbf{v}_1 - \mathbf{v}_0).\end{aligned}$$

Finally, the angle θ *is independent of the choice of* $\mathbf{e}_2, \mathbf{e}_3$ *giving the positive frame.*

The degenerate point $\mathbf{p} \in \mathrm{aff}\{\mathbf{v}_0, \mathbf{v}_1\}$ is excluded from the lemma. Nevertheless, it too has a cylindrical coordinate representation of the form $\mathbf{p} = \mathbf{v}_0 + h(\mathbf{v}_1 - \mathbf{v}_0)$ (with $r = 0$). Only uniqueness fails, because every θ gives the same representation.

Remark 2.78 The reader should carefully note the indexing of the vectors in the orthonormal frame as it appears in the cylindrical coordinate system. This book breaks with tradition by making h the coefficient of the frame vector $\mathbf{e}_1$ (rather than $\mathbf{e}_3$) and makes a corresponding change in spherical coordinates. This nontraditional order is better suited to the definition of dihedral angle, the arguments of which are grouped in pairs $\mathrm{dih}_V(\{\mathbf{v}_0, \mathbf{v}_1\}, \{\mathbf{v}_2, \mathbf{v}_3\})$ to emphasize the symmetries $\mathbf{v}_0 \leftrightarrow \mathbf{v}_1$ and $\mathbf{v}_2 \leftrightarrow \mathbf{v}_3$. Under this pairing of arguments, the axis of the dihedral angle is the line $\mathrm{aff}\{\mathbf{v}_0, \mathbf{v}_1\}$, which gives the direction $\mathbf{v}_1 - \mathbf{v}_0$ of the cylinder.

Definition 2.79 (azim) Define $\mathrm{azim}(\mathbf{v}_0, \mathbf{v}_1, \mathbf{v}_2, \mathbf{v}_3)$, the *azimuth* angle (or *longitude*), to be the uniquely determined angle θ given by the previous lemma for the points $\mathbf{p}_1 = \mathbf{v}_2$ and $\mathbf{p}_2 = \mathbf{v}_3$. By convention, let the azimuth angle be 0 in the degenerate cases when $\{\mathbf{v}_0, \mathbf{v}_1, \mathbf{v}_2\}$ or $\{\mathbf{v}_0, \mathbf{v}_1, \mathbf{v}_3\}$ is collinear.

The azimuth and dihedral angles are closely related (Figure 2.5). The azimuth angle takes values between 0 and 2π, but the dihedral angle is never greater than π. The following lemma reveals that the azimuth angle is an oriented extension of the dihedral angle and is always equal to dih or 2π – dih.

Lemma 2.80 *Let* $\mathbf{v}_1 \ne \mathbf{v}_0$ *be a nonzero vectors in* $\mathbb{R}^3$. *Assume that* $\mathbf{v}_2$ *and* $\mathbf{v}_3$

do not lie in the line $\mathrm{aff}\{\mathbf{v}_0, \mathbf{v}_1\}$. *Let*

$$\gamma = \mathrm{dih}_V(\{\mathbf{v}_0, \mathbf{v}_1\}, \{\mathbf{v}_2, \mathbf{v}_3\}).$$

Then

$$\cos(\mathrm{azim}(\mathbf{v}_0, \mathbf{v}_1, \mathbf{v}_2, \mathbf{v}_3)) = \cos\gamma.$$

Proof For simplicity, take $\mathbf{w}_i = \mathbf{v}_i - \mathbf{v}_0$. Let $\bar{\mathbf{w}}_i = (\mathbf{w}_1 \cdot \mathbf{w}_1)\mathbf{w}_i - (\mathbf{w}_1 \cdot \mathbf{w}_i)\mathbf{w}_1$. From the assumptions of the lemma, $\bar{\mathbf{w}}_2 \neq 0$. Set $\mathbf{e}_2 = \bar{\mathbf{w}}_2/\|\bar{\mathbf{w}}_2\|$. Choose a unit vector $\mathbf{e}_3$ so that $(\mathbf{e}_2 \times \mathbf{e}_3) \cdot \mathbf{w}_1 > 0$ and $\mathbf{e}_2 \cdot \mathbf{e}_3 = \mathbf{w}_1 \cdot \mathbf{e}_3 = 0$. Write $\mathbf{w}_i$ in cylindrical coordinates as

$$\begin{aligned} \mathbf{w}_2 &= \quad r_1\mathbf{e}_2 \quad + h_1\mathbf{w}_1 \\ \mathbf{w}_3 &= r_2\cos\theta\,\mathbf{e}_2 + r_2\sin\theta\,\mathbf{e}_3 + h_2\mathbf{w}_1. \end{aligned}$$

The definition of azim gives $\mathrm{azim}(\mathbf{w}_0, \mathbf{w}_1, \mathbf{w}_2, \mathbf{w}_3) = \theta$. By definition, $\cos\gamma$ is the angle between $\bar{\mathbf{w}}_2$ and $\bar{\mathbf{w}}_3$. We compute

$$\begin{aligned} \bar{\mathbf{w}}_2 &= \quad \|\bar{\mathbf{w}}_2\|\mathbf{e}_2 \\ \bar{\mathbf{w}}_3 &= (\mathbf{w}_1 \cdot \mathbf{w}_1)r_2\cos\theta\,\mathbf{e}_2 + (\mathbf{w}_1 \cdot \mathbf{w}_1)r_2\sin\theta\,\mathbf{e}_3. \end{aligned}$$

The result $\cos\theta = \cos\gamma$ is now a result of the definition of angle (Definition 2.52). □

The previous lemma identifies the cosine of the azimuth angle. The final lemma of this subsection determines the sign of its sine.

Lemma 2.81 *Write* $x \sim y$ *when there exists* $t > 0$ *such that* $x = ty$. *Then*

$$\sin(\mathrm{azim}(\mathbf{0}, \mathbf{v}_1, \mathbf{v}_2, \mathbf{v}_3)) \sim (\mathbf{v}_1 \times \mathbf{v}_2) \cdot \mathbf{v}_3.$$

Proof The relation $\sim$ is an equivalence relation. We may assume that $\{\mathbf{0}, \mathbf{v}_1, \mathbf{v}_2\}$ and $\{\mathbf{0}, \mathbf{v}_1, \mathbf{v}_3\}$ are not collinear sets, because otherwise both sides are zero. Let $(\mathbf{e}_1, \mathbf{e}_2, \mathbf{e}_3)$ be the positive frame adapted to $(\mathbf{0}, \mathbf{v}_1, \mathbf{v}_2)$. Write $\mathbf{v}_3 = r\cos\theta\,\mathbf{e}_2 + r\sin\theta\,\mathbf{e}_3 + h\,\mathbf{e}_1$ in cylindrical coordinates, where $\theta = \mathrm{azim}(\mathbf{0}, \mathbf{v}_1, \mathbf{v}_2, \mathbf{v}_3)$. Then by the explicit formulas for the positive frame,

$$\begin{aligned} (\mathbf{v}_1 \times \mathbf{v}_2) \cdot \mathbf{v}_3 &\sim (\mathbf{e}_1 \times \mathbf{v}_2) \cdot \mathbf{v}_3 \\ &\sim (\mathbf{e}_1 \times \mathbf{e}_2) \cdot \mathbf{v}_3 \\ &= \mathbf{e}_3 \cdot \mathbf{v}_3 \\ &= r\sin\theta \\ &\sim \sin\theta. \end{aligned}$$

□

2.5.2 zenith angle

The following lemma identifies the *zenith* angle ϕ. Because it is easily expressed in terms of the more basic function arc_V, there is little need to refer to it directly.

Lemma 2.82 (zenith) *Let* $(\mathbf{v}_0, \mathbf{v}_1)$ *be an ordered pair of distinct points in* $\mathbb{R}^3$. *Let* $\mathbf{v}_2 \neq \mathbf{v}_0$. *Set* $\phi = \mathrm{arc}_V(\mathbf{v}_0, \{\mathbf{v}_2, \mathbf{v}_1\}) \in [0, \pi]$. *Let* $\mathbf{e}_1$ *be the unit vector* $(\mathbf{v}_1 - \mathbf{v}_0)/\|\mathbf{v}_1 - \mathbf{v}_0\|$. *Let* $r = \|\mathbf{v}_2 - \mathbf{v}_0\|$. *Then* $\mathbf{v}_2$ *can be expressed in the form*

$$\mathbf{v}_2 = \mathbf{v}_0 + \mathbf{u} + r\cos\phi\, \mathbf{e}_1,$$

where $\mathbf{u} \cdot \mathbf{e}_1 = 0$. *The angle* ϕ *is called the zenith angle (or latitude) of* $\mathbf{v}_2$ *along* $(\mathbf{v}_0, \mathbf{v}_1)$.

Proof The lemma is a direct consequence of the definition of arc_V:

$$(\mathbf{v}_2 - \mathbf{v}_0) \cdot \mathbf{e}_1 = r\cos\phi.$$

□

Lemma 2.83 (spherical coordinates) *Assume that* $\{\mathbf{v}_0, \mathbf{v}_1, \mathbf{v}_2\} \subset \mathbb{R}^3$ *is not a collinear set. Let* $(\mathbf{e}_1, \mathbf{e}_2, \mathbf{e}_3)$ *be the positive frame adapted to* $(\mathbf{v}_0, \mathbf{v}_1, \mathbf{v}_2)$. *Then for any* $\mathbf{p}$,

$$\mathbf{p} = \mathbf{v}_0 + r\cos\theta\sin\phi\, \mathbf{e}_2 + r\sin\theta\sin\phi\, \mathbf{e}_3 + r\cos\phi\, \mathbf{e}_1, \tag{2.84}$$

where[5]

$$\begin{aligned} r &= \|\mathbf{v}_0 - \mathbf{p}\| \\ \phi &= \textit{zenith angle of } \mathbf{p} \textit{ along } (\mathbf{v}_0, \mathbf{v}_1) \\ \theta &= \mathrm{azim}(\mathbf{v}_0, \mathbf{v}_1, \mathbf{v}_2, \mathbf{p}). \end{aligned}$$

Proof Cylindrical coordinates give

$$\mathbf{p} = \mathbf{v}_0 + r'\cos\theta\, \mathbf{e}_2 + r'\sin\theta\, \mathbf{e}_3 + h\, \mathbf{e}_1,$$

for some h and $r' = \|\mathbf{p} - \mathbf{v}_0 - h\, \mathbf{e}_1\| \geq 0$. The zenith angle puts $\mathbf{p}$ in the form

$$\mathbf{p} = \mathbf{v}_0 + r'\cos\theta\, \mathbf{e}_2 + r'\sin\theta\, \mathbf{e}_3 + r\cos\phi\, \mathbf{e}_1,$$

where

$$\begin{aligned} r^2 &= \|\mathbf{p} - \mathbf{v}_0\|^2 \\ &= \|\mathbf{p} - \mathbf{v}_0 - h\, \mathbf{e}_1\|^2 + \|h\, \mathbf{e}_1\|^2 \\ &= (r')^2 + r^2\cos^2\phi, \end{aligned}$$

[5] This book follows the variable naming conventions (θ, ϕ) of American calculus textbooks, which reverses the international scientific notation.

Since $\sin\phi$, r, and r' are nonnegative, it follows that $r' = r\sin\phi$, as desired. □

Definition 2.85 (spherical coordinates) Equation (2.84) is called the spherical coordinate representation of $\mathbf{p}$ with respect to $(\mathbf{v}_0, \mathbf{v}_1, \mathbf{v}_2)$.

2.6 Cycle

The azimuth angle of the spherical coordinate system determines a cyclic permutation, called the azimuth cycle, on a finite set of points in $\mathbb{R}^3$, ordered according to increasing azimuth angle. The basic properties of that permutation are developed.

2.6.1 polar cycle

Let $V = \{\mathbf{v}_1, \ldots, \mathbf{v}_k\}$ be a finite set of nonzero points in the plane, with polar coordinates $\mathbf{v}_i = (r_i \cos\theta_i, r_i \sin\theta_i)$. It is useful to order the set of points according to increasing angle. To deal with degenerate cases when some points have exactly the same angle, order the points with the lexicographic order on their polar coordinates. We write $\mathbf{v}_i < \mathbf{v}_j$ for the total lexicographical order on points: $\theta_i < \theta_j$ or both $\theta_i = \theta_j$ and $r_i < r_j$. (The degenerate case of two equal angles does not occur in this book, but by defining a total order, there is no need to revisit the issue.) See Figure 2.7.

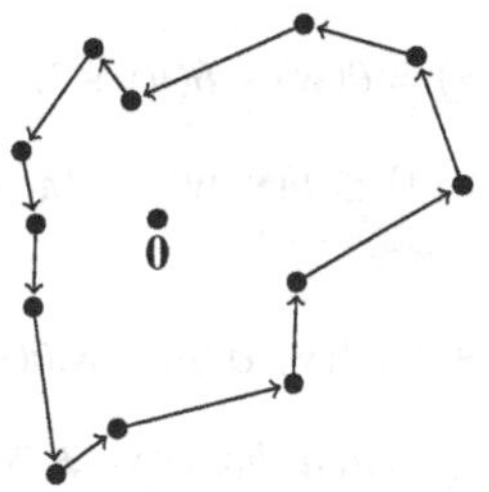

Figure 2.7 The polar cycle is the cyclic permutation of a finite set of nonzero points in the plane in a counterclockwise direction.

Definition 2.86 (polar cycle) A cyclic permutation $\sigma : V \to V$ sends $\mathbf{v} \in V$ to the next larger element with respect to this order or back to the first element if $\mathbf{v}$ is the largest. We call σ the *polar cycle* of the set V.

For $\psi \in \mathbb{R}$, let $T : \mathbb{R}^2 \to \mathbb{R}^2$ be the rotation of the plane:

$$(x, y) \mapsto (x\cos\psi + y\sin\psi, -x\sin\psi + y\cos\psi). \qquad (2.87)$$

Let σ' be the polar cycle for $T(V)$. Then

$$\sigma'(T\mathbf{v}) = T(\sigma\mathbf{v}), \qquad \text{for } \mathbf{v} \in V.$$

Lemma 2.88 *Let θ_i be real numbers such that $0 \leq \theta_i < 2\pi$ for $i = 1, 2$. Let*

$$\theta_{ji} = \theta_i - \theta_j + 2\pi k_{ji},$$

where integers k_{ij} satisfy $0 \leq \theta_{ji} < 2\pi$. Then

$$\theta_{12} + \theta_{21} = \begin{cases} 2\pi, & \text{if } \theta_i \neq \theta_j \\ 0, & \text{if } \theta_i = \theta_j. \end{cases}$$

Proof The proof is elementary. □

The next lemma gives a precise form to the observation that given a finite number of rays emanating from the origin in the plane, the sum of the included angles is 2π. In precise form, the polar cycle is used to place a cyclic order on the rays. There is a degenerate case when there is at most one ray.

Lemma 2.89 *Let $V \subset \mathbb{R}^2$ be a set of cardinality n that does not contain 0. Let σ be the polar cycle on V. In polar coordinates,*

$$\mathbf{v} = (\, r(\mathbf{v}) \cos\theta(\mathbf{v}),\ r(\mathbf{v}) \sin\theta(\mathbf{v})\,),$$

for $\mathbf{v} \in V$, with $0 \leq \theta(\mathbf{v}) < 2\pi$. Write

$$\theta(\mathbf{v}, \mathbf{w}) = \theta(\mathbf{w}) - \theta(\mathbf{v}) + 2\pi k_{pq},$$

for some integers k_{pq} that give $0 \leq \theta(\mathbf{v}, \mathbf{w}) < 2\pi$. Then for all $\mathbf{v} \in V$ and all $0 \leq i \leq j < n$,

$$\theta(\mathbf{v}, \sigma^i(\mathbf{v})) + \theta(\sigma^i(\mathbf{v}), \sigma^j(\mathbf{v})) = \theta(\mathbf{v}, \sigma^j(\mathbf{v})).$$

Moreover, if there exist $\mathbf{v}, \mathbf{w} \in V$ such that $\theta(\mathbf{v}) \neq \theta(\mathbf{w})$,

$$\sum_{i=0}^{n-1} \theta(\sigma^i\mathbf{v}, \sigma^{i+1}\mathbf{v}) = 2\pi.$$

(If $\theta(\mathbf{v}) = \theta(\mathbf{w})$ for all $\mathbf{v}, \mathbf{w} \in V$, then all the summands are zero.)

Proof Fix $\mathbf{v} \in V$. For $0 \leq i < n$, define θ_i by $\theta_0 = \theta(\mathbf{v})$ and

$$\theta_i = \theta(\sigma^i(\mathbf{v})) + 2\pi\ell_i,$$

where ℓ_i satisfies $\theta_0 \leq \theta_i < \theta_0 + 2\pi$. It follows from the definition of the polar cycle that $\theta_i \le \theta_j$ for $0 \le i \le j < n$. Then $\theta(\sigma^i \mathbf{v}, \sigma^j \mathbf{v}) = \theta_j - \theta_i$. The first conclusion of the lemma reduces to

$$(\theta_i - \theta_0) + (\theta_j - \theta_i) = (\theta_j - \theta_0).$$

The second conclusion reduces to

$$\sum_{i=0}^{n-2}(\theta_{i+1} - \theta_i) + \theta(\sigma^{n-1}\mathbf{v}, \mathbf{v}) = \theta(\mathbf{v}, \sigma^{n-1}\mathbf{v}) + \theta(\sigma^{n-1}\mathbf{v}, \mathbf{v}).$$

By the previous lemma, this is 0 or 2π. □

2.6.2 azimuth cycle

As already defined, the polar cycle is a cyclic permutation on a set of vectors in the plane that traverses them in order of increasing angle. What follows is the corresponding construction in three dimensional space. There is a cyclic permutation, called the *azimuth cycle*, on a set V of vectors in space that traverses them in order of increasing azimuth angle. Most of the work for this construction has already been done in the subsection on polar cycle, because the azimuth cycle may be constructed as the polar cycle on the projection of V to a plane. However, a nondegeneracy condition must be imposed on V to ensure that the projection to the plane is one-to-one. The following definition captures this nondegeneracy condition.

Definition 2.90 (cyclic set) Let $(\mathbf{v}_0, \mathbf{v}_1)$ be an ordered pair of distinct points in $\mathbb{R}^3$. Let V be a finite set of points in $\mathbb{R}^3$. We say that V is *cyclic* with respect to $(\mathbf{v}_0, \mathbf{v}_1)$ if the following two conditions hold.

1. If $\mathbf{u} = \mathbf{w} + h(\mathbf{v}_1 - \mathbf{v}_0)$, with $\mathbf{u}, \mathbf{w} \in V$ and $h \in \mathbb{R}$, then $\mathbf{u} = \mathbf{w}$.
2. The line through $\mathbf{v}_0$ and $\mathbf{v}_1$ does not meet V.

A cyclic set V has a well-defined azimuth cycle (Figure 2.8).

Definition 2.91 (azimuth cycle) Let $\mathbf{v}_0$ and $\mathbf{v}_1$ be distinct points in $\mathbb{R}^3$. Let V be a finite set of points in $\mathbb{R}^3$ that is cyclic with respect to $(\mathbf{v}_0, \mathbf{v}_1)$. Select $\mathbf{p} \in \mathbb{R}^3$ such that $\{\mathbf{v}_0, \mathbf{v}_1, \mathbf{p}\}$ is not collinear and let $\{\mathbf{e}_1, \mathbf{e}_2, \mathbf{e}_3\}$ be the corresponding positive, adapted, frame. Let f be the projection map:

$$\mathbf{v}_0 + x\,\mathbf{e}_2 + y\,\mathbf{e}_3 + z\,\mathbf{e}_1 \mapsto (x, y).$$

Let σ' be the polar cycle on $f(V)$. We define $\sigma : V \to V$ by $f\sigma(\mathbf{u}) = \sigma' f(\mathbf{u})$ and call σ the *azimuth cycle* on V with respect to $(\mathbf{v}_0, \mathbf{v}_1)$.

Because facts about the polar cycle lift to facts about the azimuth cycle, the next few lemmas follow naturally.

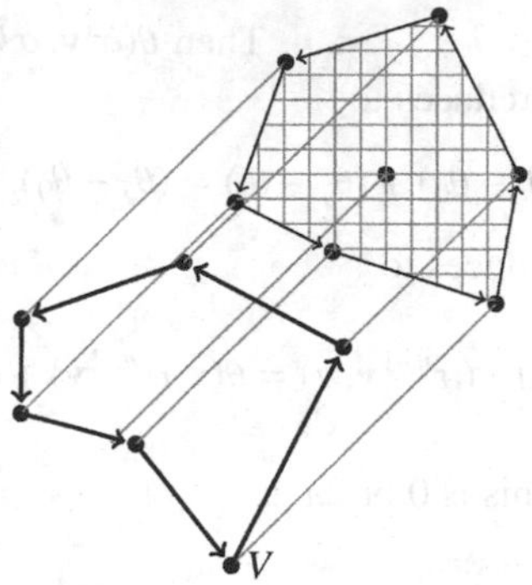

Figure 2.8 The azimuth cycle is a cyclic permutation of a finite set V of points in $\mathbb{R}^3$ that projects orthogonally to the polar cycle in the plane.

Lemma 2.92 *The azimuth cycle $\sigma : V \to V$ on a cyclic set V with respect to $(\mathbf{v}_0, \mathbf{v}_1)$ does not depend on the choice of $\mathbf{p} \in \mathbb{R}^3$ such that $\{\mathbf{v}_0, \mathbf{v}_1, \mathbf{p}\}$ is noncollinear.*

Proof The lemma follows from independence of σ' from rotations in the $\{\mathbf{e}_2, \mathbf{e}_3\}$ plane in (2.87). □

Lemma 2.93 *Let $(\mathbf{v}_0, \mathbf{v}_1)$ be an ordered pair of points in $\mathbb{R}^3$, with $\mathbf{v}_0 \neq \mathbf{v}_1$. Assume that $\{\mathbf{v}_2, \mathbf{v}_3\}$ is cyclic with respect to $(\mathbf{v}_0, \mathbf{v}_1)$. Then*

$$\mathrm{azim}(\mathbf{v}_0, \mathbf{v}_1, \mathbf{v}_2, \mathbf{v}_3) + \mathrm{azim}(\mathbf{v}_0, \mathbf{v}_1, \mathbf{v}_3, \mathbf{v}_2) = \begin{cases} 2\pi, & \text{if } \mathrm{azim}(\mathbf{v}_0, \mathbf{v}_1, \mathbf{v}_2, \mathbf{v}_3) \neq 0, \\ 0, & \text{if } \mathrm{azim}(\mathbf{v}_0, \mathbf{v}_1, \mathbf{v}_2, \mathbf{v}_3) = 0. \end{cases}$$

Proof The lemma follows immediately from Lemma 2.88. □

Lemma 2.94 *Let $(\mathbf{v}_0, \mathbf{v}_1)$ be an ordered pair of points in $\mathbb{R}^3$, with $\mathbf{v}_0 \neq \mathbf{v}_1$. Let V be a finite set in $\mathbb{R}^3$ of cardinality n that is cyclic with respect to $(\mathbf{v}_0, \mathbf{v}_1)$, with azimuth cycle σ. Then for all $\mathbf{u} \in V$, and all $0 \le i \le j < n$,*

$$\mathrm{azim}(\mathbf{v}_0, \mathbf{v}_1, \mathbf{u}, \sigma^i(\mathbf{u})) + \mathrm{azim}(\mathbf{v}_0, \mathbf{v}_1, \sigma^i(\mathbf{u}), \sigma^j(\mathbf{u})) = \mathrm{azim}(\mathbf{v}_0, \mathbf{v}_1, \mathbf{u}, \sigma^j(\mathbf{u})).$$

Moreover, if there exists $\mathbf{w} \in V$ such that $\mathrm{azim}(\mathbf{v}_0, \mathbf{v}_1, \mathbf{u}, \mathbf{w}) \neq 0$, then

$$\sum_{i=0}^{n-1} \mathrm{azim}(\mathbf{v}_0, \mathbf{v}_1, \sigma^i\mathbf{u}, \sigma^{i+1}\mathbf{u}) = 2\pi.$$

(If $\mathrm{azim}(\mathbf{v}_0, \mathbf{v}_1, \mathbf{u}, \mathbf{w}) = 0$ for all $\mathbf{w} \in V$, then all the summands are zero.)

Proof This follows immediately from Lemma 2.89. □

2.6.3 spherical triangle inequality

The geodesic length between two points $\mathbf{u}, \mathbf{v}$ on a unit sphere centered at $\mathbf{v}_0$ is $\mathrm{arc}_V(\mathbf{v}_0, \{\mathbf{u}, \mathbf{v}\})$. The following lemma is part of the verification that the function $d(\mathbf{u}, \mathbf{v}) = \mathrm{arc}_V(\mathbf{v}_0, \{\mathbf{u}, \mathbf{v}\})$ is a metric on the unit sphere. The lemma excludes the degenerate case when points on the sphere are antipodal.

Lemma 2.95 *Let $\{\mathbf{v}_0, \mathbf{v}_1, \mathbf{v}_2, \mathbf{v}_3\}$ be a set of four points in $\mathbb{R}^3$. Assume that $\mathbf{v}_0$ is not collinear with any pair of other points. Then*

$$\mathrm{arc}_V(\mathbf{v}_0, \{\mathbf{v}_1, \mathbf{v}_3\}) \le \mathrm{arc}_V(\mathbf{v}_0, \{\mathbf{v}_1, \mathbf{v}_2\}) + \mathrm{arc}_V(\mathbf{v}_0, \{\mathbf{v}_2, \mathbf{v}_3\}).$$

Equality occurs if and only if $\mathbf{v}_2 \in \mathrm{aff}_+(\mathbf{v}_0, \{\mathbf{v}_1, \mathbf{v}_3\})$.

Proof Let $\mathbf{v}_2'$ be the projection of $\mathbf{v}_2$ to the plane $\mathrm{aff}\{\mathbf{v}_0, \mathbf{v}_1, \mathbf{v}_3\}$. By the spherical law of cosines in the special case of a right triangle,

$$\cos\psi = \cos\beta \cos\alpha \le \cos\beta,$$

where $\psi = \mathrm{arc}_V(\mathbf{v}_0, \{\mathbf{v}_1, \mathbf{v}_2\})$, $\beta = \mathrm{arc}_V(\mathbf{v}_0, \{\mathbf{v}_1, \mathbf{v}_2'\})$, $\alpha = \mathrm{arc}_V(\mathbf{v}_0, \{\mathbf{v}_2, \mathbf{v}_2'\})$. Thus, $\mathrm{arc}_V(\mathbf{v}_0, \{\mathbf{v}_1, \mathbf{v}_2'\}) = \beta \le \psi = \mathrm{arc}_V(\mathbf{v}_0, \{\mathbf{v}_1, \mathbf{v}_2\})$. Similarly,

$$\mathrm{arc}_V(\mathbf{v}_0, \{\mathbf{v}_2', \mathbf{v}_3\}) \le \mathrm{arc}_V(\mathbf{v}_0, \{\mathbf{v}_2, \mathbf{v}_3\}).$$

Thus, it is enough to show that

$$\mathrm{arc}_V(\mathbf{v}_0, \{\mathbf{v}_1, \mathbf{v}_3\}) \le \mathrm{arc}_V(\mathbf{v}_0, \{\mathbf{v}_1, \mathbf{v}_2'\}) + \mathrm{arc}_V(\mathbf{v}_0, \{\mathbf{v}_2', \mathbf{v}_3\}).$$

The points $\mathbf{v}_0, \mathbf{v}_1, \mathbf{v}_3, \mathbf{v}_2'$ are coplanar. By the additivity of angles (Lemma 2.89), if $\mathbf{v}_2' \in \mathrm{aff}_+(\mathbf{v}_0, \{\mathbf{v}_1, \mathbf{v}_3\})$, then

$$\mathrm{arc}_V(\mathbf{v}_0, \{\mathbf{v}_1, \mathbf{v}_3\}) = \mathrm{arc}_V(\mathbf{v}_0, \{\mathbf{v}_1, \mathbf{v}_2'\}) + \mathrm{arc}_V(\mathbf{v}_0, \{\mathbf{v}_2', \mathbf{v}_3\}),$$

and otherwise,

$$\mathrm{arc}_V(\mathbf{v}_0, \{\mathbf{v}_1, \mathbf{v}_3\}) = \|\, \mathrm{arc}_V(\mathbf{v}_0, \{\mathbf{v}_1, \mathbf{v}_2'\}) - \mathrm{arc}_V(\mathbf{v}_0, \{\mathbf{v}_2', \mathbf{v}_3\}) \|.$$

The inequality ensues.

Further inspection shows that equality occurs exactly when $\alpha = 0$ and $\mathbf{v}_2' \in \mathrm{aff}_+(\mathbf{v}_0, \{\mathbf{v}_1, \mathbf{v}_3\})$. Equivalently, $\mathbf{v}_2' = \mathbf{v}_2 \in \mathrm{aff}_+(\mathbf{v}_0, \{\mathbf{v}_1, \mathbf{v}_3\})$. □

Lemma 2.96 *Let $\{\mathbf{v}_0, \mathbf{u}_0, \mathbf{u}_1, \mathbf{u}_2, \ldots, \mathbf{u}_r\}$ be a set of points in $\mathbb{R}^3$. Assume that no triple $\{\mathbf{v}_0, \mathbf{u}_i, \mathbf{u}_{i+1}\}$ is collinear. Assume that $\{\mathbf{v}_0, \mathbf{u}_0, \mathbf{u}_r\}$ is not collinear. Then*

$$\mathrm{arc}_V(\mathbf{v}_0, \{\mathbf{u}_0, \mathbf{u}_r\}) \le \sum_{i=0}^{r-1} \mathrm{arc}_V(\mathbf{v}_0, \{\mathbf{u}_i, \mathbf{u}_{i+1}\}).$$

Proof The proof is an easy induction on r with base case given by Lemma 2.95.

□

2.7 Chapter Summary

We give a brief chapter summary. The trigonometric functions cos, sin, arctan, arccos are defined in the standard way. The function $\arctan_2(x, y)$ is an extension of $\arctan(y/x)$ to every point (x, y) in the plane. It is the polar coordinate angle of (x, y).

$\mathbb{R}^N$ is the vector space of functions from the finite set N to $\mathbb{R}$. If $n \in \mathbb{R}$, then by convention, $\mathbb{R}^n = \mathbb{R}^N$, where $N = \{0, \ldots, n-1\}$. A bold face $\mathbf{u}, \mathbf{v}, \mathbf{p}, \mathbf{q}$ is used for points in $\mathbb{R}^N$. Vector space operations, the dot product $\mathbf{u} \cdot \mathbf{v}$, and the norm $\|\mathbf{u}\|$ are defined in the standard way.

We write $\text{aff}(S)$ for the affine hull of a set and $\text{conv}(S)$ for the convex hull of a set. The notation is extended to allow inequality constraints:

$$\text{aff}_{\pm}(V, V') = \{t_1\mathbf{v}_1 + \cdots t_n\mathbf{v}_n \ : \ t_1 + \cdots + t_n = 1, \pm t_j \geq 0, \text{ for } j > k\},$$
$$\text{aff}^0_{\pm}(V, V') = \{t_1\mathbf{v}_1 + \cdots t_n\mathbf{v}_n \ : \ t_1 + \cdots + t_n = 1, \pm t_j > 0, \text{ for } j > k\}.$$

Lines, planes, rays, cones, half-planes, half-spaces, and convex hulls can all be represented in this compact notation.

The function $\text{arc}_V(\mathbf{u}, \{\mathbf{v}, \mathbf{w}\})$ gives the angle at point $\mathbf{u}$ of a triangle with vertices $\mathbf{u}, \mathbf{v}, \mathbf{w}$. The function $\text{arc}(a, b, c)$ is the angle opposite c of a triangle with sides a, b, c. The function dih_V is the dihedral angle of a simplex, expressed as a function of its four vertices. The function dih is the dihedral angle of a simplex, expressed as a function of its six edges. The polynomials υ and Δ, which appear in formulas for volume, area, and angle, depend on three and six variables, respectively.

The cylindrical coordinates of a point in $\mathbb{R}^3$ are (r, θ, h). The spherical coordinates are (r, θ, ϕ). The angle θ is called the azimuth angle and is determined by four points $\mathbf{v}_0, \mathbf{v}_1, \mathbf{v}_2, \mathbf{v}_3$. The angle ϕ is the zenith angle. This book follows a nonstandard convention for the labeling of the coordinate axes in cylindrical and spherical coordinates: the central line of the cylinder and the line through the poles of the coordinate sphere lie in the direction of the first unit vector $\mathbf{e}_1$.

The cyclic permutation of a finite set of points in the plane, ordered by increasing angle in polar coordinates is called the polar cycle. The cyclic permutation σ of a finite set of points in three-dimensional space, ordered by increasing azimuth angle is called the azimuth cycle.

3

Volume

Summary. *In this chapter a general working knowledge of measure in three-dimensional Euclidean space is assumed. Nowhere does this book directly need integration. Measure alone suffices. Volume formulas for various classical solids are stated. Most of the volumes considered in this chapter (such as that of the ball, a rectangle, a tetrahedron, and a frustum) have been known since antiquity. This chapter describes the volumes of every three-dimensional solid that appears anywhere in the book.*

To keep the presentation as simple as possible, we avoid surface integrals. As an elementary substitute for surface integration on a sphere, this book systematically replaces subsets of the unit sphere with three-dimensional solids, called radial sets. This chapter presents basic properties of radial sets.

Finally, Section 3.3 uses the volume formula for cubes to estimate the number of integer lattice points in a ball of large radius.

3.1 Background in Measure

This book uses the concepts of null set, measurable set, and volume in three dimensions; the existence of these concepts with stated properties is assumed without proof. [1]

Definition 3.1 (vol, measurable, null) Let vol be the Lebesgue measure on

[1] Harrison has already implemented gauge integration, which is much more than what this book requires, inside the proof assistant HOL Light [24].

Euclidean space $\mathbb{R}^3$. A *null set* is a Lebesgue measurable subsets of $\mathbb{R}^3$ of measure zero. A *measurable* set in this book is a subset of $\mathbb{R}^3$ that is bounded and Lebesgue measurable.

Remark 3.2 The Lebesgue measure may be replaced with various alternatives (for example, Riemann or gauge) of measure in this definition. A list of three-dimensional solids (a rectangle, a ball, a tetrahedron, a frustum, and so forth), called *primitive regions*, is provided by Definition 3.22. These primitive regions are measurable (and have the same volume) with respect to almost any normalized translation invariant measure.

Lemma 3.3 (null set) *Null sets have the following properties.*

1. *A measurable subset of a null set is a null set.*
2. *A union of two null sets is a null set.*
3. *A plane is a null set.*
4. *A sphere is a null set.*
5. *A circular cone is a null set; that is, a union of all lines through a fixed point* $\mathbf{p}$ *and at a fixed angle to a given line through* $\mathbf{p}$.

Lemma 3.4 (measurable) *Measurability has the following properties.*

1. *Null sets are measurable.*
2. *Primitive regions are measurable (Definition 3.22).*
3. *The union of two measurable sets is measurable.*
4. *The intersection of two measurable sets is measurable.*
5. *The difference of two measurable sets is measurable.*

Lemma 3.5 (volume) *Volume has the following properties.*

1. (NONNEGATIVE) *The volume is defined for every measurable set. It is a nonnegative real number.*
2. (NULL SET) *The volume of a null set is zero.*
3. (NULL DIFFERENCE) *If X and Y are measurable, and if the symmetric difference of X and Y is contained in a null set, then X and Y have the same volume.*
4. (UNION) *If X and Y are measurable sets, and if $X \cap Y$ is contained in a null set, then*

$$\mathrm{vol}(X \cup Y) = \mathrm{vol}(X) + \mathrm{vol}(Y).$$

5. (LINEAR STRETCH) *For* $X \subset \mathbb{R}^3$, *and* $\mathbf{v} \in \mathbb{R}^3$, *set*

$$T_{\mathbf{v}}(X) = \{(v_1u_1, v_2u_2, v_3u_3) \ : \ \mathbf{u} \in X\}.$$

   If X *and* $T_{\mathbf{v}}(X)$ *are measurable, then* $\mathrm{vol}(T_{\mathbf{v}}(X)) = |v_1v_2v_3|\mathrm{vol}(X)$.
6. (TRANSLATION) *If* $X \subset \mathbb{R}^3$ *and* $\mathbf{v} \in \mathbb{R}^3$, *then let* $X + \mathbf{v} = \{\mathbf{p} + \mathbf{v} \ : \ \mathbf{p} \in X\}$. *If* X *is measurable, then* $X + \mathbf{v}$ *is as well, and* $\mathrm{vol}(X) = \mathrm{vol}(X + \mathbf{v})$.
7. (PRIMITIVES) *If* X *is a primitive region, then* $\mathrm{vol}(X)$ *is given by the formulas of Lemma 3.23 and Lemma 3.25.*

3.2 Primitive Volume

This book accepts certain elementary volume calculations as assumed background. These volumes are called primitive volumes. All further volumes calculations are obtained from these through the basic properties of measure.

To fix a convention, this book prefers to take the volume of open sets whenever that can be arranged. It begins with a description of some of the primitive regions.

3.2.1 radial set

Surface integrals are not required in this book. Although the *solid angle* is traditionally defined as a surface integral, it is simpler to give an alternative definition based on volume.

Definition 3.6 (ball) The open ball $B(\mathbf{v}, r)$ with center $\mathbf{v}$ and radius r is the set

$$\{\mathbf{q} \in \mathbb{R}^3 \ : \ \|\mathbf{v} - \mathbf{q}\| < r\}.$$

The closed ball $\bar{B}(\mathbf{v}, r)$ is

$$\{\mathbf{q} \in \mathbb{R}^3 \ : \ \|\mathbf{v} - \mathbf{q}\| \leq r\}.$$

Definition 3.7 (radial) A set C is *r-radial* at center $\mathbf{v}$ if the two conditions $C \subset B(\mathbf{v}, r)$ and $\mathbf{v} + \mathbf{u} \in C$ imply $\mathbf{v} + t\mathbf{u} \in C$ for all t satisfying $0 < \|\mathbf{u}\|t < r$. A set C is *eventually radial* at center $\mathbf{v}$ if $C \cap B(\mathbf{v}, r)$ is r-radial at center $\mathbf{v}$ for some $r > 0$.

Lemma 3.8 (radial intersection) *If* C *and* C' *are r-radial at* $\mathbf{v}$, *then* $C \cap C'$ *is also r-radial at* $\mathbf{v}$.

Lemma 3.9 (eventually radial) *If C is r-radial for some $r > 0$ then it is eventually radial.*

Lemma 3.10 (radial scale) *Assume that C is measurable and r-radial at* $\mathbf{v}$. *Let $0 \le r' < r$. Then $C \cap B(\mathbf{v}, r')$ is measurable and* $\mathrm{vol}(C \cap B(\mathbf{v}, r')) = \mathrm{vol}(C)(r'/r)^3$.

Proof The radial set C transforms into $C \cap B(\mathbf{v}, r')$ by a sequence of translations and linear stretches. □

Definition 3.11 (solid angle) When C is measurable and eventually radial at center $\mathbf{v}$, define the *solid angle* of C at $\mathbf{v}$ to be

$$\mathrm{sol}(\mathbf{v}, C) = 3\mathrm{vol}(C \cap B(\mathbf{v}, r))/r^3,$$

where r is as in the definition of eventually radial. By Lemma 3.10, this definition is independent of any such r. When the center $\mathbf{v}$ is clear from the context, write $\mathrm{sol}(C)$ for $\mathrm{sol}(\mathbf{v}, C)$.

Lemma 3.12 (solid angle) *If C is measurable and r-radial at* $\mathbf{v}$, *then the volume of C satisfies*

$$\mathrm{vol}(C) = \mathrm{sol}(\mathbf{v}, C)r^3/3.$$

Moreover, if C is bounded away from $\mathbf{v}$, *then C is eventually radial at* $\mathbf{v}$, *and* $\mathrm{sol}(\mathbf{v}, C) = 0$.

Proof These properties follow immediately from the definitions. □

3.2.2 wedge

Definition 3.13 (wedge) Assume that $\mathbf{v}_0 \ne \mathbf{v}_1$ and that $\mathbf{w}_1$ and $\mathbf{w}_2$ do not lie on the line $\mathrm{aff}\{\mathbf{v}_0, \mathbf{v}_1\}$.

$$W^0(\mathbf{v}_0, \mathbf{v}_1, \mathbf{w}_1, \mathbf{w}_2) = \{\mathbf{p} \;:\; 0 < \mathrm{azim}(\mathbf{v}_0, \mathbf{v}_1, \mathbf{w}_1, \mathbf{p}) < \mathrm{azim}(\mathbf{v}_0, \mathbf{v}_1, \mathbf{w}_1, \mathbf{w}_2)\},$$
$$W(\mathbf{v}_0, \mathbf{v}_1, \mathbf{w}_1, \mathbf{w}_2) = \{\mathbf{p} \;:\; 0 \le \mathrm{azim}(\mathbf{v}_0, \mathbf{v}_1, \mathbf{w}_1, \mathbf{p}) \le \mathrm{azim}(\mathbf{v}_0, \mathbf{v}_1, \mathbf{w}_1, \mathbf{w}_2)\}.$$

Both sets are called *wedges*.

Definition 3.14 (lune) The set $\mathrm{aff}^0_+(\{\mathbf{v}_0, \mathbf{v}_1\}, \{\mathbf{v}_2, \mathbf{v}_3\})$, called a *lune*, is given in Definition 2.35. It is the intersection of two open half-spaces

$$\mathrm{aff}^0_+(\{\mathbf{v}_0, \mathbf{v}_1\}, \{\mathbf{v}_2, \mathbf{v}_3\}) = \mathrm{aff}^0_+(\{\mathbf{v}_0, \mathbf{v}_1, \mathbf{v}_2\}, \mathbf{v}_3) \cap \mathrm{aff}^0_+(\{\mathbf{v}_0, \mathbf{v}_1, \mathbf{v}_3\}, \mathbf{v}_2).$$

A lune is a special case of a wedge, restricted to have an azimuth angle less than π. To specify a lune, the ordering of $\{\mathbf{v}_2, \mathbf{v}_3\}$ is irrelevant, just as it is in the dihedral angle $\mathrm{dih}(\{\mathbf{v}_0, \mathbf{v}_1\}, \{\mathbf{v}_2, \mathbf{v}_3\})$. To specify a wedge, the ordering of $(\mathbf{v}_0, \mathbf{v}_1, \mathbf{w}_1, \mathbf{w}_2)$ does matter. A reversal of $(\mathbf{w}_1, \mathbf{w}_2)$ complements the wedge (up to the shared boundary). The next lemma shows that a lune is indeed a special case of a wedge.

Lemma 3.15 (lune wedge) *Let* $\{\mathbf{v}_0, \mathbf{v}_1, \mathbf{v}_2, \mathbf{v}_3\}$ *be a set of four points in* $\mathbb{R}^3$. *Assume that the set is not coplanar. Assume that* $\mathrm{azim}(\mathbf{v}_0, \mathbf{v}_1, \mathbf{v}_2, \mathbf{v}_3) < \pi$. *Then,*

$$W^0(\mathbf{v}_0, \mathbf{v}_1, \mathbf{v}_2, \mathbf{v}_3) = \mathrm{aff}_+^0(\{\mathbf{v}_0, \mathbf{v}_1\}, \{\mathbf{v}_2, \mathbf{v}_3\}).$$

3.2.3 primitive types

Definition 3.16 (tetrahedron) A tetrahedron is a set of the form

$$\mathrm{TET}\{\mathbf{v}_1, \mathbf{v}_2, \mathbf{v}_3, \mathbf{v}_4\} = \mathrm{aff}_+^0(\emptyset, \{\mathbf{v}_1, \mathbf{v}_2, \mathbf{v}_3, \mathbf{v}_4\}).$$

Definition 3.17 (solid triangle) The *solid triangle* $\mathrm{TRI}(\mathbf{v}_0, V, r)$ is determined by $\mathbf{v}_0 \in \mathbb{R}^3$, a set $V = \{\mathbf{v}_1, \mathbf{v}_2, \mathbf{v}_3\} \subset \mathbb{R}^3$, and a radius $r \geq 0$:

$$\mathrm{TRI}(\mathbf{v}_0, V, r) = B(\mathbf{v}_0, r) \cap \mathrm{aff}_+(\mathbf{v}_0, V).$$

Definition 3.18 (rcone) Define the following *right-circular cones*. If $\mathbf{v}$ and $\mathbf{w}$ are points in $\mathbb{R}^3$, and $h \in \mathbb{R}$, then set

$$\begin{aligned}
\mathrm{rcone}(\mathbf{v}, \mathbf{w}, h) &= \{\mathbf{p} \;:\; (\mathbf{p} - \mathbf{v}) \cdot (\mathbf{w} - \mathbf{v}) \geq \|\mathbf{p} - \mathbf{v}\| \, \|\mathbf{w} - \mathbf{v}\| h\}, \\
\mathrm{rcone}^0(\mathbf{v}, \mathbf{w}, h) &= \{\mathbf{p} \;:\; (\mathbf{p} - \mathbf{v}) \cdot (\mathbf{w} - \mathbf{v}) > \|\mathbf{p} - \mathbf{v}\| \, \|\mathbf{w} - \mathbf{v}\| h\}.
\end{aligned}$$

Definition 3.19 (frustum, FR) The frustum $\mathrm{FR}(\mathbf{v}_0, \mathbf{v}_1, h', h, a)$ is specified by an apex $\mathbf{v}_0 \in \mathbb{R}^3$, heights $0 \leq h' \leq h$, a vector $\mathbf{v}_1 - \mathbf{v}_0$ giving its direction, and $a \in [0, 1]$. The set FR is given as

$$\{\mathbf{p} \in \mathrm{rcone}^0(\mathbf{v}_0, \mathbf{v}_1, a) \;:\; h' \|\mathbf{v}_1 - \mathbf{v}_0\| < (\mathbf{p} - \mathbf{v}_0) \cdot (\mathbf{v}_1 - \mathbf{v}_0) < h \|\mathbf{v}_1 - \mathbf{v}_0\|\}.$$

That is, the frustum is the part of a right-circular cone between two parallel planes that cut the axis of the cone at a right angle. When $h' = 0$, the frustum extends to the apex of the cone. When $h' = 0$, it is dropped from the notation: $\mathrm{FR}(\mathbf{v}_0, \mathbf{v}_1, h, a) = \mathrm{FR}(\mathbf{v}_0, \mathbf{v}_1, 0, h, a)$.

Definition 3.20 (conic cap) The conic cap $\mathrm{CAP}(\mathbf{v}_0, \mathbf{v}_1, r, a)$ is specified by an apex $\mathbf{v}_0 \in \mathbb{R}^3$, radius $r \geq 0$, nonzero vector $\mathbf{v}_1 - \mathbf{v}_0$ giving direction, and constant a. The conic cap is the intersection of the ball $B(\mathbf{v}_0, r)$ with a solid right-circular cone:

$$\mathrm{CAP}(\mathbf{v}_0, \mathbf{v}_1, r, a) = \{\mathbf{p} \in B(\mathbf{v}_0, r) \ : \ (\mathbf{p} - \mathbf{v}_0) \cdot (\mathbf{v}_1 - \mathbf{v}_0) > \|\mathbf{p} - \mathbf{v}_0\| \, \|\mathbf{v}_1 - \mathbf{v}_0\| \, a\}.$$

Remark 3.21 (quadratic solids) By Tarski arithmetic, a tetrahedron can also be described as the intersection of four open half-spaces. Moreover, all of the three-dimensional solids that have just been defined are described by linear and quadratic constraints.

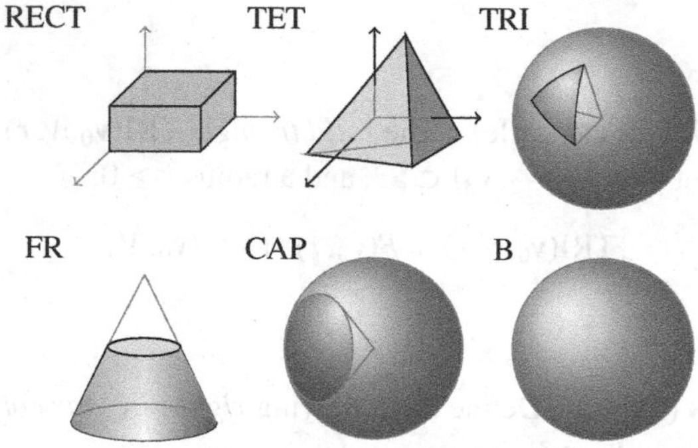

Figure 3.1 All solids in this book can be constructed from the rectangle, tetrahedron, solid spherical triangle, and wedges of a frustum, conic cap, and ball.

Definition 3.22 (primitive) A *primitive region* (Figure 3.1) is any of the following:

1. A rectangle

$$\mathrm{RECT}(a, b) = \{\mathbf{p} \ : \ a_i < p_i < b_i, \ \text{for } i = 1, 2, 3.\}.$$

2. A tetrahedron $\mathrm{TET}\{\mathbf{v}_0, \mathbf{v}_1, \mathbf{v}_2, \mathbf{v}_3\}$.
3. A solid triangle $\mathrm{TRI}(\mathbf{v}_0, \{\mathbf{v}_1, \mathbf{v}_2, \mathbf{v}_3\}, r)$.

4. A frustum or a wedge of a frustum:

$$\mathrm{FR}(\mathbf{v}_0, \mathbf{v}_1, h_1, h_2, a) \cap W^0(\mathbf{v}_0, \mathbf{v}_1, \mathbf{v}_2, \mathbf{v}_3).$$

5. A conic cap or a wedge of a conic cap:

$$\mathrm{CAP}(\mathbf{v}_0, \mathbf{v}_1, r, c) \cap W^0(\mathbf{v}_0, \mathbf{v}_1, \mathbf{v}_2, \mathbf{v}_3).$$

6. A ball or a wedge of a ball

$$B(\mathbf{v}_0, r) \cap W^0(\mathbf{v}_0, \mathbf{v}_1, \mathbf{v}_2, \mathbf{v}_3).$$

The set of primitives is not minimal. In particular, the rectangle is a union of tetrahedra, and the wedge of a ball is a union of solid triangles (up to a null set).

3.2.4 volume calculations

Lemma 3.23 (primitive volume)

1. *A rectangle* RECT(a, b) *has volume zero unless* $a_i < b_i$ *for all* i*. In this case, the volume is*

$$(b_3 - a_3)(b_2 - a_2)(b_1 - a_1).$$

2. *A tetrahedron* TET$\{\mathbf{v}_1, \mathbf{v}_2, \mathbf{v}_3, \mathbf{v}_4\}$ *has volume*

$$\sqrt{\Delta(x_{12}, x_{13}, x_{14}, x_{34}, x_{24}, x_{23})}/12,$$

where $x_{ij} = \|\mathbf{v}_i - \mathbf{v}_j\|^2$.

3. *Let* $\mathbf{v}_1, \mathbf{v}_2, \mathbf{v}_3$ *be unit vectors. A solid triangle* TRI$(\mathbf{v}_0, \{\mathbf{v}_1, \mathbf{v}_2, \mathbf{v}_3\}, r)$ *has volume*

$$(\alpha_1 + \alpha_2 + \alpha_3 - \pi)r^3/3,$$

where $\alpha_i = \mathrm{dih}_V(\{\mathbf{v}_0, \mathbf{v}_i\}, \{\mathbf{v}_j, \mathbf{v}_k\})$.

4. *The frustum* FR$(\mathbf{v}_0, \mathbf{v}_1, h, a)$ *(with* $h' = 0$*) has volume*

$$\pi(t^2 - h^2)h/3, \quad h = ta.$$

5. *The conic cap* CAP$(\mathbf{v}_0, \mathbf{v}_1, r, a)$ *has volume*

$$2\pi(1 - a)r^3/3.$$

6. *The ball* $B(\mathbf{v}_0, r)$ *has volume*

$$4\pi r^3/3, \text{ when } r \geq 0,$$

We do not give a proof of Lemma 3.23. It is part of the background in measure theory assumed at the beginning of the chapter. (The enumeration of background properties of volume in Lemma 3.5 makes a forward reference to Lemma 3.23.) The measurability of these primitive regions is also part of the background assumptions. An incomplete sketch of the lemma follows.

Proof sketch The volume of a tetrahedron is

$$|\det(\mathbf{v}_2 - \mathbf{v}_1, \mathbf{v}_3 - \mathbf{v}_1, \mathbf{v}_4 - \mathbf{v}_1)|/6.$$

Cayley and Menger evaluated the square of this determinant (Section 2.3.3). The square is $\Delta/4$, with $\Delta \geq 0$.

The formula for the volume of a solid triangle is $r^3/3$ times its solid angle. Girard's formula

$$\alpha_1 + \alpha_2 + \alpha_3 - \pi \tag{3.24}$$

for the area of a spherical triangle is classical (Figure 3.2). Euler's formula (Lemma 2.73) gives another formula for the area, which is sometimes more convenient.

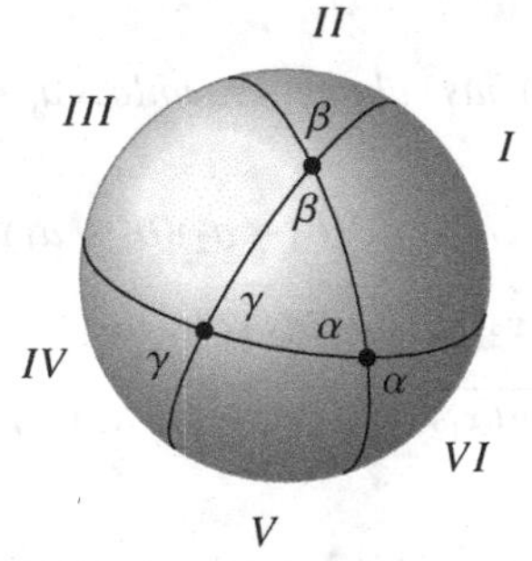

Figure 3.2 The area of a spherical triangle T is calculated by inclusion-exclusion: six lunes with areas $2\alpha, 2\alpha, 2\beta, 2\beta, 2\gamma, 2\gamma$ cover both T and the congruent antipodal triangle three times and the rest of the unit sphere once. This gives the equation $6\,\mathrm{area}(T) + (4\pi - 2\,\mathrm{area}(T)) = 4\alpha + 4\beta + 4\gamma$, or $\mathrm{area}(T) = \alpha + \beta + \gamma - \pi$.

The volume of a right-circular cone, $1/3$ its base area times height, has been known from antiquity.

The conic cap volume is $r^3/3$ times its solid angle, which computed as a surface of revolution for the curve $x^2 + y^2 = 1$ (Figure 3.3):

$$2\pi \int_a^1 |y|\sqrt{1 + (y')^2}\,dx = 2\pi \int_a^1 |y|\sqrt{1/y^2}\,dx = 2\pi \int_a^1 dx = 2\pi(1 - a). \quad \square$$

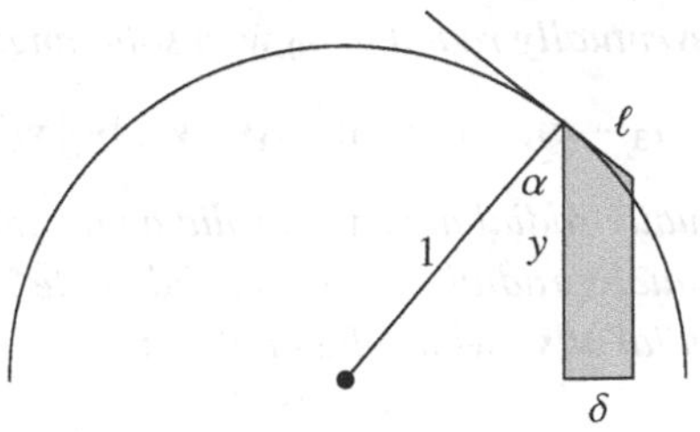

Figure 3.3 To first order approximation in δ, the surface area $2\pi y\,\ell$ of rotation of a slice of width δ of a unit sphere equals the area $2\pi y\,(\delta \sec\alpha) = 2\pi\,\delta$ of the surface area of the cone tangent to the sphere, which is independent of y and α. It follows that the surface area of the part of a unit sphere between two parallel planes depends only on the separation of the two planes.

When a region is realized by revolution along an axis $\text{aff}\{\mathbf{v}_0, \mathbf{v}_1\}$, we can also give the volume of the intersection of the region with a wedge $W^0 = W^0(\mathbf{v}_0, \mathbf{v}_1, \mathbf{v}_2, \mathbf{v}_3)$. These intersections are measurable by the last statement of Lemma 3.23 (with r sufficiently large). In the following lemma, let

$$\theta = \text{azim}(\mathbf{v}_0, \mathbf{v}_1, \mathbf{v}_2, \mathbf{v}_3).$$

Lemma 3.25 (wedge volume) *Let C be*

$$\text{CAP}(\mathbf{v}_0, \mathbf{v}_1, r, a), \quad B(\mathbf{v}_0, r), \quad \textit{or} \quad \text{FR}(\mathbf{v}_0, \mathbf{v}_1, h, a).$$

Let m be the volume of C. Then $C \cap W^0$ has volume $m\,\theta/(2\pi)$.

This lemma is also part of the assumed background material on volumes, cited at the beginning of the chapter in Lemma 3.5. Of course, these are elementary integrals.

All of the primitive regions (except the rectangle) are eventually radial at the natural base point $\mathbf{v}_0$ and have a solid angle. In fact, the intersection of any primitive region (again except the rectangle) with a small ball at $\mathbf{v}_0$ is again a primitive region. Thus, the earlier volume calculations immediately yield solid angle formulas as well.

Lemma 3.26 (primitive solid angle)

1. $\text{TET}\{\mathbf{v}_0, \mathbf{v}_1, \mathbf{v}_2, \mathbf{v}_3\}$ *is eventually radial at* $\mathbf{v}_0$ *with solid angle*

$$(\alpha_1 + \alpha_2 + \alpha_3 - \pi), \quad \alpha_i = \text{dih}_V(\{\mathbf{v}_0, \mathbf{v}_i\}, \{\mathbf{v}_j, \mathbf{v}_k\}).$$

2. $\text{TRI}(\mathbf{v}_0, \{\mathbf{v}_1, \mathbf{v}_2, \mathbf{v}_3\})$ *is eventually radial at* $\mathbf{v}_0$ *with solid angle*

$$(\alpha_1 + \alpha_2 + \alpha_3 - \pi), \quad \alpha_i = \text{dih}_V(\{\mathbf{v}_0, \mathbf{v}_i\}, \{\mathbf{v}_j, \mathbf{v}_k\}).$$

3. $\text{FR}(\mathbf{v}_0, \mathbf{v}_1, h, a)$ *is eventually radial at* $\mathbf{v}_0$ *with solid angle* $2\pi(1 - a)$.
4. $\text{CAP}(\mathbf{v}_0, \mathbf{v}_1, r, a)$ *is eventually radial at* $\mathbf{v}_0$ *with solid angle* $2\pi(1 - a)$.
5. $B(\mathbf{v}_0, r)$ *is eventually radial at* $\mathbf{v}_0$ *with solid angle* 4π.

Proof In every case, the intersection of the region with $B(\mathbf{v}_0, r')$, for $r' > 0$ sufficiently small, is a ball, a conic cap, or a solid triangle. These volumes have already been calculated. This gives the results as stated. □

Lemma 3.27 (wedge solid angle) *Let C be either* $\text{CAP}(\mathbf{v}_0, \mathbf{v}_1, r, a)$, $B(\mathbf{v}_0, r)$, *or* $\text{FR}(\mathbf{v}_0, \mathbf{v}_1, h, a)$. *Then C and $C \cap W^0$ are eventually radial at $\mathbf{v}_0$. Furthermore, $C \cap W^0$ has solid angle $s\,\theta/(2\pi)$, where s is the solid angle of C.*

Proof The intersection of $C \cap W^0$ is one of the primitive regions with a solid angle that has already been calculated in Lemma 3.25. □

3.3 Finiteness and Volume

Previous sections have developed all of the volume calculations that are needed in this book. This chapter concludes with some elementary estimates based on the volumes of cubes and balls.

Lemma 3.28 *For all $\mathbf{p} \in \mathbb{R}^3$ and all $r \geq 0$, the set $\mathbb{Z}^3 \cap B(\mathbf{p}, r)$ is finite of cardinality at most* $4\pi(r + \sqrt{3})^3/3$.

Proof If $\mathbf{v} \in \mathbb{Z}^3 \cap B(\mathbf{p}, r)$, then the ith coordinate v_i of $\mathbf{v}$ must lie in the finite range

$$p_i - r \leq v_i \leq p_i + r.$$

Hence, there are only finitely many possibilities for $\mathbf{v}$.

Place an open unit cube at each point of $\mathbb{Z}^3 \cap B(\mathbf{p}, r)$. The cubes are measurable, disjoint, and contained in $B(\mathbf{p}, r + \sqrt{3})$. Thus, the combined volume of the cubes, which is $\text{card}(\mathbb{Z}^3 \cap B(\mathbf{p}, r))$, is no greater than the volume of the containing ball. The result ensues. □

Lemma 3.29 *For all $\mathbf{p} \in \mathbb{R}^3$ and all $r \geq \sqrt{3}$, the set $\mathbb{Z}^3 \cap B(\mathbf{p}, r)$ is finite of cardinality at least* $4\pi(r - \sqrt{3})^3/3$.

Proof Lemma 3.28 establishes finiteness. Place a closed unit cube at each point of $\mathbb{Z}^3 \cap B(\mathbf{p}, r)$. The cubes are measurable and cover $B(\mathbf{p}, r - \sqrt{3})$. Thus, the combined volume of the cubes is at least the volume of the covered ball. The result ensues. □

Lemma 3.30 (lattice shell) *For all* $\mathbf{p} \in \mathbb{R}^3$ *and all* $r_0, r_1 > 0$, *there exists a* C *such that for all* $r \geq r_1$,

$$\text{card}(\mathbb{Z}^3 \cap (B(\mathbf{p}, r + r_0) \setminus B(\mathbf{p}, r - r_1))) \leq Cr^2.$$

Proof When $r \geq r_1 + \sqrt{3}$, the previous two lemmas show that the cardinality is at most $4\pi/3$ times

$$(r + r_0 + \sqrt{3})^3 - (r - r_1 - \sqrt{3})^3 \leq C' r^2$$

for some C'. Similarly, if $r_1 \leq r \leq r_1 + \sqrt{3}$, the cardinality is at most some fixed constant C''. The result ensues. □

4

Hypermap

Summary. *A planar graph, which is a graph that admits a planar embedding, has too little structure for our purposes because it does not specify a particular embedding. A plane graph carries a fixed embedding, which gives it a topological structure where combinatorics alone should suffice. A hypermap gives just the right amount of structure. It is a purely combinatorial object, but carries information that the planar graph lacks by encoding the relations among nodes, edges, and faces. This chapter is about hypermaps.*

In the original proof of the Kepler conjecture, the basic combinatorial structure was that of a planar map, as defined by Tutte [47]. Although planar maps appear throughout that proof, they are lightweight objects, in the sense that no significant structural results are needed about them.

Gonthier makes hypermaps the fundamental combinatorial structure in his formal proof of the four-color theorem [16]. His formal proof eliminates topological arguments such as the Jordan curve theorem in favor of purely combinatorial arguments. When I learned of Gonthier's work, I significantly reorganized the proof by replacing planar maps with hypermaps, making them heavyweight objects, in the sense that significant structural results about them are needed.

As a result of these changes, many parts of the proof that were originally done topologically can now be done combinatorially, a change that significantly reduces the effort required to formalize the proof. These changes also make it possible to treat rigorously what was earlier done by geometric intuition. For example, the original proof made implicit use of the equivalence of two different notions of a planar map: a combinatorial notion that was used in the computer algorithms and a topological notion of an equivalence class of embeddings of a graph into a sphere. Hypermaps make this equivalence explicit and rigorous.

Put simply a hypermap is a finite set together with two permutations on that set. It is therefore useful to start the chapter with a brief review of permutations. The second section develops the basic terminology of hypermaps. The next section describes various transformations of hypermaps called walkup transformations. These transformations can be viewed as corresponding to operations such as contracting an edge in a graph or deleting a node of a graph. Next, properties of planar hypermaps are developed. The final sections prove the correctness of an algorithm to generate all planar hypermaps with given properties. The algorithm can be described heuristically in terms of drawing graphs in pencil and pen on a sheet of paper. This algorithm has been implemented as a computer program. The output from this program is an essential part of the proof of the Kepler conjecture.

4.1 Background on Permutations

This section reviews the theory of permutations, as presented in standard textbooks.

Definition 4.1 (permutation) A *permutation* f on a set D is a bijection $f : D \to D$.

For example, the identity map I_D on a set D is a permutation:

$$I_D(x) = x \text{ for all } x \in D.$$

If $f : D \to D$ is a permutation, then its inverse function $f^{-1} : D \to D$ is also a permutation. It satisfies

$$ff^{-1} = f^{-1}f = I_D.$$

(This chapter uses product notation fg for the composition $f \circ g$ of maps.) If D is a finite set, and two maps $f, g : D \to D$ satisfy $fg = I_D$ on D, then f and g are permutations and are inverses of one another:

$$fg = gf = I_D.$$

Natural number powers f^k of a permutation $f : D \to D$ are defined recursively by

$$f^0 = I_D, \quad \text{and} \quad f^{k+1} = ff^k.$$

Integer powers f^m of a permutation are defined as

$$f^m = f^i(f^{-1})^j,$$

where $m = i - j$. This is well-defined. The usual rule of exponents holds:

$$f^{m+n} = f^m f^n.$$

If $f : D \to D$ is a permutation on a finite set D, then there is a smallest positive integer k such that $f^k = I_D$. The integer k is the *order* of the permutation f. If $f^m = I_D$ for some m, then $m = ki$ for some integer i, where k is the order of f. The inverse $f^{-1} = f^k f^{-1} = f^{k-1}$ can be written as a nonnegative power of f.

A permutation f of a finite set D is *cyclic* if for every $x, y \in D$, there exists an integer i such that $f^i x = y$.

The set of all permutations of the set $\{0, 1, 2, \ldots, k-1\}$ is written Sym(k). The set Sym(k) is finite and has cardinality $k!$.

4.2 Definitions

A hypermap, presented in the next definition, is the main subject of this chapter.

Definition 4.2 (hypermap, dart) A *hypermap* is a finite set D, together with three functions $e, n, f : D \to D$ that compose to the identity

$$enf = I_D.$$

The elements of D are called *darts*. The functions e, n and f are called the *edge map*, the *node map*, and the *face map*, respectively.

Remark 4.3 (plane graphs as hypermaps) A hypermap is an abstraction of the concept of plane graph. Place a dart at each angle of a plane graph (Figure 4.1). One function, f, cycles counterclockwise around the angles of each face. Another function, n, rotates counterclockwise around the angles at each node. A third function, e, pairs angles at opposite ends of each edge. The hypermap extracts the data (D, e, n, f) from the plane graph and discards the rest.

This construction of hypermaps from plane graphs is our primary reason to study hypermaps. Many of the lemmas and proofs in this chapter have are standard results about plane graphs, translated into the language of hypermaps.

It follows from the background on permutations that e, n, f are all permutations on D. A hypermap satisfies

$$enf = nfe = fen = I_D. \tag{4.4}$$

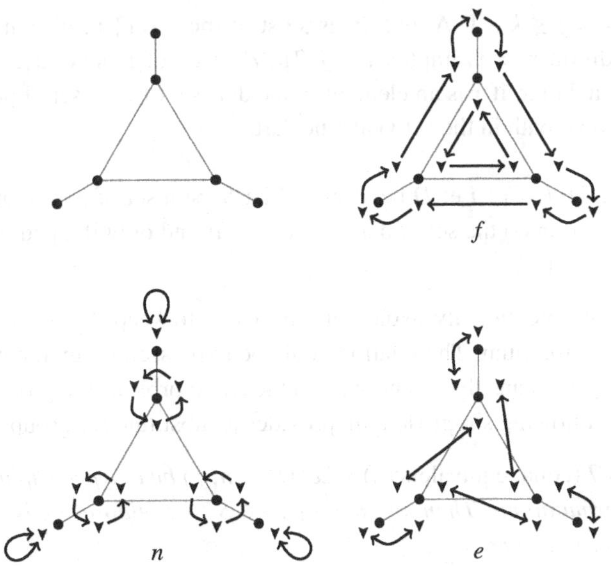

Figure 4.1 A plane graph is given by the gray edges and circular nodes as shown. Twelve darts mark the angles of the plane graph. Darts may be permuted about faces (f), nodes (n), and edges (e) of the plane graph to form a hypermap.

This *triality relation* shows that if (e, n, f) give a hypermap, then so do (n, f, e) and (f, e, n). Because of these symmetries in the defining relation, there are multiple versions of theorems about hypermaps, all obtained from one proof by symmetry. Alternatively, it is possible to define a hypermap as a finite set with two permutations e, n, leaving f to be derived from $enf = I_D$; however, the triality symmetry would be lost in such a definition.

Inverted, this triality becomes

$$n^{-1}e^{-1}f^{-1} = (fen)^{-1} = I_D.$$

This inversion is the abstract form of the duality between nodes and faces in a plane graph.

Definition 4.5 (path, list, sublist, dart set, visit) Let D be a set of darts and let S be a set of permutations of D. A *path* with *steps* in S from x_0 to x_{k-1} is a *list*[1] of darts $[x_0; x_1; \ldots; x_{k-1}]$ such that for each i, $x_{i+1} = h_i x_i$ for some $h_i \in S$. A *sublist* of a list is a consecutive subsequence $[x_i; x_{i+1}; x_{i+1}; \ldots; x_j]$,

[1] The empty path [] seems to have an ancient origin: "This is the path made known to me when I had learned to remove all darts." –The Dhammapada

with $0 \le i \le j \le k-1$. A *unit list* is a list of the form $[x]$. A path is *injective* if the condition $x_i = x_j$ implies $i = j$. The *dart set* of L is $\{x_0, x_1, \ldots, x_{k-1}\}$. A path *visits* a dart x, if x is an element of the dart set of L. A set of paths *visits* a dart x, if some path in the set visits the dart.

Definition 4.6 ($\sim_S$) Let D be a set and let S be a set of permutations on D. Define a relation on the set of darts by $x \sim_S y$ if and only if a path runs from x to y with steps in S.

This book intentionally avoids group theory to keep the theoretical background to a minimum. The relation could be expressed group theoretically by saying $x \sim_S y$ means that x and y lie in the same orbit of the group generated by S. The following simple lemma provides a substitute for group theory.

Lemma 4.7 (equal equivalences) *Let (D, e, n, f) be a hypermap and let S be a set of permutations. Then for each $h_1, h_2 \in S$, the relation $\sim_S$ is the same as the relation $\sim_T$, where*

$$T = S \cup \{h_1 h_2\}.$$

Moreover, for each $h \in S$, the relation $\sim_S$ is the same as the relation $\sim_T$, where

$$T = S \cup \{h^{-1}\}.$$

Also, the relation $\sim_S$ (that is, $\sim_T$) is an equivalence relation.

Proof If $x \sim_S y$, then clearly $x \sim_T y$. Conversely, if $x \sim_T y$, where $T = S \cup \{h_1 h_2\}$, select a path P from x to y with steps in T that contains the fewest $h_1 h_2$-steps.

P does not contain[2] *any $h_1 h_2$-steps.* Otherwise, a sublist $[\ldots; u; h_1 h_2 u; \ldots]$ of P can be expanded to a path $[\ldots; u; h_2 u; h_1 u; \ldots]$ that contradicts the minimal property of P. This claim gives the first conclusion of the lemma.

Fix h in a set of permutations R. By an induction that uses the first conclusion, for all i, $\sim_R$ equals the relation $\sim_{R(h,i)}$, where $R(h, i) = R \cup \{h, h^2, \ldots, h^i\}$. If $h \in S$ is an element of order k, and $T = S \cup \{h^{-1}\}$, then the second conclusion follows because the following sets give the same relation:

$$S, \quad S(h, k-1) = T(h, k-2), \quad T.$$

By repeated action of the previous conclusion, we obtain $(\sim_S) = (\sim_T)$, where $T = S \cup S^{-1} \cup \{I_D\}$, and where $S^{-1} = \{h^{-1} : h \in S\}$. The unit path $[x]$ yields reflexivity of $\sim_T$. Also, $T^{-1} = T$ gives the symmetry. Finally,

[2] This paragraph continues with the book's general convention of typesetting in italic a claim that is followed by its justification.

concatenation of paths gives transitivity. Thus, $\sim_T$ (i.e., $\sim_S$) is an equivalence relation. □

Several of the following definitions – components, connected, node, edge, face – have been appropriated from planar graph theory. These definitions are intended to capture the intuitive notions of the nodes and edges of a graph, faces of a plane graph, and connectivity.

Definition 4.8 (combinatorial component, connected) A *combinatorial component* of a hypermap (D, e, n, f) is an equivalence class of the relation $\sim_S$, where $S = \{e, f, n\}$. (See Lemma 4.7 for other sets that define the same equivalence classes.) Let $\#c$ denote the number of combinatorial components of (D, e, n, f). The hypermap is *connected* if $\#c = 1$.

Definition 4.9 (orbit, node, face, edge) The *orbit* of $x \in D$ under a permutation h on a set D is a set of the form $\{h^i x \; : \; i \in \mathbb{N}\}$. A *node* of a hypermap (D, e, n, f) is the orbit of a dart $x \in D$ under n. A *face* is an orbit under f. An *edge* is an orbit under e.

Plane graphs are conventionally illustrated in such a way that nodes, edges, and faces have an entirely different appearance: nodes as point, edges as curves, and faces as polygons. In an abstract hypermap, the triality symmetry shows that the nodes, edges, and faces have equal footing. Nevertheless, to retain the intuition of plane graphs, this book often depicts hypermaps with the darts in a node arranged in a small cluster around a point, the darts in a edge along a curve, and the darts in a face around a polygon.

Let $\#h$ denote number of orbits of a permutation h on D.

Lemma 4.10 (orbit) *Let D be a finite set. The orbit of $x \in D$ of a permutation $h : D \to D$ is the equivalence class of x under the relation $\sim_S$, when $S = \{h\}$.*

Definition 4.11 (plain) A hypermap (D, e, n, f) is *plain* (carefully note[3] the spelling) when e is an involution on D (that is, $e^2 = I_D$).

Definition 4.12 (degenerate) A dart in a hypermap (D, e, n, f) is *degenerate* if it is a fixed point of one of the maps e, n, f; otherwise it is *nondegenerate*.

Definition 4.13 (simple) A hypermap is *simple* if the intersection of each face with each node contains at most one dart.

Lemma 4.14 (nodal fixed point) *Let (D, e, n, f) be a simple plain hypermap such that every face has at least three darts. Then n has no fixed point.*

[3] This is a deliberate and dangerous play on the homophonous *plane* that makes this topic unspeakable. Although plane graphs are planar, not all plain hypermaps are planar.

Proof For a contradiction, let $x \in D$ be a fixed point of n.

The darts ex and fx lie in the same node and face and are therefore equal in the simple hypermap. Indeed, they lie in the same node because $n(fx) = e^{-1}x = ex$ and they lie in the same face because

$$f^2(ex) = f(fenx) = fx.$$

So $ex = fx$.

Thus, $f^2(ex) = fx = ex$, and ex lies on a face with at most two darts. This contradicts what is given.   □

4.3 Walkup

This section describes various operations to transform one hypermap to another. The simplest of these operations is the *walkup* transformation that deletes one dart from a hypermap and constructs permutations that skip past the deleted dart. More complex transformations of hypermaps can be constructed as a sequence of walkup transformations and correspond to standard operations on graphs such as the contraction or deletion of an edge.

To focus attention on a dart x in a hypermap, it can be useful to draw a hexagon around x and place the six darts $ex, fx, e^{-1}x, nx, f^{-1}x, ex, n^{-1}x$ at its corners as shown in Figure 4.2. Some of these seven darts may be equal to one another, even if the figure draws them apart. Figure 4.3 shows the layout of a degenerate dart.

$$fx \rightarrow e^{-1}x$$
$$\nwarrow \swarrow$$
$$n^{-1}x \rightarrow x \longrightarrow nx$$
$$\nwarrow \swarrow \nwarrow \swarrow$$
$$ex \quad f^{-1}x$$

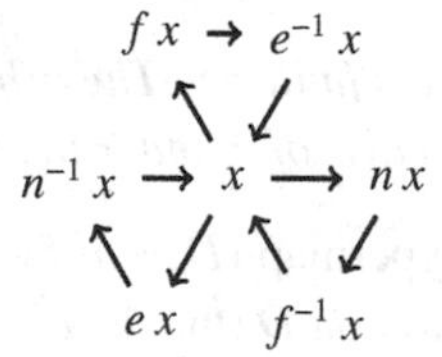

Figure 4.2 A fragment of a hypermap showing a dart x and its entourage. The node map is given by the horizontal arrows, the edge map by arrows descending towards the lower left, and the face map by arrows rising towards the upper left.

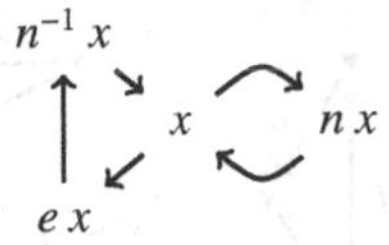

Figure 4.3 A fragment of a hypermap showing a dart x and its entourage when x is fixed by the permutation f. The vertical arrow on the left has type f. The two arrows pointing to the right along the top of the diagram have type n. The two arrows pointing to the left along the bottom of the diagram have type e.

4.3.1 single

A walkup deletes a dart from a hypermap and reforms the edge, node, and face maps to produce a hypermap on the reduced set of darts. Walkups come in three varieties: edge walkups, face walkups, and node walkups.

Definition 4.15 (walkup, degenerate) The edge *walkup* W_e at a dart $x \in D$ of a hypermap (D, e, n, f) is the hypermap (D', e', n', f'), where $D' = D \setminus \{x\}$ and the maps skip over x:

$$\begin{aligned} f'y &= \text{ if } (fy = x) \text{ then } fx \text{ else } fy \\ n'y &= \text{ if } (ny = x) \text{ then } nx \text{ else } ny \\ e' &= (n'f')^{-1}. \end{aligned}$$

A walkup at x is said to be *degenerate* if the dart x is degenerate.

Figure 4.4 shows the result of an edge walkup on the hexagon around a dart x. The triality symmetry 4.4, applied to the definition of edge walkups, yields the definition of face walkup W_f and node walkup W_n.

At a degenerate dart x, all three walkups are equal: $W = W_e = W_n = W_f$ (Figure 4.5).

Definition 4.16 (merge, split) Let (D, e, n, f) be a hypermap and let $h = n, e$, or f. Let $O(h, x)$ denote the orbit of $x \in D$ under h. Let (D', e', n', f') be the hypermap obtained from (D, e, n, f) by the walkup W_h at $x \in D$. Let $h' = e', n', f'$, respectively, according to the choice of h. The walkup W_h at x *merges* when the walkup joins the orbit of h through x with another orbit. That is, the orbit O of some $y \in D'$ under $h' : D' \to D'$ has the form

$$O \cup \{x\} = O(h, x) \cup O(h, y),$$

where $y \notin O(h, x)$. It *splits* when the walkup splits the orbit at x into two orbits. That is, there are distinct orbits O_1, O_2 under h' in the hypermap (D', e', n', f')

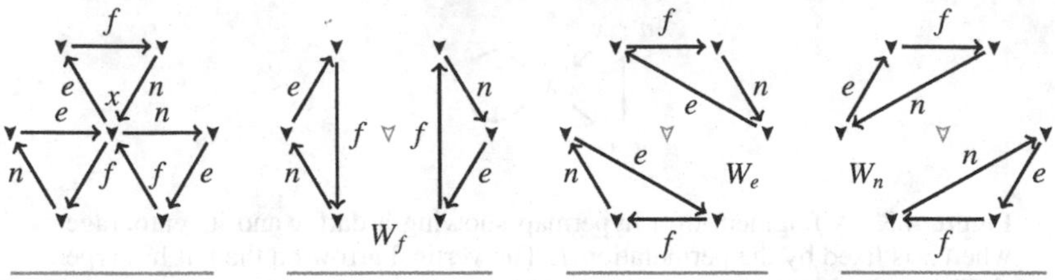

Figure 4.4 A fragment of a hypermap in its original state, and three modified fragments after applying a face walkup W_f, edge walkup W_e, and node walkup W_n at the dart x at the center. Each walkup eliminates the dart x. The other darts are the same as in the original, but undergo modified permutations e, n, and f.

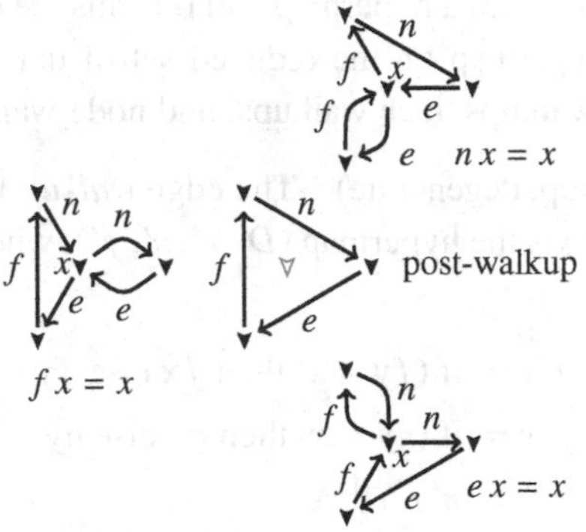

Figure 4.5 All three walkup transformations have the same effect at a dart x that is degenerate, regardless of whether the degeneracy is a face-, edge-, or node-degeneracy. The three different kinds of degenerate darts are shown in the hypermap fragments on the outside of the figure, and the hypermap fragment after applying a walkup is shown at the center of the diagram.

such that

$$\{x\} \cup O_1 \cup O_2 = O(h, x).$$

Lemma 4.17 (merge-split) *Let (D, e, n, f) be a hypermap and let W_h be a nondegenerate walkup at a dart x. Then W_h merges or splits. Moreover, it merges if and only if x and y lie in distinct h-orbits, where $(h, y) = (f, ex)$, (e, nx), or (n, fx).*

Proof The walkup W_f splits if and only if fx (or x) and ex lie in the same f-orbit before the split. Figure 4.6 makes this clear. The other cases $h = e, n$ hold by triality. □

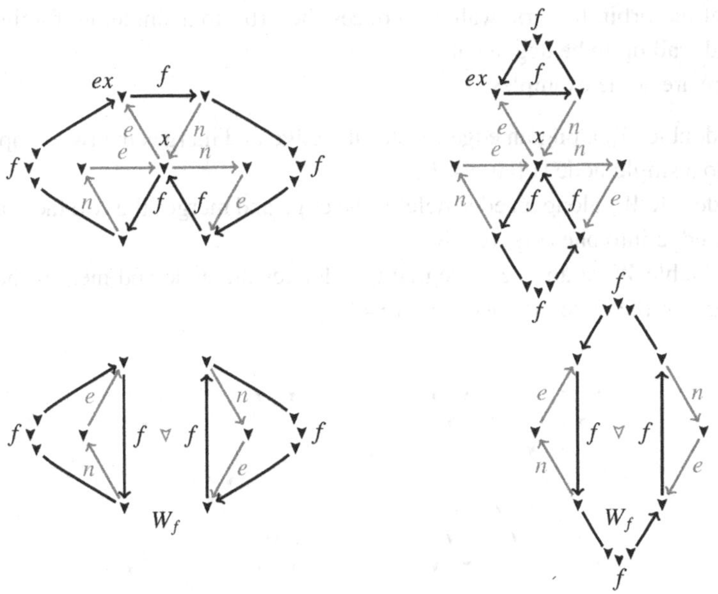

Figure 4.6 When darts x and $e\,x$ lie in the same orbit of f (upper left), the face walkup W_f at x splits the orbit into two (lower left). But when the darts x and $e\,x$ lie in different orbits of f (upper right), the edge walkup W_f at x merges the two orbits into one (lower right).

The following is a useful way to tell if a walkup merges.

Lemma 4.18 (merge criterion)*Suppose, in a simple plain hypermap (D, e, n, f), that an edge $\{x, y\}$ consists of two nondegenerate darts. Then the walkup W_f at x merges.*

Proof The darts fx and ex lie in the same node: $n(fx) = e^{-1}x = ex$. If they are also in the same face of a simple hypermap, then $fx = ex = y$. So

$$ny = nfx = nfey = y,$$

and y is a fixed point of n, and hence degenerate, contrary to assumption. Thus, fx and ex are in different faces, and the walkup merges by Lemma 4.17. □

4.3.2 double

A double walkup is the composite of two walkups of the same type. The two darts for the two walkups are to be the members of an orbit of cardinality two (under n, e, or f). By choosing the type of the walkups to be different from the type of the orbit, the first walkup reduces the orbit to a singleton, forcing the second walkup to be degenerate.

Here are some examples.

1. A double W_n along an edge deletes the edge and merges the two endpoints into a single node (Figure 4.7).
2. A double W_f along an edge deletes the edge and merges the two faces along the edge into one (Figure 4.8).
3. A double W_e at a node of degree two deletes the node and merges the two edges at the node into one (Figure 4.9).

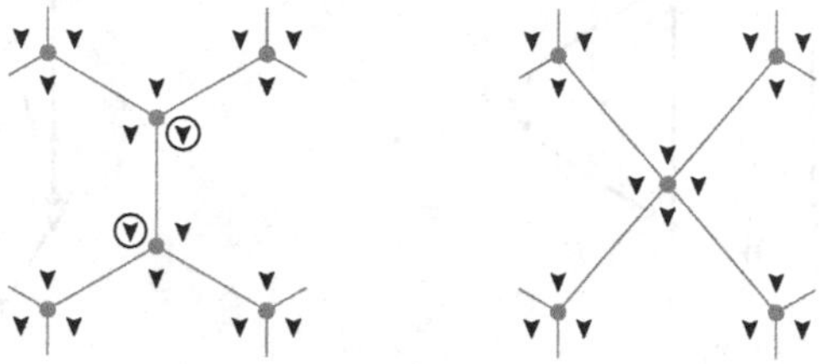

Figure 4.7 When a hypermap comes from a plane graph, the double *node* walkup at the two darts of an edge contracts the corresponding edge of the graph.

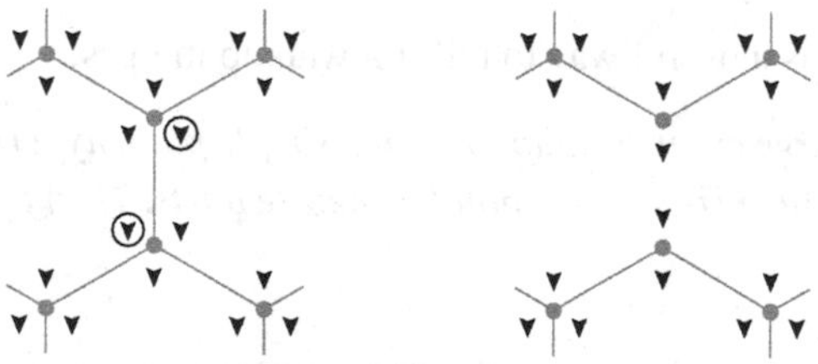

Figure 4.8 When a hypermap comes from a plane graph, the double *face* walkup at the two darts of an edge deletes the corresponding edge from the graph.

Figure 4.9 When a hypermap comes from a plane graph, the double *edge* walkup at the two darts of a node deletes the corresponding node from the graph.

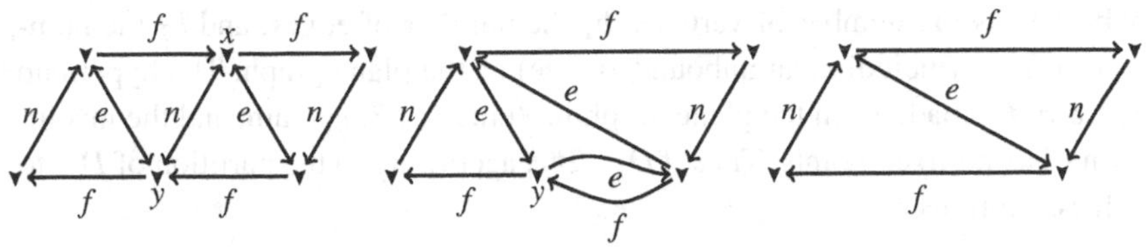

Figure 4.10 A double edge walkup of a hypermap at the two darts of a node preserves plainness, depicted as bidirectional e-arrows. The three frames show a fragment of a plain hypermap, the fragment after the first edge walkup at x, and after the second edge walkup at y.

Lemma 4.19 (plain walkup) *The three preceding double walkups carry plain hypermaps into plain hypermaps.*

Proof The walkups W_n and W_f preserve the orbit structure of edges, except for dropping one dart. By dropping both darts from the same edge, one edge is lost and all others edges remain unchanged.

Figure 4.10 illustrates the double W_e at $\{x, y\}$. The two edges $\{x, ex\}$, $\{y, ey\}$ meeting the node are fused by the double walkup into $\{ex, ey\}$, which is still an edge of cardinality two. □

4.4 Planarity

Just as a graph can be planar or nonplanar, so too can a hypermap. A plane graph satisfies Euler's formula, which relates the number of vertices, edges, and faces of the graph. Euler's formula can be expressed in purely combinatorial terms, then generalized to hypermaps. A hypermap for which Euler's formula holds is defined to be planar.

Definition 4.20 (planar) A hypermap is *planar* (note the spelling!) when the Euler relation holds:

$$\#n + \#e + \#f = \#D + 2\#c.$$

Remark 4.21 (Eulerian relation) The Euler relation for plane graphs can be translated into the language of hypermaps. Consider a connected plane graph that satisfies the Euler relation for the alternating sum of Betti numbers:

$$b_0 - b_1 + b_2 = 2,$$

where b_0 is the number of vertices, b_1 the number of edges, and b_2 the number of faces (including an unbounded face) of the plane graph. The hypermap (D, e, n, f), made from the plane graph in Remark 4.3, is plain, and the involution e has no fixed points. Thus, $\#D = 2\#e$, according to the partition of D into edges. Moreover,

$$\begin{aligned} b_0 &= \#n \\ b_1 &= \#e \\ b_2 &= \#f \\ 2b_1 &= \#D \\ 1 &= \#c \\ b_0 - b_1 + b_2 &= \#n + (\#e - \#D) + \#f = 2\#c. \end{aligned}$$

Thus, the hypermap is also planar.

The Euler relation for graphs has many consequences, one of which is that in a triangulation of the sphere, the number of edges is equal to 3/2 times the number of faces because each face has three edges and each edge borders two faces. Here is another simple consequence, expressed in the language of hypermaps.

Lemma 4.22 (dart bound) *Let H be a connected plain planar hypermap such that every edge has cardinality two. Assume that there are at least three darts in every node. Then*

$$\#D \leq (6\#f - 12).$$

Proof In a connected plain planar hypermap, the Euler relation becomes

$$6\#f - 12 = \#D + 2(\#D - 3\#n),$$

so it is enough to show that

$$\#D \geq 3\#n.$$

The inequality follows directly by assumption: the set of darts can be partitioned into nodes, with at least three darts per node. □

Definition 4.23 (planar index) The *planar index* of a hypermap is

$$\iota = \#f + \#e + \#n - \#D - 2\#c.$$

The index measure the departure of a hypermap from planarity; a hypermap with index zero is planar.

Lemma 4.24 (walkup index) *Let x be a nondegenerate dart of a hypermap (D, e, n, f). Let (D', e', n', f') be the result of the face walkup W at x. The walkup changes the cardinality of some orbits.*

$$
\begin{aligned}
\#f' &= \#f + \mathrm{split}_f \\
\#e' &= \#e \\
\#n' &= \#n \\
\#D' &= \#D - 1 \\
\#c' &= \#c + \mathrm{split}_c \\
\iota' &= \iota + 1 + \mathrm{split}_f - 2\mathrm{split}_c,
\end{aligned}
$$

where

$$
\mathrm{split}_f = \begin{cases} 1, & \text{if } W \text{ splits} \\ -1, & \text{if } W \text{ merges} \end{cases}
$$

and $\mathrm{split}_c = 1$ *if ex and $f^{-1}x$ belong to different combinatorial components after the walkup W, and* $\mathrm{split}_c = 0$ *otherwise. Moreover, a walkup at a degenerate dart preserves the planar index.*

Proof The proof is evident from Figures 4.4, 4.5, and 4.6. □

Lemma 4.25 (index inequality) *Let ι be the index of a hypermap (D, e, n, f) and let ι' be the index after a walkup W_h at a dart x. Then $\iota \le \iota'$.*

Proof Without loss of generality, by triality symmetry, the walkup is a face walkup. If $\mathrm{split}_c = 0$, then the inequality is immediate by Lemma 4.24. If $\mathrm{split}_c = 1$, then ex and $f^{-1}x$ lie in different components after the walkup and in different faces as well. Thus, the walkup splits by Lemma 4.17, and $\mathrm{split}_f = 1$. The result follows by Lemma 4.24. □

The following lemma is a hypermap analogue of the fact that the Euler characteristic of a surface graph is never greater than the Euler characteristic of a plane graph.

Lemma 4.26 (nonpositive index) *The planar index of a hypermap is never positive.*

Proof An face walkup never decreases the index. A sequence of face walkups leads to the empty hypermap, which has index zero. □

Lemma 4.27 (planar walkup) *Walkups take planar hypermaps to planar hypermaps.*

Proof A planar hypermap has maximum index. The walkup can only increase the index, but never beyond its maximum. Thus, the index remains at its maximum value. □

4.5 Path

This section develops the basic properties of paths in hypermaps.

4.5.1 contour

We make a distinction between injective paths, which never repeat a dart, and loops that return to the initial dart in the trajectory. In the definition of loop, we wish to remove any dependence on an initial dart. A loop can start in at an arbitrary dart in the trajectory. If a path is a certain kind of list, then a loop is a certain kind of cyclic list, as given in the definition that follows.

Definition 4.28 (cyclic list) A *cyclic list* $[\![x_0; \ldots; x_{k-1}]\!]$ is an equivalence class of lists under the transitive closure of the relation:

$$[x_0; x_1; x_2; \ldots; x_{k-1}] \sim [x_1; x_2; \ldots; x_{k-1}; x_0].$$

A *sublist* of a cyclic list is a sublist of some representative of the equivalence class.

Definition 4.29 (contour path, contour loop) A *contour path* from x_0 to x_{k-1} is a path $[x_0; x_1; \ldots; x_{k-1}]$ such that $x_{i+1} = n^{-1}x_i$ or fx_i for each $i < k$. (That is, each step in the path is a clockwise step around a node or a counterclockwise step around a face.) A *contour loop* is an injective cyclic list $[\![x_0; x_1; \ldots; x_{k-1}]\!]$ such that for every i, there exists $h_i \in \{f, n^{-1}\}$ such that $x_{i+1} = h_i x_i$, where the subscripts are read modulo k.

If we consider steps $x \mapsto n^{-1}x$ and $x \mapsto fx$ as positive and other steps as negative, then a contour path can be intuitively imagined as a path that carries an intrinsic positive orientation. That is, a contour path is an oriented path of sorts.

Remark 4.30 (contour path illustration) Figure 4.1 constructs a hypermap from a plane graph by drawing darts next to each angle. In this representation, the darts along a contour path lie to the left of the corresponding plane graph edges. For that reason, a shaded region to the left of a curve depicts a contour path.

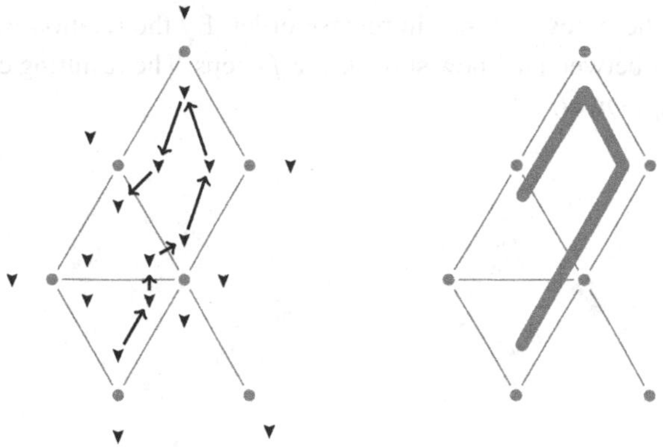

Figure 4.11 When a hypermap is constructed from a plane graph, a contour path can be depicted by a sequence of arrows between darts that follow alongside the edges of the graph. The contour path can be reconstructed from a shaded path alongside the edges of the graph. The direction of the path is such that the edges of the graph remain to the right of motion.

Lemma 4.31 (injective path) *An injective contour path from x to y can be constructed from an arbitrary contour path from x to y by dropping some darts from the path.*

Proof Repeatedly replace $[\ldots; a; b; \ldots; b; c; \ldots]$ with $[\ldots; a; b; c; \ldots]$. □

Lemma 4.32 (contours-components) *Let H be a hypermap. If x and y are darts in the same combinatorial component of H if and only if there exists a contour path from x to y.*

Proof Combinatorial components are defined by an equivalence relation $\sim_S$, where $S = \{e, n, f\}$. By Lemma 4.7, this is the same equivalence relation as $\sim_T$, where $T = \{n^{-1}, f\}$. By the definition of the equivalence relation T, $x \sim_T y$ if and only if some contour path runs from x to y. □

Definition 4.33 (complement) Let (D, e, n, f) be a plain hypermap. Let $P = [\![x; y; \ldots]\!]$ be a contour loop that does not return to visit any node a second time. (In particular, the dart set of P intersected with a node is the dart set of a maximal sublist $[z; n^{-1}z; \ldots; n^{-k}z]$ of n^{-1} steps.) Replace each maximal sublist of n^{-1}-steps

$$[z; n^{-1}z; \ldots; n^{-k}z]$$

with the sublist

$$[n^{-(k+1)}z; n^{-(k+2)}z; \ldots; nz].$$

Concatenate these new sublists in reverse order. By the relation $nf = f^{-1}n^{-1}$, the transitions between the new sublists are f-steps. The resulting contour loop P^c is the *complement*.

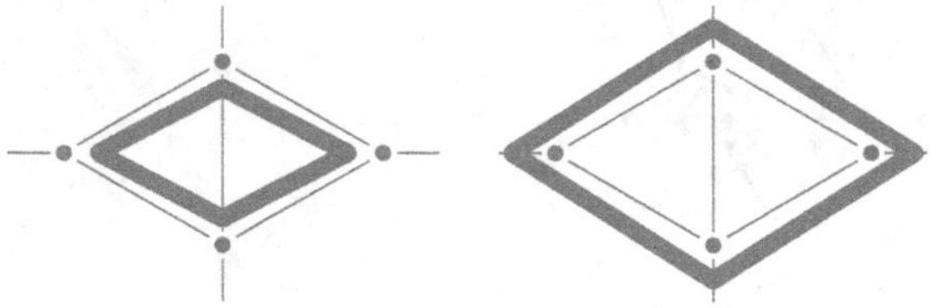

Figure 4.12 The complement contour traces the remaining darts at the same nodes as the original contour loop. The shaded path uses the conventions of Figure 4.11.

4.5.2 Möbius

The rest of this section develops the basic properties of Möbius contours. A Möbius contour should be thought of as the simplest kind of nonplanar contour path. Möbius contours turn out to be extremely useful because in practice the best way to certify that a given hypermap is nonplanar is to produce a Möbius contour in the hypermap (Lemma 4.38).

Definition 4.34 (Möbius contour) A Möbius contour in a hypermap (D, e, n, f) is an injective contour path $P = [x_0; \ldots]$ that satisfies

$$x_j = nx_0, \quad x_k = nx_i \tag{4.35}$$

for some $0 < i \leq j < k$ (Figure 4.13).

Remark 4.36 (Four-color theorem) Gonthier devised the notion of Möbius contour as a way to prove the four-color theorem without appeal to topology. (The Appel–Haken proof of the four-color theorem ultimately relies on the Jordan curve theorem.) This chapter uses a significant amount of material from [15].

Remark 4.37 (Möbius strip) Heuristically, a Möbius contour is a combinatorial Möbius strip that twists one side to the other (Figure 4.15). A planar hypermap has no such contour.

Lemma 4.38 (planar-non-Möbius) *A planar hypermap does not have a Möbius contour.*

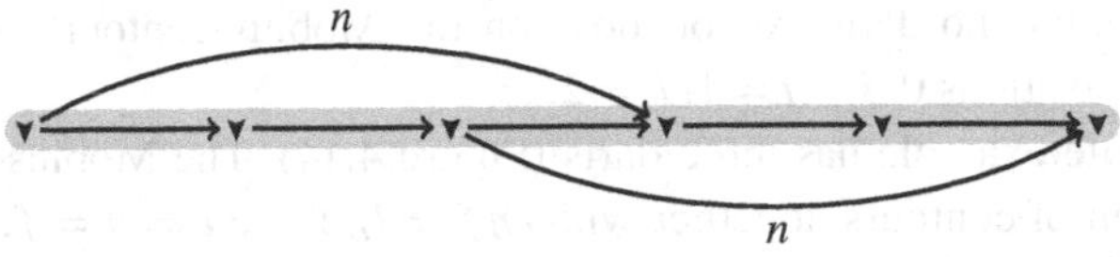

Figure 4.13 The horizontal contour path is a Möbius contour.

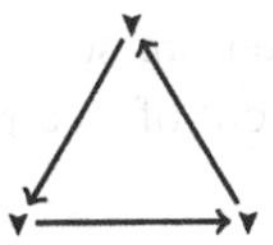

Figure 4.14 For every set D of cardinality three and cyclic permutation of that set, there is a hypermap with dart set D and $e = f = n$ all equal to the given cyclic permutation. This hypermap has a Möbius contour $[x; fx; f^2x]$, for each $x \in D$.

Proof For a contradiction, assume that there exist planar hypermaps with Möbius contours. To simplify the counterexample and reduce the number of darts, we may use walkups that transform a planar hypermap with a Möbius contour into another planar hypermap with a Möbius contour. An edge walkup at a dart that is not on the Möbius contour transforms a counterexample in this way. A walkup of the right sort at a dart that is not at position 0, i, j, k also transforms a counterexample into another counterexample. (See Möbius condition 4.35.) Eventually, a counterexample with the smallest possible number

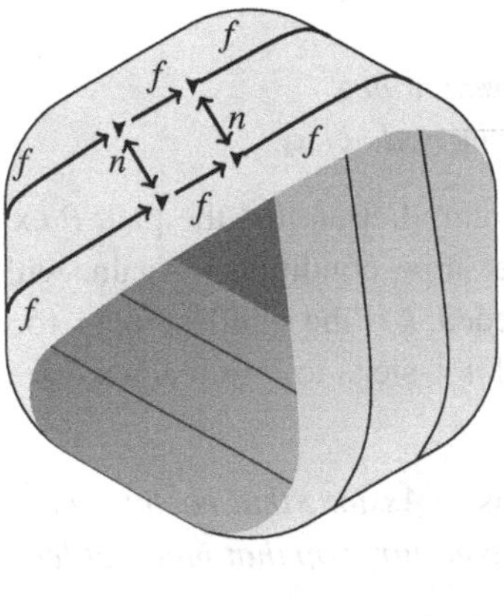

Figure 4.15 A Möbius contour of a hypermap embedded in a Möbius strip.

of darts contains no darts except those on the Möbius contour, and its only darts are at positions 0, $i = j = 1$, $k = 2$.

This counterexample has three darts (Figure 4.14). The Möbius condition, the definition of contours, together with $enf = I_D$ force $e = n = f$, which are all permutations of order three. This hypermap is not planar:

$$\#e + \#n + \#f = 3 \neq 5 = \#D + 2\#c. \qquad \square$$

The final results in this section are somewhat technical lemmas that are needed later in the correctness proof of an algorithm that generates planar hypermaps.

Lemma 4.39 (step coherence) *Suppose that a hypermap has no Möbius contours. Let L be a contour loop. Let P be any injective contour path with at least three darts, that starts and ends on L, but visits no other darts of L. Then the first and last steps of P are both of the same type (n^{-1} or f).*

Proof The proof shows the contrapositive. Suppose $P = [nx; fnx; \ldots; ny; y]$. The successor of nx on L is x. Starting at x, follow L to y, and on to nx. Follow P back to ny. This is a Möbius contour $x \ldots y \ldots nx \ldots ny$.

Suppose $P = [nx; x; \ldots; f^{-1}y; y]$. Starting at x, follow P to y, then follow L to nx, and on to ny. This is also a Möbius contour.[4] □

Lemma 4.40 (loop separation) *Suppose that a hypermap has no Möbius contours. Let L be a contour loop. Then there does not exist a contour path $[x_0; \ldots; x_k]$ for $k \geq 1$ with the following properties.*

1. *x_i lies on L if and only if $i = 0$.*
2. $x_1 = fx_0$.
3. *x_0 and x_k lie in different nodes.*
4. *Some dart of L is at the node of x_k.*

Proof Assume for a contradiction that the path P exists. Some sublist is injective and satisfies the same conditions. Again, without loss of generality, shrinking the path if needed, k is the smallest index for which the last two conditions are met. Append n^{-1}-steps to P to reach a dart of L. This is contrary to Lemma 4.39. □

Lemma 4.41 (three darts) *Assume that each face of a hypermap has at least three darts. Then every contour loop that meets at least two nodes has at least three darts.*

4 The second statement can also be deduced from the first statement by the duality $(D, e, n, f) \leftrightarrow (D, e^{-1}, f^{-1}, n^{-1})$ that swaps f-steps with n^{-1}-steps in a path.

Proof Let $P = [\![x; y]\!]$ be a contour loop meeting two nodes. Then $y = fx$ and $x = fy$, so that the face has cardinality two. □

4.6 Subquotient

This section develops the properties of *subquotient* hypermaps. Each dart in a subquotient hypermap is an equivalence class of darts in the originating hypermap. It is a subquotient rather than a quotient because the equivalence relation is only defined on a *subset* of the darts of the originating hypermap.

Subquotients are used in the algorithm to generate planar hypermaps, described later in the chapter. The originating hypermap is the one we wish to construct and the subquotient is the partially constructed hypermap of the unfinished algorithm. The material in this section has the narrow purpose of providing a descriptive language for the algorithm. This material is not used outside this chapter.

4.6.1 definition

An isomorphism is a structure preserving map.

Definition 4.42 (isomorphism) Two hypermaps (D, e, n, f) and (D', e', n', f') are *isomorphic* when there is a bijection $G : D \to D'$ such that

$$h' \circ G = G \circ h$$

for $(h, h') = (e, e'), (f, f'), (n, n')$.

The faces of a subquotient are to be traced out by a collection of contour loops in the originating hypermap. To get a well-defined subquotient, the collection of contour loops must be normal in the following sense.

Definition 4.43 (normal family) Let (D, e, n, f) be a hypermap. Let $\mathcal{L}$ be a family of contour loops. The family $\mathcal{L}$ is *normal* if the following conditions hold of its loops.

1. No dart is visited by two different loops.
2. Every loop visits at least two nodes.
3. If a loop visits a node, then every dart at that node is visited by some loop.

A normal family determines a new hypermap. A dart in the new set D' of darts is a maximal sublist $[x; n^{-1}x; n^{-2}x; \ldots; n^{-k}x]$ of n^{-1} steps appearing in some loop in $\mathcal{L}$. The map f' takes the maximal path $[x; n^{-1}x; \ldots; y]$ to the maximal path (in the same contour loop) starting at fy. The map n'^{-1} takes the maximal path $[\dots; y]$ to the maximal sequence (in some other contour loop) starting $[n^{-1}y; \dots]$. Equivalently, n' takes the maximal path $[x; \dots]$ to the maximal path ending $[\dots; nx]$. The map e' is defined by $e'n'f' = I_{D'}$.

Definition 4.44 (subquotient) The hypermap (D', e', n', f') constructed from the normal family $\mathcal{L}$ of $H = (D, e, n, f)$ is called the *subquotient* of H by $\mathcal{L}$ and is denoted $H/\mathcal{L}$. If x is a dart visited by some loop in $\mathcal{L}$, then the maximal path $[\dots; x; \dots]$ is called the *quotient dart* of x.

Intuitively, the subquotient hypermap is represented as a graph with cycles under f' that are precisely the contour loops in the normal family (Figure 4.16).

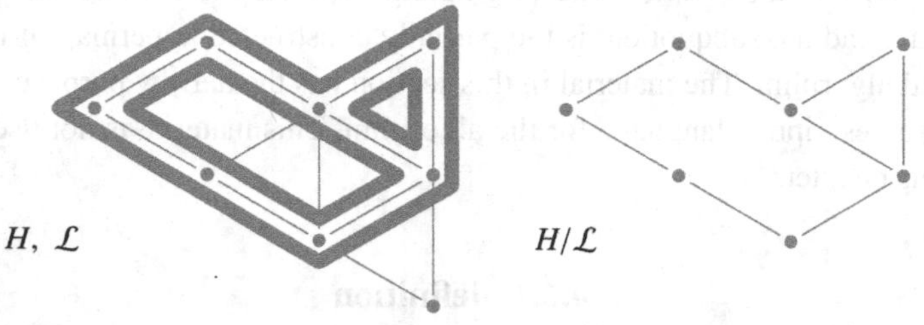

Figure 4.16 When the hypermap H comes from a plane graph, we may depict the normal family $\mathcal{L}$ as a collection of shaded loops alongside the edges of the graph (Figure 4.11). The subquotient is the hypermap associated with the traversed edges of the plane graph.

4.6.2 properties

This subsection explores some of the properties of a subquotient hypermap. The first two lemmas describe the faces and the nodes of the subquotient in terms of the combinatorics of the normal family.

Lemma 4.45 (subquotient face, **F**) *Let $\mathcal{L}$ be a normal family of the hypermap H. Then $\mathcal{L}$ is in natural bijection with the set of faces of the subquotient $H/\mathcal{L}$. If $x' = [x; \dots; n^{-k}x]$ is a maximal path of n^{-1} steps in the contour loop $L \in \mathcal{L}$, then the corresponding face* $\mathbf{F}(L)$ *of $H/\mathcal{L}$ is the one containing the quotient dart x'.*

Proof This is left as an exercise for the reader. □

Lemma 4.46 (subquotient node) *Let H be a hypermap and let $\mathcal{L}$ be a normal family of H. Then there is a natural bijection between the set of nodes of $H/\mathcal{L}$*

and the set of nodes of H that are visited by some contour loop in $\mathcal{L}$. The bijective function sends the node in $H/\mathcal{L}$ of the dart $x' = [x; n^{-1}x; \ldots; n^{-k}x]$ to the node of x in H.

Proof The proof is an elementary verification. Let $H/\mathcal{L} = (D', e', n', f')$. *The function described in the lemma is well-defined.* Indeed,

$$(n')^{-1}x' = [n^{-(k+1)}x; \ldots]$$

is also sent to the node of x in H.

This function is onto. Indeed, If $L \in \mathcal{L}$ visits x, and x' is the quotient dart of x in D', then the node of x' clearly maps to the node of x.

Finally, the function is one-to-one. Indeed, if the nodes of two quotient darts x', y' map to the same node of H, then $x' = [x; \ldots]$ and $y' = [y; \ldots]$, where $n^j x = y$ for some j. It follows by the definition of the node map on the subquotient that x' and y' belong to the same node. □

The next two lemmas look at properties of the subquotient that are inherited from the original hypermap.

Lemma 4.47 (plain subquotient) *Let H be a plain hypermap and let $\mathcal{L}$ be a normal family. Then $H/\mathcal{L}$ is a plain hypermap.*

Proof Write $H = (D, e, n, f)$ and $H/\mathcal{L} = (D', e', n', f')$. Write $[\ldots; x]$ for the node in the subquotient ending in dart $x \in D$ and $[x; \ldots]$ for the node in the subquotient starting with dart $x \in D$. Plainness gives $e^2 x = x$, so that for any dart $[\ldots x]$ in the subquotient

$$\begin{aligned} e'^{-2}[\ldots; x] &= n'f'n'f'[\ldots; x] = n'f'n'[fx; \ldots] \\ &= n'f'[\ldots; nfx] = n'[fnf; \ldots] = [\ldots; nfnfx] \\ &= [\ldots; e^{-2}x] = [\ldots; x]. \end{aligned}$$

Thus, e' has order two on the subquotient. □

Definition 4.48 (no double joins) A hypermap H has no *double joins* if for every two nodes (possibly equal to each other), at most one edge of H meets both of them.

Lemma 4.49 (subquotient-no-double-joins) *Let H be a plain hypermap with no double joins and let $\mathcal{L}$ be a normal family of H. Then $H/\mathcal{L}$ has no double joins.*

Proof By Lemma 4.47, the subquotient $H/\mathcal{L}$ is plain. Let $\{x', e'x'\}$ and $\{y', e'y'\}$ be edges with the property that x' and y' lie at one node of $H/\mathcal{L}$ and $e'x'$ and $e'y'$ lie at a second node. Write $x' = [\ldots; x]$ and $y' = [\ldots; y]$. Then $e'x' =$

$[\dots; ex]$ and $e'y' = [\dots; ey]$. According to Lemma 4.46, the map from nodes of $H/\mathcal{L}$ to nodes of H is one-to-one. It follows that x and y belong to the same node and that ex and ey belong to the same node. By the assumption that H has no double joins, it follows that $\{x, ex\} = \{y, ey\}$. Hence also $\{x', e'x'\} = \{y', e'y'\}$, and $H/\mathcal{L}$ has no double joins. □

Lemma 4.50 (nodal fixed point) *Let $H = (D, e, n, f)$ be a hypermap in which the edge map has no fixed points. Let $\mathcal{L}$ be a normal family of H, with subquotient $H/\mathcal{L} = (D', e', n', f')$. Then the following are equivalent conditions.*

1. *n' has a fixed point in D'.*
2. *The dart set of some $L \in \mathcal{L}$ contains a node.*

Proof If $x' = [x; n^{-1}x; \dots; n^{-k}x]$ is a dart in D', then $(n')^{-1}x'$ is $[n^{-(k+1)}x; \dots]$. The dart x' is a fixed point if and only if $x = n^{-(k+1)}x$. This holds if and only if the dart set of x' is an entire node. □

4.6.3 example

Example 4.51 (maximal normal family) *Assume that $H = (D, e, n, f)$ is a hypermap. Assume that every face meets at least two nodes. Then the set of all faces defines a normal family of contour loops in which each contour loop follows f around a face $[x; fx; \dots]$. If e acts without fixed points, then each dart of the subquotient is just a unit path consisting of a single dart of H, and the subquotient is isomorphic to H itself.*

Example 4.52 (minimal normal family) *Assume that $H = (D, e, n, f)$ is a plain hypermap. Let $F = \{x, fx, \dots\}$ be a face that visits at least three nodes and that meets each node in at most one dart. Let $\mathcal{L}$ be the family with two contour loops: $[\![x; fx; \dots]\!]$ and its complement $L^c = [\![n^{-1}x; \dots]\!]$. The family $\mathcal{L}$ is normal. The subquotient hypermap $H/\mathcal{L}$ has two faces: F and a back side F' of the same cardinality k.*

Example 4.53 (dihedral) *There is a hypermap* Dih_{2k} *with a dart set of cardinality $2k$. The permutations f, n, e have orders k, 2, and 2 respectively, and $enf = I$. The set of darts is given by*

$$\{x, fx, f^2x, \dots, f^{k-1}x\} \cup \{nx, nfx, nf^2x, \dots, nf^{k-1}x\}$$

for any dart x. If a hypermap is isomorphic to Dih_{2k} *for some k, then it is dihedral.*[5] *In particular, the hypermap constructed in the previous example is dihedral.*

[5] The three permutations generate the dihedral group of order $2k$, acting on a set of $2k$ darts under the left action of the group upon itself.

Lemma 4.54 *A hypermap H is isomorphic to* Dih_{2k} *if and only if the hypermap is connected, the dart set has cardinality $2k$, and the permutations f, n, e have orders k, 2, and 2, respectively.*

Proof Let y be any dart of H, and x any dart of $\mathrm{Dih}_{2k} = (D, e', n', f')$. By the connectedness of H, and the relations between f and n, every dart of H is equal to one of the following: $f^i y$, $n f^i y$ for $i = 0, \ldots, k-1$. By the cardinality assumption, these darts are all distinct. The bijection $f'^i x \mapsto f^i y$, $n' f'^i x \mapsto n f^i y$ is an isomorphism of hypermaps. □

Lemma 4.55 *Let (D, e, n, f) be a connected plain hypermap with more than two darts. Assume that the hypermap has no double joins. Assume that the two darts of each edge lie at different nodes. Then every face has at least three darts.*

Proof Otherwise, if some face is a singleton $\{x\}$, then both darts of the edge $\{x, ex\}$ lie at the same node, which is contrary to assumption.

Furthermore, if some face has only two darts, then the two edges meeting the face, $\{x, ex\}$ and $\{e^{-1} f^{-1} x, f^{-1} x\}$, which join the same two nodes, must actually be equal. That is, $x = fex = n^{-1} x$, so that x is a fixed point of the node map. Similarly, fx is a fixed point of the node map. Then $\{x, fx\}$ is a combinatorial component, which is contrary to the assumption that the hypermap is connected with more than two darts. □

4.7 Generation

This final section of the chapter, which presents an algorithm that generates all simple, plain, planar hypermaps satisfying certain general conditions (Definition 4.56), is more technical than other sections. This material may be skipped without disrupting the flow of the book because there is no need to return to the description of the algorithm, although the book relies on the output of the algorithm's execution in Chapter 8. The algorithm proceeds by adding more and more edges and nodes to a dihedral hypermap.

The algorithm itself is elementary to describe in intuitive terms. Imagine that a biconnected plane graph G has been drawn on a sheet of paper in pencil and that the purpose of the algorithm is to retrace the edges of graph in pen.[6] We

[6] Historically, this algorithm was developed in precisely this way. After drawing many plane graphs by hand while working on the Kepler conjecture, I started to generate the graphs by computer by automating the manual process in 1994. The code was first implemented in *Mathematica*, then later in *Java*, and more recently by Bauer and Nipkow in *ML* and *Isabelle/HOL* [33].

start the algorithm by selecting any face F_0 of the graph and tracing its edges in pen. Next, we select any node v on F_0 at an edge that is still in pencil. We can find a face F_1 of G that shares an edge with F_0 with endpoint v, and that has an edge in pencil at v. The second step of the algorithm pens the edges of the simple arc to complete the simple closed curve around F_1. Continue in this manner, adding a simple arc in pen to the penciled lines to add a face of G, until all the edges of G have been traced in pen.

It is remarkable (and rather unfortunate) that it requires a technical section to put these simple pencil and pen drawings into a rigorous form that function as a blueprint for a formal proof. A hypermap gives rigorous form to the pencil drawing, and various subquotients are the inked drawings at various stages of the algorithm.

We do not attempt to work in the greatest possible generality. As a matter of convenience, we impose a large number of conditions on the class of hypermaps that the algorithm generates. In particular, we assume that the hypermaps are restricted, as defined below. This definition is idiosyncratic, tailored to our needs, and matched to our particular proof methods.

Definition 4.56 (restricted) A *restricted hypermap* $H = (D, e, n, f)$ is one with the following properties.

1. The hypermap H has no double joins, and is nonempty, connected, planar, plain and simple.
2. The edge map e has no fixed points.
3. The node map n has no fixed points.
4. The cardinality of every face is at least three.

Remark 4.57 (step type) The assumption that $ex \neq x$ implies that $fx \neq n^{-1}x$ so that f-steps of a path can be distinguished from n^{-1}-steps.

4.7.1 flag

The algorithm marks certain faces as "true." Roughly, this means that the face cannot be modified at any later stage of the algorithm. That is, the edges of the faces are in ink and no pencil lines lie within its interior. When all of its faces are true, the hypermap stands in final form. The function that marks each face as true or false is a *flag*. For the algorithm to work properly, it is necessary to impose some constraints, as captured in Definition 4.59.

Under the bijection $\mathbf{F}$ between a normal family $\mathcal{L}$ and the set of faces of a subquotient $H/\mathcal{L}$, any function on $\mathcal{L}$ can be identified with a function on the set of faces of $H/\mathcal{L}$.

Definition 4.58 (canonical function) Let H be a hypermap with normal family $\mathcal{L}$. The *canonical function* $\check{\varphi}_{can}$ is the boolean-valued function on the set of faces of $H/\mathcal{L}$ that is true on $\mathbf{F}(L)$ exactly when the dart set of L maps bijectively to the face $\mathbf{F}(L)$ of $H/\mathcal{L}$, under $x \mapsto [x]$. That is, each dart of $\mathbf{F}(L)$ is a unit path in H. A face $\mathbf{F}(L)$ (or contour loop L) is said to be canonically true or false, according to the value of the canonical function.

In other words, the face in the subquotient is canonically true, exactly when the corresponding contour loop $L \in \mathcal{L}$ has no n^{-1} steps. The dart set of such a contour loop L is a face of H.

Definition 4.59 (flag) Let S be a set of darts in a hypermap H. An S-*flag* on H is a boolean-valued function $\check{\varphi}$ on the set of faces that satisfies the following two constraints.

1. If darts x, y belong to true faces, then some contour path runs from x to y that remains in true faces.
2. Each edge of the hypermap meets a true face or S.

An $\emptyset$-flag is simply called a flag.

Example 4.60 (dihedral hypermap flag) *The dihedral hypermap of Example 4.53 carries a flag that marks one face true and the other false.*

Example 4.61 (maximal subquotient flag) *Let H be a connected hypermap and let $\mathcal{L}$ be the example of Example 4.51. Then the canonical map takes value* true *on every face. This is a flag. In fact, Lemma 4.32 provides the contour paths that are required in the definition of flag.*

There is a standard way of constructing the sets S of darts that are used in S-flags.

Definition 4.62 (S) Let H be a hypermap, L a contour loop of the hypermap, and x an element of the dart set of L. If L is canonically true, then let $S = \emptyset$. Otherwise, let $m \geq -1$ to be the largest m such that

$$[x; fx; f^2x; \ldots; f^{m+1}x]$$

is a sublist of L, and set $S(H, L, x) = \{f^i x \; : \; 1 \leq i \leq m\}$.

Lemma 4.63 (flag subquotient) *Let H be a hypermap in which e acts without fixed points, L a contour loop, and x and element of the dart set of L. Let $\mathcal{L}$ be a normal family of H that contains L. Then $S(H, L, x)$ maps bijectively to a set S' of darts in the subquotient $H/\mathcal{L}$.*

Proof The darts of the subquotient are maximal sublists $[y; n^{-1}y; \ldots; n^{-k}y]$ of contour loops $L' \in \mathcal{L}$ made entirely of n^{-1} steps. Each $y \in S(H, L, x)$ is preceded by an f-step and is followed by an f-step in L. Hence, the maximal sublist of L containing y is a unit path $[y]$. The bijection ensues. □

4.7.2 markup

In the heuristics at the beginning of this section, a graph is drawn in pencil on a sheet of paper and various edges are retraced in pen. In the following definition, H represents the pencil drawing, $H/\mathcal{L}$ represents the ink drawing, L and x guide the tip of the pen to the next edge to be retraced in pen.

Definition 4.64 (marked hypermap) Let $(H, \mathcal{L}, L, x)$ be a tuple consisting of

1. a hypermap $H = (D, e, n, f)$ with no Möbius contours in which e acts without fixed points;
2. a normal family $\mathcal{L}$;
3. a contour loop $L \in \mathcal{L}$; and
4. a dart x visited by L.

Such a tuple is a *marked hypermap* if the following conditions hold.

1. The subquotient $H' = H/\mathcal{L} = (D', e', n', f')$ is simple.
2. n' has no fixed points on D'.
3. x is followed by an f-step in the loop: $L = [\![x; fx; \ldots]\!]$.
4. The contour loop $L' \in \mathcal{L}$ that visits[7] nfx is canonically true.
5. $\check{\varphi}_{can}$ is an S'-flag on H', where S' is the image of $S(H, L, x)$ in D'.

Example 4.65 (illustration) *This example illustrates the markup (Figure 4.17). In the figure, the hypermap H is represented as a plane graph. The contour loops are represented by gray loops alongside edges of the graph. The darts in each shaded face also form a contour loop under the face map. The edges of the plane graph that are flanked in gray give the edges of a plane graph representing the subquotient. The polygons that are fully shaded are true, together with the unbounded face. Two polygons in the subquotient are false. A dart of H in a false contour loop L in $\mathcal{L}$ is marked x. By inspection, $S(H, L, x) = \{fx, f^2x, f^3x\}$. By inspection, $\check{\varphi}_{can}$ is an S'-flag. (In fact, the darts in true faces form a connected region. Every edge in the subquotient meets a true face, except the edges through darts $f'x'$, f'^2x', and f'^3x', which meet S'.)*

[7] L visits fx. By the definition of normality, some contour loop in $\mathcal{L}$ visits the dart nfx at the same node.

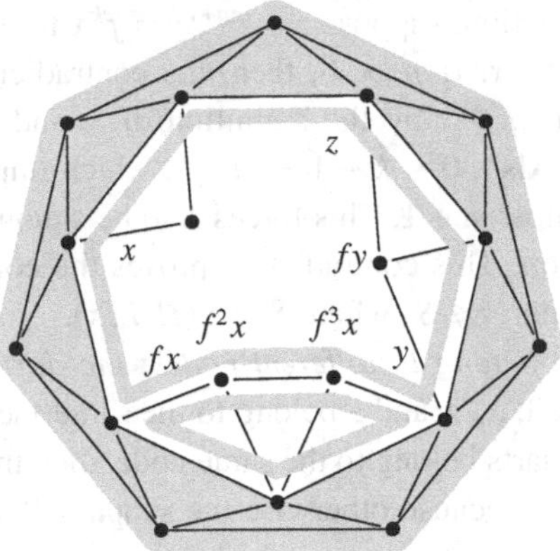

Figure 4.17 This marked hypermap is described in (Example 4.65). As usual, the angles of a plane graph are the darts of the hypermap, and the permutations e, n, f are derived from the structure of the graph. The constants (m, p, q) are $(3, 1, 5)$, because $y = f^{m+1}x = f^{3+1}x$, $z = f^{p+1}y = f^{1+1}y$, and the contour path from x to z makes $q + 1 = 5 + 1$ steps of type f.

Definition 4.66 (m, p, q, y, z) Let $(H, \mathcal{L}, L, x)$ be a marked hypermap. Set $y = f^{m+1}x$, where $m = \mathrm{card}(S(H, L, x))$. Set $z = f^{p+1}y$, where p is the smallest natural number such that some contour loop in $\mathcal{L}$ visits $f^{p+1}y$. Let x' and z' be the images of x and z respectively in $H/\mathcal{L} = (D', e', n', f')$. Let q be the smallest natural number such that $z' = (f')^{q+1}x'$.

In the pencil-pen analogy, the dart y marks the tip of the pen as it is about to draw a simple arc. The dart z marks the endpoint of that arc. The integers m, p, q track how far the pen has progressed in tracing the edges of a face, how many nodes appear on the arc about to be drawn, and how z sits in relation to y in the current pen sketch.

The following lemma gives the existence of q, by showing that x' and z' lie in the same face $\mathbf{F}(L)$ of $H/\mathcal{L}$. (The existence of q is trivial when L is canonically true.) In terms of the pencil and pen drawing, the following lemma expresses the planarity of the drawing: a continuous stroke of the pen that starts in one connected component of the complement of Jordan curve and that does not cross the Jordan curve must end in the same connected component.

Lemma 4.67 (loop confinement) *Let $(H, \mathcal{L}, L, x)$ be a marked hypermap, where H is restricted. Assume that L is canonically false. Let the natural numbers m and p and darts y and z be given by Definition 4.66. Then, L visits z, and $z \neq f^k x$ when $0 < k \leq m + 1$. Furthermore, the darts x and y belong to different nodes; the darts y and z belong to different nodes.*

Proof For a contradiction, suppose $z = f^{p+1}y = f^k x$ for some $0 < k \leq m + 1$. Then also, $f^p y = f^{k-1}x$. If $p > 0$, then this contradicts the minimality of p. (Note that L visits $f^{k-1}x$ by the definition of S and m.) So $p = 0$, and $y = f^{k-1}x = f^{m+1}x$. Also, $0 \leq k - 1 < m + 1$, which implies that the face of x has cardinality at most $m + 1$. This forces L to be canonically true, which is contrary to assumption. This contradiction proves the conclusion $z \neq f^k x$ of the lemma. In particular, $z \notin S$, where $S = S(H, L, x)$.

The darts x and y belong to different nodes; the darts y and z belong to different nodes. Indeed, x, y, and z belong to the same face. By the simplicity of H, if two of these darts belong to the same node, then they are equal to each other. However, $x \neq y$ because otherwise the subpath $P = [x; fx; \ldots; f^{m+1}x]$ gives a canonically true contour loop, which is contrary to the assumption that L is canonically false. Also, $y \neq z$ because otherwise the face of y is equal to $\{y, fy, f^2y, \ldots, f^p y\}$. It follows that $x = f^k y$ is visited by L for some $1 \leq k \leq p$. This contradicts the defining minimality property of p.

Let L' be the contour loop of $\mathcal{L}$ that visits z. For a final contradiction, assume that $L' \neq L$.

L' is false. Otherwise, L' is true with respect to the canonical flag and is therefore a loop consisting entirely of f-steps. In particular, L' visits z, x, and y. This is contrary to the assumption that the contour loop containing x is false.

Let $H' = (D', e', n', f') = H/\mathcal{L}$ and let S' be the image of S in D'. Let $z' = [\ldots; u] \in D'$ be the image in D' of z. As $z \notin S$, we also have $z' \notin S'$. By the definition of S'-flag, the dart $e'z'$ lies in a true face or $e'z' \in S'$. This disjunction splits the proof into two cases.

1. In the case that $e'z'$ lies in a true face,

$$e'z' = f'^{-1}n'^{-1}[\ldots, u] = f'^{-1}[n^{-1}u; \ldots] = [f^{-1}n^{-1}u],$$

so that $n^{-1}u$ is visited by a true contour loop. Consider the contour path in H, obtained by concatenating

$$[y; fy; \ldots; z], \quad [n^{-1}z; \ldots; u], \text{ and } [n^{-1}u; \ldots; n^{-1}x].$$

The first segment consists of f-steps, the second of n^{-1}-steps, and the third segment exists within true contour loops of $\mathcal{L}$ by the connectedness of true faces (by properties of flags). This path satisfies all the properties of Lemma 4.40. In particular, the second segment avoids L by Lemma 4.39. The lemma asserts that the path does not exist.

2. In the case that $e'z' \in S'$, it follows that $f^{-1}n^{-1}u \in S$ and L visits $n^{-1}u$ at the node of z. Consider the following path of f-steps in H:

$$[y; fy; \ldots; z].$$

This path satisfies all the enumerated properties of Lemma 4.40. The lemma asserts that the path does not exist.

□

Lemma 4.68 (parameters) *Let $(H, \mathcal{L}, L, x)$ be a marked hypermap, where H is restricted. Assume that L is canonically false. Let m, p, and q be the natural numbers and let x, y, and z be the darts given by Definition 4.66. Let* $r = \text{card}(\mathbf{F}(L))$. *Then*

$$0 \le p, \quad 0 \le m < q < r, \quad m + 1 < p + q.$$

Proof Let $H/\mathcal{L} = (D', e', n', f')$. Let y' and z' be the images of y and z in D', respectively. Let S' be the image of $S(H, L, x)$ in D'. By definition $m = \text{card}(S(H, L, x))$. Both m and p are natural numbers, so $0 \le p$ and $0 \le m$.

$q < r$. Indeed, by definition, $z' = (f')^{q+1}x'$ and no smaller natural number has this property. Also, x' and y' belong to the face $\mathbf{F}(L)$. If $q \ge r$, then $(f')^{q+1} = (f')^{q-r+1}$, which contradicts the minimality of q.

We claim $m < q$. Indeed, if $0 \le k < m$, then

$$(f')^{k+1}x' = [f^{k+1}x] \in S', \quad \text{and} \quad (f')^{q+1}x' = z' \notin S',$$

by Lemma 4.63 and Lemma 4.67. Thus, $m \le q$. Also, $q \ne m$ because otherwise $z' = (f')^{m+1}x' = y'$. This implies that z and y lie in the same node, which has been proved impossible. This completes the proof that $m < q$.

We claim $m + 1 < p + q$. Indeed, the inequalities $0 \le p$ and $m < q$ imply that either $m + 1 < p + q$ or $p = 0 \wedge m + 1 = q$. The second disjunct cannot hold because otherwise $z' = (f')^{q+1}x' = f'y'$. Write $y' = [y; \ldots; u]$. This is not a unit path by the definitions of $S(H, L, x)$ and m, so $y \ne u$; however, y and u lie in the same node. Also from $p = 0$, it follows that $z = f^{p+1}y = fy$. So, ey and eu both lie in the node of z'. The existence of two edges, $\{y, ey\}$ and $\{u, eu\}$, between the same nodes contradicts the hypothesis on H of no double joins. This proves the claim and the lemma. □

4.7.3 transform

Definition 4.69 (transform) From one marked hypermap $(H, \mathcal{L}, L, x)$ in which L is canonically false, we construct a new tuple

$$T(H, \mathcal{L}, L, x) = (H, \mathcal{M}, L_1, x),$$

called the *transform* of $(H, \mathcal{L}, L, x)$. As the notation indicates, the hypermap H and the dart x are the same for both tuples. The data $\mathcal{M}$ and L_1 are specified in the following paragraph.

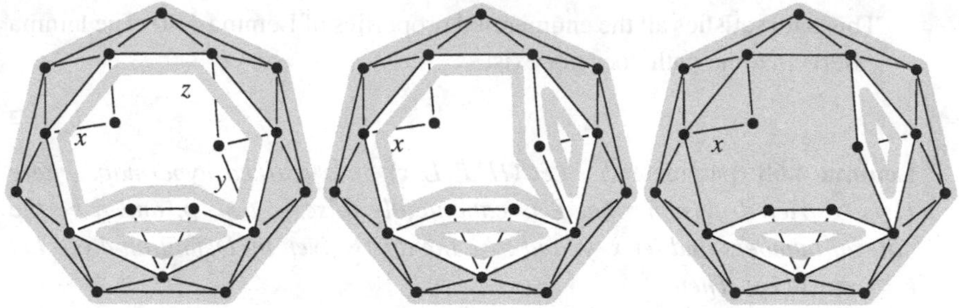

Figure 4.18 Starting in the first frame with the marked hypermap described in (Example 4.65), we take its transform to obtain the second frame, and the transform again to obtain the third frame. Each transform replaces the contour loop through the dart x with two contour loops. A shaded face indicates each contour loop that is confined to a face. A gray loop indicates a multi-face contour loop.

In the pencil-pen heuristic, the transform is the act of drawing a single simple arc in pen (Figure 4.18). Let m, p, y, and z be given by Definition 4.66. Let L_1 be the contour loop in H that follows L from x to y, takes f-steps from y to z, and then continues along L back to x. Let L_2 be the contour loop in H that follows L from $n^{-1}y$ to nz, and then complements the path of L_1 from y to z, traveling instead from nz to $n^{-1}y$. Let

$$\mathcal{M} = (\mathcal{L} \setminus \{L\}) \cup \{L_1, L_2\}. \tag{4.70}$$

Remark 4.71 (canonical compatibility) There is a canonical boolean function on $\mathcal{L}$ and one on $\mathcal{M}$. The canonical boolean functions agree on the intersection $\mathcal{L} \cap \mathcal{M}$. This means there is a well defined boolean-valued function on $\mathcal{L} \cup \mathcal{M}$. There is no ambiguity.

Lemma 4.72 (markup transform) *Let $(H, \mathcal{L}, L, x)$ be a marked hypermap in which H is a restricted hypermap and L is canonically false. Then the transform $(H, \mathcal{M}, L_1, x)$ is also a marked hypermap.*

In the pencil-pen heuristic, the marking of the hypermap records the state of the pen. The transform draws a single simple arc on paper in pen. The lemma asserts that in the act of drawing a simple arc, we retain a record of the state of the pen.

Proof Let

$$H = (D, e, n, f),\ H' = (D', e', n', f') = H/\mathcal{L},\ H'' = (D'', e'', n'', f'') = H/\mathcal{M}.$$

Let S' be the image of $S(H, L, x)$ in D'. Let y and z be the darts constructed in

Definition 4.66 from the marked hypermap $(H, \mathcal{L}, L, x)$. The dart z is not at the same node as y (by Lemma 4.67).

The proof can be organized into independent claims (typeset as usual in italic), according to the separate properties of a marked hypermap. The first part of the proof establishes that $\mathcal{M}$ is a normal family.

(NORMAL-1) *No dart is visited by two different loops.* Indeed by construction, the sets of darts of L_1 and L_2 are disjoint from each other and disjoint from the sets of darts of $L' \in \mathcal{L} \setminus \{L\}$. The result now follows from the normality of $\mathcal{L}$.

(NORMAL-2) *Every loop visits at least two nodes.* Indeed, this is true for L_1 and L_2 because they visit the nodes of y and z. It is also true of the other loops because they belong to the normal family $\mathcal{L}$.

(NORMAL-3) *If a loop visits a node, then every dart at that node is visited by some loop.* Indeed, the nodes that are visited by some loop in $\mathcal{M}$ are precisely those visited by some loop in $\mathcal{L}$, together with the *new* nodes; that is, the nodes of $fy, \ldots, f^p y$. The set of darts that are visited by some loop of $\mathcal{M}$ is the union of the set visited by loops in $\mathcal{L}$, together with the darts at the new nodes. As L_1 and L_2 are complementary at each new node, and as $\mathcal{L}$ itself is normal, the claim ensues. It follows that $\mathcal{M}$ is normal.

Next we check the properties of a marked hypermap (page 98).

(SIMPLE) To prove the simplicity of the subquotient, it is enough to show that none of the contour loops in $\mathcal{M}$ ever return to a node after leaving it. (In particular, the dart set of any $L' \in \mathcal{M}$ intersected with a node is the dart set of a maximal sublist $[z; n^{-1}z; \ldots; n^{-k}z]$ of n^{-1} steps.) This is true of $L' \in \mathcal{L} \setminus \{L\}$ by assumption and true of L_1 and L_2 by construction. Simplicity ensues.

(FIXED-POINT FREE) By Lemma 4.50, to prove that n'' does not have a fixed-point, it is enough to show that no loop in $\mathcal{M}$ has a dart set containing a node. It is sufficient to consider the loops L_1 and L_2. The set of darts of L_1 and L_2 at the old nodes (that is, those not meeting $\{fy, \ldots, f^p y\}$) are subsets of the set of darts of L at those nodes. As the dart set of L does not contain an old node, neither do L_1 and L_2. At the new nodes, L_1 and L_2 both have at least one dart, so neither contains the entire node. It follows that n'' is fixed-point free.

The third and fourth properties of a marked hypermap are routine verifications. Consider the flag property (page 97).

We claim that $e'y' \notin S'$, where y' is the image of y in H'. Otherwise, write $y' = [y; n^{-1}y; \ldots; u]$, a sublist of L, and select k such that $e'y' = [f^k x] \in S'$. Then

$$(n')^{-1}y' = f'e'y' = f'[f^k x] = [f^{k+1}x; \ldots].$$

By the construction of $S(H, L, x)$, we know that L visits $f^{k+1}x$. Hence, y' and $(n')^{-1}y'$ both lie in the same node and in the same face $\mathbf{F}(L)$. By the simplicity of H', it follows that $y' = (n')^{-1}y'$. That is, y' is a fixed point of n'. This is contrary to assumption. The claim ensues.

We claim that $e'y'$ lies in a true face of H'. Indeed, since φ_{can} is an S'-flag on H', the edge $\{y', e'y'\}$ meets a true face or S'. However, $y' \notin S'$, because each dart of S' is a unit path but y' is not. Also, $e'y' \notin S'$, by the previous paragraph. The dart y' lies in the false face $\mathbf{F}(L)$. The only remaining possibility is that $e'y'$ lies in a true face.

(FLAG-1) *The true faces of H'' are connected.* Indeed, L_1 is connected to a true face by the contour path $[x; n^{-1}x]$ because $n^{-1}x$ lies in the same face as nfx, which is a true face by assumption. If $L' \in \mathcal{M} \setminus \{L_1, L_2\}$ is true, then $L' \in \mathcal{L}$, and it connects with the true faces of $\mathcal{L}$ as before. If L_2 is true, then the proof requires more argument. Write $y' = [y; \ldots; u]$ as above. (The dart $e'y'$ is naturally identified with a dart $[eu]$ on in H'' because the face is true.) The dart $u'' = [n^{-1}y; \ldots; u]$ of H'' lies in the face $\mathbf{F}(L_2)$. Also,

$$(n'')^{-1}u'' = [n^{-1}u; \ldots] = [feu; \ldots] = f''[eu] = f''e'y'.$$

Thus, $(n'')^{-1}$ steps from $u'' \in \mathbf{F}(L_2)$ into a true face, and from there any true face may be reached.

(FLAG-2) *Each edge of H'' meets a true face or S'', where S'' is the image of $S(H, L_1, x)$ in D''.* Indeed, the function φ_{can} is an S'-flag on H'. Let $S_1 = \{fx, \ldots, f^{m+p+1}x\}$. Then $S(H, L, x) \subset S_1$ and

$$S_1 \subset \begin{cases} \mathbf{F}(L_1) & \text{if } \mathbf{F}(L_1) \text{ is true} \\ S(H, L_1, x) & \text{otherwise.} \end{cases}$$

The following four cases exhaust all edges of H''.

1. Edges of H'' that are identical to an edge of H', which is verified using the S'-flag on H';
2. edges that meet S_1;
3. the edge $\{u'', e''u''\}$, which is discussed and treated above;
4. the edge $\{z'', e''z''\}$, where $e'[z; \ldots] = [ez'']$, with z'' in a true face.

The other verifications are routine. □

4.7.4 algorithm

The aim in this chapter is to prove that every restricted hypermap with a given bound on the cardinality of the dart set is generated by a particular algorithm. The proof, a long induction argument, starts by showing how to go from one partially constructed hypermap to another more fully constructed one. The hypermap H represents the fully constructed one and the subquotient $H/\mathcal{L}$ represents a partially constructed one.

The data structures used to implementation of algorithm in computer are less abstract than the structures we have described in this chapter. This subsection describes how hypermaps have been implemented in code.

Definition 4.73 (listing) Let $H = (D, e, n, f)$ be a hypermap. Write $O(h, x)$ for the orbit of a dart x under a permutation h of D. In particular, $O(n, x)$ is the node of x. The *listing* of a dart $x \in D$ is the list

$$\ell_H(x) = [O(n, x); O(n, fx); \ldots ; O(n, f^{k-1}x)],$$

where k is the cardinality of the face of x. If ℓ is a list, let $[\![\ell]\!]$ denote the cyclic list of ℓ (Definition 4.28). The listing of H is the set

$$\{\ell_H(x) \;:\; x \in D\}.$$

To describe the computer implementation of the algorithm, we require some basic operators on lists.

Definition 4.74 (set, carrier, map, proper isomorphism) If ℓ is a list, let set(ℓ) be the set whose elements are the entries of ℓ. If $\ell = [x_0; \ldots ; x_{k-1}]$ is a list and ϕ is a function, we define the operator *map* by

$$\text{map}(\phi, \ell) = [\phi(x_0); \ldots ; \phi(x_{k-1})],$$

which applies ϕ to each entry of the list. Let g be a set of lists. The *carrier* of g is the set

$$\bigcup \{\text{set}(\ell) \;:\; \ell \in g\}.$$

Let g_1 and g_2 be two sets of lists. We say that they are *properly isomorphic*, if there is a bijection ϕ from the carrier of g_1 to the carrier of g_2, such that

$$\{[\![\text{map}(\phi, \ell)]\!] \;:\; \ell \in g_1\} = \{[\![\ell]\!] \;:\; \ell \in g_2\}.$$

Lemma 4.75 *Let x be a dart of a restricted hypermap $H = (D, e, n, f)$. Then the entries of $\ell_H(x)$ are all distinct.*

Proof If $O(n, f^i x) = O(n, f^j x)$, then by the simplicity of the hypermap, we have $f^i x = f^j x$. By construction, the exponents i and j are nonnegative and less than the cardinality of the face of x. This forces $i = j$. □

Lemma 4.76 *Let x and y be darts of a restricted hypermap $H = (D, e, n, f)$. If $x \neq y$, then $\ell_H(x) \neq \ell_H(y)$. In fact, the first two entries of the listings are sufficient to distinguish darts.*

Proof If the first two entries of $\ell_H(x)$ equal the first two entries of $\ell_H(y)$, then $O(n, x) = O(n, y)$ and $O(n, fx) = O(n, fy)$. The edges $\{x, nfx\}$ and $\{y, nfy\}$ of the hypermap run between the same nodes. By the no double joins condition of restricted hypermaps (Definition 4.48), the two edges are equal:

$$\{x, nfx\} = \{y, nfy\}.$$

The two nodes $O(n, x)$ and $O(n, fx)$ are distinct by Lemma 4.75. This forces $x = y$. □

Lemma 4.77 *Let x be a dart of a restricted hypermap $H = (D, e, n, f)$. If $\ell_H(x) = [x_0; \ldots, x_{k-1}]$, then $\ell_H(fx) = [x_1; \ldots; x_{k-1}; x_0]$. In particular, $[\![\ell_H(x)]\!] = [\![\ell_H(fx)]\!]$. As y runs over the face of x, $\ell_H(y)$ runs over the set of representatives of the cyclic list $[\![\ell_H(x)]\!]$.*

Proof This is an elementary verification. □

If we write ρ for the rotation operator on lists:

$$\rho[x_0; \ldots, x_{k-1}] = [x_1; \ldots; x_{k-1}; x_0],$$

then the first statement of the lemma takes the form

$$\ell_H(fx) = \rho\, \ell_H(x).$$

Lemma 4.78 *Let H_1 and H_2 be restricted hypermaps that have properly isomorphic listings. Then H_1 and H_2 are isomorphic hypermaps.*

Proof Write $H_1 = (D_1, e_1, n_1, f_1)$ and $H_2 = (D_2, e_2, n_2, f_2)$. Set $\ell_i = \ell_{H_i}$. Let ϕ be the bijection ϕ from the set of nodes of H_1 to the set of nodes of H_2 that realizes the proper isomorphism between listings. By the definition of proper isomorphism, for every $x \in D_1$, there exists $y \in D_2$ such that

$$[\![\mathrm{map}(\phi, \ell_1(x))]\!] = [\![\ell_2(y)]\!].$$

By Lemma 4.77, there is a dart z in the face of y such that

$$\mathrm{map}(\phi, \ell_1(x)) = \ell_2(z).$$

By Lemma 4.76, z is uniquely determined by this condition. Define $\psi : D_1 \to D_2$ by $\psi x = z$. By Lemma 4.76 again, from ϕ being one-to-one, we find that ψ is one-to-one.

Next we check the equivariance of ψ with respect to face permutations:

$$\begin{aligned}\ell_2(\psi f_1 x) &= \mathrm{map}(\phi, \ell_1(f_1 x))\\ &= \rho\, \mathrm{map}(\phi, \ell_1(x))\\ &= \rho\, \ell_2(\psi x)\\ &= \ell_2(f_2 \psi x).\end{aligned}$$

By Lemma 4.76, it follows that $\psi f_1 x = f_2 \psi x$.

Next we check the equivariance of ψ with respect to edge permutations. Recall that in a restricted hypermap, we have $e^2 = I_D$. This implies

$$O(n, ex) = O(n, fx), \quad O(n, fex) = O(n, x),$$

so that $\ell_H(ex) = [O(n, fx); O(n, x); \ldots]$. We compute

$$\begin{aligned}\ell_2(\psi e_1 x) &= \mathrm{map}(\phi, \ell_1(e_1 x))\\ &= [\phi(O(n_1, f_1 x)); \phi(O(n_1, x)); \ldots]\\ &= [O(n_2, \psi f_1 x); O(n_2, \psi x); \ldots]\\ &= [O(n_2, f_2 \psi x); O(n_2, \psi x); \ldots]\\ &= \ell_2(e_2 \psi x).\end{aligned}$$

It follows that $\psi e_1 x = e_2 \psi x$.

By the hypermap triality relation $enf = I_D$, the equivariance of ψ with respect to f and e give the equivariance with respect to n as well.

The map ψ is one-to-one between finite sets D_1 and D_2. Proper isomorphism is a symmetric relation. This means that there exists a one-to-one mapping from D_2 to D_1. Thus, D_1 and D_2 have the same cardinality, and ψ is a bijection. This proves that ψ is an isomorphism from H_1 to H_2. □

We give a description of a hypermap generating algorithm in very broad terms. In the algorithm, the primary data structure is a *graph record*, which has the general form of a tuple $(r, \ldots)$, where r is a list of ordered pairs (a, b), where a itself is a list of integers and b takes values true or false.

The algorithm depends on a parameter p, which gives the size of the largest face of the graphs generated by the algorithm. The algorithm starts with a particular graph record, called the pth seed:

$$\begin{aligned}\mathrm{seed}_p = ([&\\ &([0; 1; 2; \ldots; p-1], \mathrm{true});\\ &([p-1; \ldots; 2; 1; 0], \mathrm{false})\\ ],& \ldots).\end{aligned}$$

Every step of the algorithm has input a graph record, and output a finite set of graph records. We use the notation $g_1 \dashrightarrow_p g_2$ for the relation asserting that g_2 is among the graph records output, when the input is g_1. We write $\dashrightarrow_p^*$ for the reflexive transitive closure of the relation $\dashrightarrow_p$. That is, $\dashrightarrow_p^*$ is the smallest set of ordered pairs that contains $\dashrightarrow_p$ and that is both reflexive and transitive.

A graph record $([(a_0, b_0); \ldots], \ldots)$ is *final*, if all of the terms b_i are true. The algorithm consists of iterating $\dashrightarrow_p$ on seed_p, to obtain all graph records g such that

1. $\text{seed}_p \dashrightarrow_p^* g$, and
2. g is final.

We call graph records that satisfy these two properties *algorithmically planar* (with parameter p).

We now show in broad terms how to relate this algorithm to the classification of restricted hypermaps.

Definition 4.79 (record) Let $(H, \mathcal{L})$ be a pair consisting of a restricted hypermap H and a normal family $\mathcal{L}$ of H. We say that a graph record

$$([(a_0, b_0); \ldots; (a_{k-1}, b_{k-1})], \ldots)$$

is a *record* of $(H, \mathcal{L})$ provided that there is an proper isomorphism ϕ between the listing of $H/\mathcal{L}$ and the set

$$\{a_0, \ldots, a_{k-1}\}$$

such that for all darts x in the hypermap $H/\mathcal{L}$ and all indices i, if

$$[\![\text{map}(\phi, \ell_{H/\mathcal{L}}(x))]\!] = [\![a_i]\!],$$

then the canonical flag on the face of x takes value b_i.

Theorem 4.80 *Let H be a restricted hypermap. Let p be the cardinality of the largest face of H. Let $\mathcal{L}$ be the maximal normal family of Example 4.51. Then there exists an algorithmically plane graph record (with parameter p) that records $(H, \mathcal{L})$.*

The rest of this chapter sketches a proof of this theorem. The theorem shows that the hypermap generating algorithm captures all restricted planar hypermaps up to isomorphism. Specifically, a restricted hypermap H is isomorphic to its subquotient by the maximal normal family $\mathcal{L}$. If two hypermaps H_1 and H_2 have properly isomorphic records with respect to their maximal normal families, then they are isomorphic, by Lemma 4.78.

There is a more refined version of this theorem that produces records for each step of the algorithm. The following lemma treats the initialization step.

Lemma 4.81 *Let H be a restricted hypermap. Let p be the cardinality of the largest face of H. Let $\mathcal{L}$ be any face with p darts. Let $\mathcal{L}$ be the minimal normal family of Example 4.52 associated with F. Then* seed_p *records* $(H, \mathcal{L})$.

Proof This follows from the explicit description of seed_p and Examples 4.53 and 4.60, which identify $H/\mathcal{L}$ with the dihedral hypermap. □

Lemma 4.82 *Let H be a restricted hypermap. Let p be the cardinality of the largest face of H. Suppose that $\mathcal{L}$ is a normal family of H that is not maximal, and such that $H/\mathcal{L}$ is a simple hypermap on which the node map acts without fixed points. Suppose that the canonical function is a flag on $H/\mathcal{L}$. Suppose that g_1 is a graph record that records $(H, \mathcal{L})$. Then there exists a normal family $\mathcal{M}$ of H and a graph record g_2 with the following properties.*

1. *$H/\mathcal{M}$ is simple.*
2. *The node map acts without fixed points on $H/\mathcal{M}$.*
3. *The canonical function on $H/\mathcal{M}$ is a flag.*
4. *The set of darts visited by $\mathcal{M}$ has larger cardinality than the set of darts visited by $\mathcal{L}$.*
5. *g_2 records $(H, \mathcal{M})$.*
6. *$g_1 \dashrightarrow_p g_2$.*

Remark 4.83 A given graph record may record many different pairs $(H, \mathcal{L})$. In general, the relation $\dashrightarrow_p$ is multi-valued. Because of multiplicities of choices, the lemma only asserts the existence of $\mathcal{M}$ and g_2, but makes no uniqueness claims.

Proof sketch We give a few more details about the input-output relation $\dashrightarrow_p$ on a graph record

$$g_1 = ([(a_0, b_0); \ldots], \ldots).$$

The assumption that $\mathcal{L}$ is not maximal, implies that some term b_i is false. The algorithm selects some list a_i such that b_i is false. The claim of the lemma is independent of how this choice is made, and we disregard the details of the choice. The algorithm also selects an entry c_j of the list a_i. Because g_1 records $(H, \mathcal{L})$, there is a contour loop L in $\mathcal{L}$ that is canonically false that corresponds to a_i. Similarly, there is a node of $H/\mathcal{L}$ that corresponds to the entry c_j. Under the injection from nodes of $H/\mathcal{L}$ to nodes of H (Lemma 4.46), we obtain a node of H visited by L. Because the subquotient $H/\mathcal{L}$ is simple, the set of darts visited by L at the node has the form $\{n^{-k}x, \ldots, n^{-1}x, x\}$ for some uniquely determined dart x of H.

The rules about how to choose c_i are such that the resulting dart x has the property $S(H, L, x) = \emptyset$, and the loop of $\mathcal{L}$ visiting nfx is true. This means that $(H, \mathcal{L}, L, x)$ is a marked hypermap (Definition 4.64) with canonically false L.

By Lemma 4.72, the transform $(H, \mathcal{M}_1, L_1, x)$ of $(H, \mathcal{L}, L, x)$ is marked. We repeatedly take the tranform

$$T^i(H, \mathcal{L}, L, x) = (H, \mathcal{M}_i, L_i, x),$$

until L_i is canonically true. Set $\mathcal{M} = \mathcal{M}_i$.

We claim that there exists g_2 such that $g_1 \dashrightarrow_p g_2$, and such that g_2 records $(H, \mathcal{M})$. In fact, without elaborating on details, this is not surprising, because it has been arranged to be true by construction: the algorithm is just the implementation at the level of lists of integers the iterated transform of marked hypermaps. Specifically, it implements the transform (Definition 4.69) in lists. But there is one important distinction. The iterated transform starts with a specific marked hypermap H and results in a single normal family $\mathcal{M}$. However, $\dashrightarrow_p$ works without prior assumptions about the eventual end hypermap H, and generates all possible iterated transforms, subject to Lemma 4.68. For this reason, in general, g_2 will be just one of many graph records g such that $g_1 \dashrightarrow_p g$. Nevertheless, the claim of the lemma holds for this particular g_2. □

Remark 4.84 According to the pencil-pen heuristic, this section describes in rigorous terms the process of retracing a pencil drawing in pen. One transform of a marked hypermap corresponds to retracing a single pencil arc in pen. The iterated transform corresponds to retracing in pen an entire face. The step $\dashrightarrow_p$ corresponds to working without the pencil tracings, and drawing in pen a single face of cardinality at most p, along a given edge in all possible ways. The relation $\dashrightarrow_p^*$ corresponds to drawing all possible sequences of faces. Theorem 4.80 asserts that all restricted hypermaps admit a sequential construction in this fashion.

Finally, we turn to the proof of Theorem 4.80.

Proof Let H be a restricted hypermap. Let p be the cardinality of the largest face of H. Let $\mathcal{L}_0$ be the minimal normal family of H. By Lemma 4.81, $g_0 = \text{seed}_p$ records $(H, \mathcal{L}_0)$. Assume inductively, that g_i records $(H, \mathcal{L}_i)$ and that $H/\mathcal{L}_i$ has simple subquotient. If $i > 0$, we assume inductively that $g_{i-1} \dashrightarrow_p g_i$. If $\mathcal{L}_i$ contains a canonically false loop, then Lemma 4.82 constructs g_{i+1} and $\mathcal{L}_{i+1}$. The process terminates when every face of $\mathcal{L}_i$ is canonically true. The process must terminate, because the number of darts visited by $\mathcal{L}_i$ is increasing in i, and is bounded from above by the cardinality of the dart set of H. When every loop $\mathcal{L}_i$ is canonically true, then $\mathcal{L}_i$ is the maximal normal family, g_i is final, and

$$g_0 \dashrightarrow_p g_1 \dashrightarrow \cdots \dashrightarrow_p g_i, \quad \text{or} \quad g_0 \dashrightarrow_p^* g_i.$$

That is, g_i is an algorithmically plane graph record with parameter p. It records $(H, \mathcal{L}_i)$, where $\mathcal{L}_i$ is the maximal normal family. □

5

Fan

Summary. *This chapter is the final foundational chapter. The main concept is that of a fan, a geometric object that is related both to sphere packings and to hypermaps. A fan determines a set V of points in $\mathbb{R}^3$, which later chapters interpret as the set of centers of a packing of congruent balls. The same set V can be interpreted as the set of nodes of a hypermap or as the set of nodes of a graph. In fact, a fan is a graph and a geometric realization of a hypermap. The main result of this chapter, an Euler formula for fans, implies that the hypermap of a fan is planar. To make the material in this chapter as self-contained as possible, the planarity results in this chapter have been carefully organized to avoid any use of the Jordan curve theorem.*

Fans are also closely related to polyhedra. This chapter associates a fan with every bounded polyhedron in $\mathbb{R}^3$ with nonempty interior. Polyhedra inherit various properties from fans, such as an Euler formula for polyhedra.

5.1 Definitions

If $S \subset \mathbb{R}^3$ is a set of points, abbreviate

$$C_\pm(S) = \mathrm{aff}_\pm(\mathbf{0}, S),$$
$$C_\pm^0(S) = \mathrm{aff}_\pm^0(\mathbf{0}, S).$$

When the subscript is absent, the subscript + is implied: $C_+(S) = C(S)$, and so forth. The parentheses around the set are frequently omitted:

$$C^0\{\mathbf{v}, \mathbf{w}\} = C^0_+(\{\mathbf{v}, \mathbf{w}\}) = \mathrm{aff}^0_+(\{\mathbf{0}\}, \{\mathbf{v}, \mathbf{w}\}).$$

The following definition gives the main object of study in this chapter (Figure 5.1). The separate defining properties of a fan are given by name because we need to make frequent reference to them.

Definition 5.1 (fan, blade) Let (V, E) be a pair consisting of a set $V \subset \mathbb{R}^3$ and a set E of unordered pairs of distinct elements of V. The pair is said to be a *fan* if the following properties hold.

1. (CARDINALITY) V is finite and nonempty.
2. (ORIGIN) $\mathbf{0} \notin V$.
3. (NONPARALLEL) If $\{\mathbf{v}, \mathbf{w}\} \in E$, then $\mathbf{v}$ and $\mathbf{w}$ are not parallel.
4. (INTERSECTION) For all $\varepsilon, \varepsilon' \in E \cup \{\{\mathbf{v}\} : \mathbf{v} \in V\}$,
$$C(\varepsilon) \cap C(\varepsilon') = C(\varepsilon \cap \varepsilon').$$

When $\varepsilon \in E$, call $C^0(\varepsilon)$ or $C(\varepsilon)$ a *blade* of the fan.

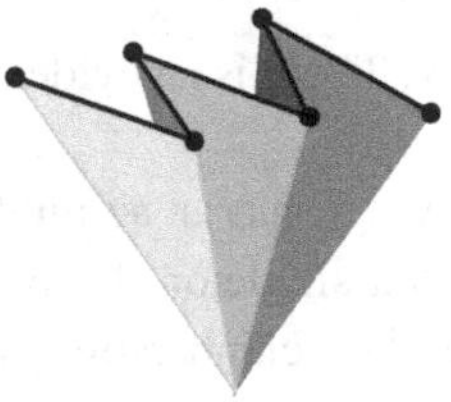

Figure 5.1 A fan with six nodes and five edges. An unbounded blade is associated with each edge.

Remark 5.2 In the mathematical literature, other objects go by the name of fan. The definition of fan given below is not the same as definitions in other mathematical contexts. In particular, a fan in this book is not a fan from the theory of toric varieties.[1]

[1] According to Fulton [13], a toric variety fan is a family Δ of rational strongly convex polyhedral cones in $N_{\mathbb{R}}$ (the vector space generated by a lattice N) such that (1) each face of a cone in the family is again in the family, and (2) the intersection of two cones in the family is a face of each cone. For our purposes, a fan determines a set of strongly convex polyhedral cones $\Delta = \{C(\varepsilon) : \varepsilon \in E \cup \{\{v\}\} \cup \{\{\mathbf{0}\}\}\}$, which satisfies conditions (1) and (2). Hence, a fan in our sense bears some relation to a toric variety fan.

5.1.1 basic properties

The rest of the chapter develops the properties of fans. We begin with a completely trivial consequence of the definition.

Lemma 5.3 *If (V, E) is a fan, then for every $E' \subset E$, (V, E') is also a fan.*

Proof This proof is elementary. □

Lemma 5.4 (fan cyclic) *Let (V, E) be a fan. For each $\mathbf{v} \in V$, the set*

$$E(\mathbf{v}) = \{\mathbf{w} \in V \,:\, \{\mathbf{v}, \mathbf{w}\} \in E\}$$

is cyclic with respect to $(\mathbf{0}, \mathbf{v})$.

Proof If $\mathbf{w} \in E(\mathbf{v})$, then $\mathbf{v}$ and $\mathbf{w}$ are not parallel. Also, if $\mathbf{w} \neq \mathbf{w}' \in E(\mathbf{v})$, then

$$C\{\mathbf{v}, \mathbf{w}\} \cap C\{\mathbf{v}, \mathbf{w}'\} = C\{\mathbf{v}\}.$$

This implies that $E(\mathbf{v})$ is cyclic. □

Remark 5.5 (easy consequences of the definition) Let (V, E) be a fan.

1. The pair (V, E) is a graph with nodes V and edges E. The set

$$\{\{\mathbf{v}, \mathbf{w}\} \,:\, \mathbf{w} \in E(\mathbf{v})\}$$

   is the set of edges at node $\mathbf{v}$. There is an evident symmetry: $\mathbf{w} \in E(\mathbf{v})$ if and only if $\mathbf{v} \in E(\mathbf{w})$.
2. Since $E(\mathbf{v})$ is cyclic, each $\mathbf{v} \in V$ has an azimuth cycle $\sigma(\mathbf{v}) : E(\mathbf{v}) \to E(\mathbf{v})$. The set $E(\mathbf{v})$ can reduce to a singleton. If so, $\sigma(\mathbf{v})$ is the identity map on $E(\mathbf{v})$. To make the notation less cumbersome, $\sigma(\mathbf{v}, \mathbf{w})$ denotes the value of the map $\sigma(\mathbf{v})$ at $\mathbf{w}$.
3. The property (NONPARALLEL) implies that the graph has no loops: $\{\mathbf{v}, \mathbf{v}\} \notin E$.
4. The property (INTERSECTION) implies that distinct sets $C^0(\varepsilon)$ do not meet. This property of fans is eventually related to the planarity of hypermaps.

Remark 5.6 (verifying the fan properties) We are often given a pair (V, E) and asked to verify that it is a fan. Here are a few tips about how to verify the fan properties in practice.

1. (CARDINALITY) If V is defined as a subset or image of a finite set, then it is evidently finite. Also, if V is a bounded subset of $\mathbb{R}^3$ and if V has no limit point, then V is finite. Lemma 6.2 gives a finiteness result when the minimum distance is 2.

2. (ORIGIN) If (V, E) is a fan, then any subset of V inherits the property $\mathbf{0} \notin V$ from V.

3. (NONPARALLEL) *If the property* (INTERSECTION) *is known, then to prove that* $\mathbf{u}$ *and* $\mathbf{v}$ *are not parallel, it is enough to show the strict form of the triangle inequality:*

$$\|\mathbf{u} - \mathbf{v}\| < \|\mathbf{u}\| + \|\mathbf{v}\|. \tag{5.7}$$

Indeed, the strict form of the triangle inequality implies that $\mathbf{0} \notin \mathrm{conv}\{\mathbf{u}, \mathbf{v}\}$. Also, the intersection property implies that $C^0\{\mathbf{v}\} \cap C^0\{\mathbf{u}\} = \varnothing$. These conditions imply that $\mathbf{u}$ and $\mathbf{v}$ are not parallel. Inequality (5.7) is equivalent to

$$\mathrm{arc}_V(\mathbf{0}, \{\mathbf{u}, \mathbf{v}\}) < \pi.$$

4. (INTERSECTION) The intersection property is generally the most difficult to verify in practice. Some geometrical reasoning based on additional facts about (V, E) is generally required to verify the intersection property. If (V, E) is a fan, then the intersection property is inherited by subsets of V and E. Also, note that

$$C(\varepsilon \cap \varepsilon') \subset C(\varepsilon) \cap C(\varepsilon')$$

always holds by elementary geometry. Hence, it is enough to check the reverse inclusion. Furthermore, if $\varepsilon = \varepsilon'$, then the intersection property is a triviality. The verification comes down to checking two cases:

$$C(\varepsilon) \cap C(\varepsilon') = \{\mathbf{0}\},$$

when $\varepsilon \cap \varepsilon' = \varnothing$, and

$$C(\varepsilon) \cap C(\varepsilon') = C\{\mathbf{v}\},$$

when $\varepsilon \cap \varepsilon' = \{\mathbf{v}\}$.

5.1.2 hypermap

One of the main uses of a fan in this book is to provide a link between sphere packings and hypermaps. Next, we associate a hypermap can be associated with a fan. We define a set of darts and permutations on that set of darts as follows.

Let (V, E) be a fan. Define a set of darts D to be the disjoint union of two sets D_1, D_2:

$$D_1 = \{(\mathbf{v}, \mathbf{w}) \ : \ \{\mathbf{v}, \mathbf{w}\} \in E\},$$
$$D_2 = \{(\mathbf{v}, \mathbf{v}) \ : \ \mathbf{v} \in V, \ E(\mathbf{v}) = \varnothing\}, \ \text{and}$$
$$D \ = D_1 \cup D_2.$$

Darts in D_2 are said to be *isolated* and darts in D_1 are *nonisolated*. Define permutations n, e, and f on D_1 by

$$n(\mathbf{v}, \mathbf{w}) = (\mathbf{v}, \sigma(\mathbf{v}, \mathbf{w})),$$
$$f(\mathbf{v}, \mathbf{w}) = (\mathbf{w}, \sigma(\mathbf{w})^{-1}\mathbf{v}),$$
$$e(\mathbf{v}, \mathbf{w}) = (\mathbf{w}, \mathbf{v}).$$

Define permutations n, e, f on D_2 by making them degenerate on D_2:

$$n(\mathbf{v}) = e(\mathbf{v}) = f(\mathbf{v}) = \mathbf{v}.$$

Set $\text{hyp}(V, E) = (D, e, n, f)$. The next lemma shows that (D, e, n, f) is indeed a hypermap.

Lemma 5.8 *Let* (V, E) *be a fan. Let* $D = D_1 \cup D_2$ *and* $\text{hyp}(V, E) = (D, e, n, f)$, *as constructed above. Then*

1. $\text{hyp}(V, E)$ *is a plain hypermap.*
2. e *has no fixed points in* D_1.
3. f *has no fixed points in* D_1.
4. *For every pair of distinct nodes, at most one edge meets both.*
5. *The two darts of an edge of* D_1 *lie at different nodes.*

Proof

$$e(n(f(\mathbf{v}, \mathbf{w}))) = e(n(\mathbf{w}, \sigma(\mathbf{w})^{-1}\mathbf{v})) = e(\mathbf{w}, \mathbf{v})$$
$$= (\mathbf{v}, \mathbf{w}).$$

So $\text{hyp}(V, E)$ is a hypermap. Plainness is an elementary calculation:

$$e(e(\mathbf{v}, \mathbf{w})) = e(\mathbf{w}, \mathbf{v}) = (\mathbf{v}, \mathbf{w}).$$

There is no fixed point in D_1 under e, because otherwise, $\mathbf{v} = \mathbf{w} \in E(\mathbf{v})$ but by construction $\mathbf{v} \notin E(\mathbf{v})$. The argument that f has no fixed points is similar.

The next step is to show that for every two distinct nodes, there is at most one edge meeting both. That is,

$$(n^i ex = en^j x) \Rightarrow (n^j x = x).$$

Let $x = (\mathbf{v}, \mathbf{w}) \in D_1$. Let $\sigma = \sigma(\mathbf{v})$. Then

$$\begin{aligned} n^j x &= (\mathbf{v}, \sigma^j \mathbf{w}) \\ e n^j x &= (\sigma^j \mathbf{w}, *) \\ e x &= (\mathbf{w}, *) \\ n^i e x &= (\mathbf{w}, *) \\ n^i e x &= e n^j x \\ &\Rightarrow (\mathbf{w} = \sigma^j \mathbf{w}) \\ &\Rightarrow (n^j x = (\mathbf{v}, \mathbf{w}) = x). \end{aligned}$$

Finally, each dart of an edge lies on a different node. That is, $ex \neq n^i x$ for $x \in D_1$. In detail:

$$\begin{aligned} e(\mathbf{v}, \mathbf{w}) &= (\mathbf{w}, *), \quad \mathbf{w} \in E(\mathbf{v}) \\ n^i(\mathbf{v}, \mathbf{w}) &= (\mathbf{v}, *), \quad \mathbf{v} \notin E(\mathbf{v}). \end{aligned}$$

The lemma ensues. □

5.2 Topology

5.2.1 background

This chapter uses some basic notions from topology such as continuity, connectedness, and compactness.

Remark 5.9 We use the term *connected* in two different senses: in the topological sense and in a combinatorial sense for hypermaps. To reduce the confusion, this book calls a connected component of a topological space a *topological component* and a connected component of a hypermap a *combinatorial component.*

We assume basic facts about the topology of Euclidean space. In particular, the set $\mathbb{R}^3$ is a metric space under the Euclidean distance function $d(\mathbf{v}, \mathbf{w}) = \|\mathbf{v} - \mathbf{w}\|$. Every subset of $\mathbb{R}^3$ is a metric space under the restriction of the metric d to the subset. A subset carries the metric space topology. In particular,

$$S^2 = \{\mathbf{v} \;:\; \|\mathbf{v}\| = 1\},$$

the unit sphere in $\mathbb{R}^3$ centered at $\mathbf{0}$, is a metric space and a topological space.

If Y is an open set in $\mathbb{R}^3$, write $[Y]$ for its set of topological components. The family of topological components of Y has the following properties: the members are pairwise disjoint, nonempty, connected open sets; and the union of the family is all of Y. Conversely, any family with these properties must be the family of topological components of Y. If two points in $\mathbb{R}^3$ can be joined by a continuous path in Y, then the two points lie in the same topological component of Y.

5.2.2 topological component and dart

The next series of definitions and lemmas introduce terminology to refer to the different geometric features of a fan. A major theme of this chapter is the correspondence between the geometric features of a fan and the combinatorial properties of the hypermap. For example, we show that the set of nodes of the hypermap is in natural bijection with the set of nodes V of the fan (V, E) (Lemma 5.13). We describe when the set of faces of the hypermap is in bijection with the topological components of the following set $Y(V, E)$.

Definition 5.10 (X, Y) Let (V, E) be a fan. Let $X = X(V, E)$ be the union of the blades

$$C(\varepsilon)$$

as ε ranges over E. Let $Y = Y(V, E)$ be the complement $Y = \mathbb{R}^3 \setminus X$.

Definition 5.11 (W^0_{dart}, W_{dart}) Let (V, E) be a fan and let $(D, e, n, f) = \text{hyp}(V, E)$ be the associated hypermap. A wedge $W^0_{\text{dart}}(x)$ and a subset $W^0_{\text{dart}}(x, \epsilon)$ are associated with each dart $x = (\mathbf{v}, \mathbf{w}) \in D$. (See Figure 5.2.) Define

$$W^0_{\text{dart}}(x) = \begin{cases} W^0(\mathbf{0}, \mathbf{v}, \mathbf{w}, \sigma(\mathbf{v}, \mathbf{w})), & \text{if } \text{card}(E(\mathbf{v})) > 1, \\ \mathbb{R}^3 \setminus \text{aff}_+(\{\mathbf{0}, \mathbf{v}\}, \mathbf{w}), & \text{if } E(\mathbf{v}) = \{\mathbf{w}\}, \\ \mathbb{R}^3 \setminus \text{aff}\{\mathbf{0}, \mathbf{v}\}, & \text{if } E(\mathbf{v}) = \varnothing. \end{cases}$$

Define

$$W_{\text{dart}}(x) = \begin{cases} W(\mathbf{0}, \mathbf{v}, \mathbf{w}, \sigma(\mathbf{v}, \mathbf{w})), & \text{if } \text{card}(E(\mathbf{v})) > 1, \\ \mathbb{R}^3, & \text{otherwise.} \end{cases}$$

($W_{\text{dart}}(x)$ is the closure of $W^0_{\text{dart}}(x)$.) For any $x = (\mathbf{v}, \ldots) \in D$, set

$$W^0_{\text{dart}}(x, \epsilon) = W^0_{\text{dart}}(x) \cap \text{rcone}^0(\mathbf{0}, \mathbf{v}, \cos \epsilon).$$

Definition 5.12 (azim) Define azim(x) as the azimuth angle of $W^0_{\text{dart}}(x)$:

$$\text{azim}(x) = \begin{cases} \text{azim}(\mathbf{0}, \mathbf{v}, \mathbf{w}, \sigma(\mathbf{v}, \mathbf{w})), & \text{if } \text{card}(E(\mathbf{v})) > 1, \\ 2\pi, & \text{otherwise.} \end{cases}$$

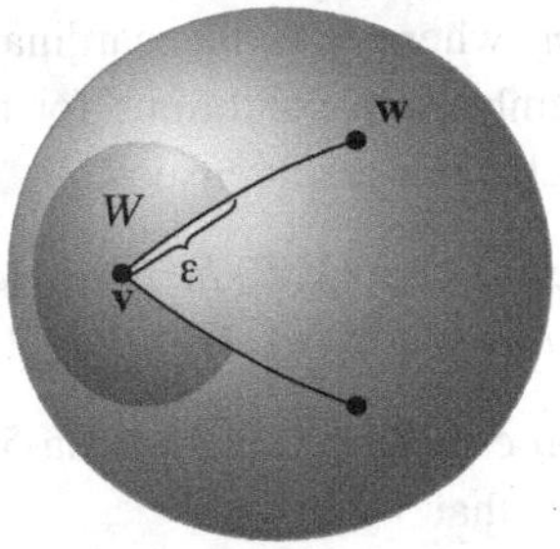

Figure 5.2 The set $W = W(x, \varepsilon)$ at the dart $x = (\mathbf{v}, \mathbf{w})$ is the cone over a wedge of a disk on the unit sphere. The geodesic radius of the disk is ε. If $\|\mathbf{v}\| = \|\mathbf{w}\| = 1$, then the disk is centered at $\mathbf{v}$, and the wedge extends counterclockwise from the geodesic arc of $(\mathbf{v}, \mathbf{w})$ until it meets the next blade of the fan through $\mathbf{v}$. The angle of this wedge at $\mathbf{v}$ equals azim(x).

Lemma 5.13 *Let (V, E) be a fan with hypermap H. There is a natural bijection between nodes of H and V that sends the node containing the dart $(\mathbf{v}, *)$ to $\mathbf{v} \in V$.*

Proof This is left as an exercise for the reader. □

Definition 5.14 (node) Write node$(x) \in V$ for the node corresponding to a dart $x \in D$, under the identification of nodes of a hypermap H with V.

Lemma 5.15 (node partition) *Let (V, E) be a fan. Let $\mathbf{v} \in V$. Then a disjoint sum decomposition of $\mathbb{R}^3$ is given by*

$$\mathbb{R}^3 = \mathrm{aff}\{\mathbf{0}, \mathbf{v}\} \cup \bigcup_{\mathrm{node}(x)=\mathbf{v}} W^0_{dart}(x) \cup \bigcup_{\{\mathbf{v},\mathbf{w}\}\in E} \mathrm{aff}^0_+(\{\mathbf{0}, \mathbf{v}\}, \mathbf{w}).$$

Proof We start the proof with the existence of the disjoint sum decomposition. First of all, $\mathbb{R}^3$ is the disjoint union of aff$\{\mathbf{0}, \mathbf{v}\}$ and its complement.

The case when card$(E(\mathbf{v})) \leq 1$ follows immediately from the definitions. Therefore, assume that card$(E(\mathbf{v})) > 1$. Fix $\mathbf{u}$ such that $\{\mathbf{v}, \mathbf{u}\} \in E$, and let σ be the azimuth cycle on $E(\mathbf{v})$. Let $\alpha(i) = \mathrm{azim}(\mathbf{0}, \mathbf{v}, \sigma^i\mathbf{u}, \sigma^{i+1}\mathbf{u})$. By Lemma 2.94, the sum of the angles $\alpha(i)$ is 2π. Every $\mathbf{p} \in \mathbb{R}^3 \setminus \mathrm{aff}\{\mathbf{0}, \mathbf{v}\}$ satisfies either

$$\sum_{i=0}^{j} \alpha(i) < \mathrm{azim}(\mathbf{0}, \mathbf{v}, \mathbf{u}, \mathbf{p}) < \sum_{i=0}^{j+1} \alpha(i).$$

or

$$\sum_{i=0}^{j} \alpha(i) = \mathrm{azim}(\mathbf{0}, \mathbf{v}, \mathbf{u}, \mathbf{p})$$

for a unique $0 \leq j < n$, where n is the cardinality of $E(\mathbf{v})$. These conditions are exactly the membership conditions for the sets $W^0_{\text{dart}}(\mathbf{v}, \sigma^j\mathbf{u})$ and $\text{aff}^0_+(\{\mathbf{0}, \mathbf{v}\}, \sigma^j\mathbf{u})$, respectively. The result ensues. □

Corollary 5.16 (disjointness) *Let (V, E) be a fan, let $x = (\mathbf{v}, \ldots)$ be a dart in the hypermap of (V, E) and let $\mathbf{w} \in E(\mathbf{v})$. Then $W^0_{dart}(x) \cap C\{\mathbf{v}, \mathbf{w}\} = \emptyset$.*

Proof The decomposition established in Lemma 5.15 is disjoint. It follows directly from the definitions that

$$C\{\mathbf{v}, \mathbf{w}\} \subset \text{aff}^0_+(\{\mathbf{0}, \mathbf{v}\}, \mathbf{w}) \cup \text{aff}\{\mathbf{0}, \mathbf{v}\}.$$

□

The next lemma gives an important map from a combinatorial structure (the set of darts) to a topological object (the set $[Y(V, E)]$). After presenting the proof, we codify the map in a definition.

Lemma 5.17 (dart and topological component) *Let (V, E) be a fan. For each dart x in the hypermap of (V, E) and for every ϵ sufficiently small and positive, $W^0_{dart}(x, \epsilon)$ is nonempty and lies in a single topological component of $Y(V, E)$.*

Proof The proof first shows that $W^0_{\text{dart}}(x, \epsilon)$ lies in Y for ϵ small. Let $x = (\mathbf{v}, \mathbf{w}) \in D_1$. Let S^2 be the unit sphere centered at $\mathbf{0}$. By making ϵ small enough, the sets[2] $W^0_{\text{dart}}(x, \epsilon) \cap S^2$ avoid the compact sets $C(\varepsilon) \cap S^2$ when $\mathbf{v} \notin \varepsilon$. Thus, $W^0_{\text{dart}}(x, \epsilon)$ also avoids $C(\varepsilon)$ when $\mathbf{v} \notin \varepsilon$. By Corollary 5.16, $W^0_{\text{dart}}(x, \epsilon)$ avoids $C(\varepsilon)$, when $\mathbf{v} \in \varepsilon$. Thus, $W^0_{\text{dart}}(x, \epsilon) \subset Y$ for ϵ small.

To complete the proof, it is enough to show that each $W^0_{\text{dart}}(x, \epsilon)$ is connected. The product of intervals

$$\{(r, \theta, \phi) \;:\; r \in (0, \infty),\ \theta \in (\theta_1, \theta_2),\ \phi \in (0, \epsilon)\}$$

is connected. The set $W^0_{\text{dart}}(x, \epsilon)$ is the image of this product under a spherical coordinate representation (Definition 2.85). It is readily verified that the spherical to Cartesian coordinate transformation is a continuous map. As the image of a connected set under a continuous map, $W^0_{\text{dart}}(x, \epsilon)$ is connected. □

Definition 5.18 (lead into) Let (V, E) be a fan. For each dart x in the hypermap of (V, E), there exists a well-defined topological component U_x of $Y(V, E)$ that contains $W^0_{\text{dart}}(x, \epsilon)$ (for all sufficiently small positive ϵ). The dart x is said to *lead into* U_x.

[2] Beware of the notational subtleties: $\epsilon \in \mathbb{R}$ is not $\varepsilon \in E$.

5.3 Planarity

One of the main results of this chapter holds that the hypermaps associated with certain fans are planar (Lemma 5.30). Recall that a hypermap is defined to be planar if Euler's formula holds for the hypermap. Every fan is a graph and is *plane* in the sense of being embedded in a sphere. The proof of hypermap planarity ultimately reduces to the Euler formula for the fan, viewed as a plane graph.

There are many proofs of Euler's formula for a graph. We may pick our favorite and translate it into the language of hypermaps. The graph has special properties that allow us to simplify the proof of Euler's formula: it is embedded in the sphere with edges formed by geodesic arcs, and the faces are all geodesically convex polygons.

We prove Euler's formula as follows. A geodesically convex polygon has a diagonal (Lemma 5.20). The diagonal, which breaks a polygon into two smaller ones, permits an induction on the number of sides for the area of each polygonal face of the hypermap, generalizing Girard's formula for the area of a triangles. An identity, which equates the sum of these areas with the total surface area of a sphere, is equivalent to Euler's formula. This proof can be recognized as a special case of the Gauss–Bonnet formula, which relates the Euler characteristic to the area of a surface of constant curvature.

The reader who is ready to believe Euler's formula for plane graphs and to accept that the hypermaps of fans are planar may safely skip this section.

5.3.1 face attributes

To simplify the proofs in this section, we generally assume that fans satisfy the following convexity condition.

Definition 5.19 (fully surrounded) A fan (V, E) is *fully surrounded*, if $\text{azim}(x) < \pi$ for all darts x in the hypermap of (V, E).

The following lemma proves the existence of a diagonal to a nontriangular face. A diagonal divides a face into two faces, each with fewer sides than the original. The existence of the diagonal occurs in induction arguments to reduce statements about a face to statements about faces with fewer sides. The following lemma is thus the key technical lemma for many of the results in this section.

Lemma 5.20 (sweep) *Let (V, E) be a fan with hypermap (D, e, n, f). Suppose that (V, E) is fully surrounded. Fix a dart $x \in D$. Let $\mathbf{v} = \text{node}(x)$,*

$\mathbf{v}_0 = \text{node}(fx)$, *and* $\mathbf{v}_1 = \text{node}(f^2x)$. *Let* $\mathbf{w}(t) = (1-t)\mathbf{v}_0 + t\mathbf{v}_1$ *for* $0 \le t \le 1$. *Then*

1. *For each* $t \in [0,1]$, $\mathbf{v}$ *and* $\mathbf{w}(t)$ *are not parallel.*
2. *If* $0 < t \le 1$, *and* $C^0\{\mathbf{v}, \mathbf{w}(t)\}$ *meets* X, *then* $t = 1$ *and* $\{\mathbf{v}, \mathbf{v}_1\} \in E$. *(See Figure 5.3.)*

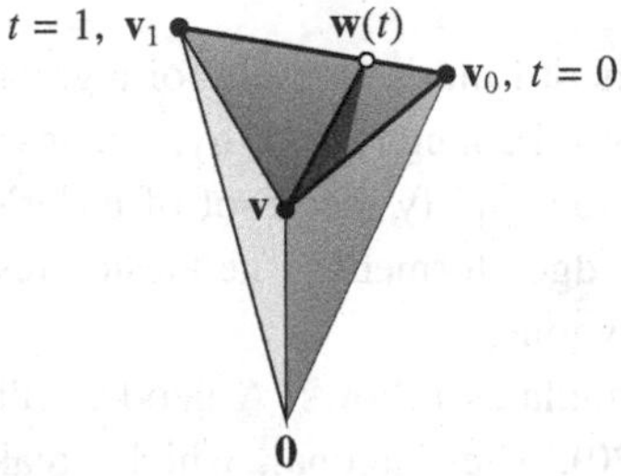

Figure 5.3 As a point $\mathbf{w}(t)$ slides from $\mathbf{v}_0$ to $\mathbf{v}_1$, the blade $C^0\{\mathbf{v}, \mathbf{w}(t)\}$ avoids the blades of the fan (V, E), when $0 < t < 1$.

Proof Abbreviate $C^0(t) = C^0\{\mathbf{v}, \mathbf{w}(t)\}$. Let $Y = Y(V, E)$ and $X = X(V, E)$. It follows from the definition of a fan that $\{\mathbf{v}, \mathbf{v}_0\} \in E$ and that $\mathbf{v}$ and $\mathbf{v}_0$ are not parallel. By continuity, $\mathbf{v}$ and $\mathbf{w}(t)$ are not parallel when t is sufficiently small and positive. Let $I \subset (0, 1]$ be any interval that contains $(0, \epsilon)$ for some sufficiently small positive ϵ, with the property that $\mathbf{v}$ and $\mathbf{w}(t)$ are not parallel for all $t \in I$.

We claim that if $t \in I$ *and* $C^0(t)$ *meets* X, *then* $t = 1$ *and* $\{\mathbf{v}, \mathbf{v}_1\} \in E$. Indeed, an inspection of possible intersections with nodes $\mathbf{u} \in V$ and blades $C^0(\varepsilon) \subset X$ shows that for $t > 0$ sufficiently small, $C^0(t)$ does not meet X; hence, $C^0(t) \subset Y$. Assuming that $C^0(t)$ meets X for some $t \in I$, let a be the smallest such $t \in I$. $C^0(a)$ cannot meet X at a node $\mathbf{u} \in V$ because $\text{azim}(y) < \pi$ whenever $\mathbf{u} = \text{node}(y)$, which means that there exists a smaller $t < a$ for which $C^0(t)$ meets a blade at $\mathbf{u}$. Thus, $C^0(a)$ first meets X along a blade $C^0(\varepsilon)$. If the intersection with this blade is transversal, again one can find a smaller t that gives an intersection with the blade. Hence, $C^0(a)$ and $C^0(\varepsilon)$ are coplanar. From the disjointness properties of blades of a fan, it follows that $\varepsilon = \{\mathbf{v}, \mathbf{v}_1\} \in E$, that $a = 1$, and that $C^0(1)$ is a blade of the fan. The claim ensues.

The vectors $\mathbf{v}$ *and* $\mathbf{w}(t)$ *are not parallel for any* $t \in (0, 1]$. Otherwise, let $b \in (0, 1]$ be the least constant for which $\mathbf{v}$ and $\mathbf{w}(b)$ are parallel. Set $I = (0, b)$. Select a such that $0 < a < b$. Then $\{\mathbf{0}, \mathbf{w}(a), \mathbf{w}(b), \mathbf{v}\}$ lie in a unique plane A. Since all $\mathbf{w}(t)$ lie in a line, $\mathbf{w}(t) \in A$ all $t \in I$. Then $\mathbf{v}_0 \in C^0(a) \cap X$, contradicting the established disjointness of X from $C^0(a)$. Thus, b does not exist, proving the claim and the first conclusion of the lemma.

Set $I = \{t : 0 < t \le 1\}$. The second conclusion of the lemma follows immediately from the claim. □

Lemma 5.21 (face to component) *Let (V, E) be a fan and let (D, e, n, f) be its hypermap. Assume that (V, E) is fully surrounded. Then for every face F of the hypermap, there exists a topological component U of $Y(V, E)$ such that for every $x \in F$, the dart x leads into U.*

This lemma strengthens the relationship between the combinatorics of hypermaps and the topology of fans by showing that there exists a well-defined map from faces to topological components. Write $F \mapsto U_F$ for this map.

Proof Fix any dart $x \in F$ and construct the set $C^0(t)$ as in the previous lemma. For all $\epsilon > 0$ sufficiently small, there exists $\delta > 0$ such that set $C^0(t)$ meets both $W^0_{\text{dart}}(x, \epsilon)$ and $W^0_{\text{dart}}(fx, \epsilon)$ for all $0 < t < \delta$. By the previous lemma, the set $C^0(t)$ lies in a single component U when t is sufficiently small and positive. Thus, x and fx lead into the same component U. By induction, for all $y \in F$, the dart y leads into U. □

The following lemma appears in induction arguments to show that a statement about one fan (V, E) can be reduced to a statement about a simpler fan (V, E').

Lemma 5.22 (fan diagonal) *Let (V, E) be a fully surrounded fan and let $\mathbf{v}, \mathbf{w} \in V$ be nonparallel. Suppose that $C^0\{\mathbf{v}, \mathbf{w}\} \subset U_F$ for some face F. Let $E' = E \cup \{\{\mathbf{v}, \mathbf{w}\}\}$. Then (V, E') is a fan.*

Proof We establish each of the defining properties of a fan in turn. The node set is unchanged, remains finite and nonempty, and does not contain $\mathbf{0}$. The property (NONPARALLEL) of E' follow from the corresponding property of E and the assumed nonparallelism for $\{\mathbf{v}, \mathbf{w}\}$.

In the verification of the intersection property

$$C(\varepsilon) \cap C(\varepsilon') = C(\varepsilon \cap \varepsilon'),$$

it is enough to consider the case $\varepsilon = \{\mathbf{v}, \mathbf{w}\}$ and $\varepsilon' \ne \varepsilon$, the other cases being trivial. Then from elementary geometry

$$C\{\mathbf{v}, \mathbf{w}\} = C^0\{\mathbf{v}, \mathbf{w}\} \cup C\{\mathbf{v}\} \cup C\{\mathbf{w}\}$$

and known facts $C^0\{\mathbf{v}, \mathbf{w}\} \subset U_F$, $C(\varepsilon') \subset X(V, E)$, and $X(V, E) \cap U_F = \emptyset$, it follows that

$$\begin{aligned} C\{\mathbf{v},\mathbf{w}\} \cap C(\varepsilon') &= (C\{\mathbf{v}\} \cup C\{\mathbf{w}\}) \cap C(\varepsilon') \\ &= C(\{\mathbf{v}\} \cap \varepsilon') \cap C(\{\mathbf{w}\} \cap \varepsilon') \\ &= C(\{\mathbf{v},\mathbf{w}\} \cap \varepsilon'). \end{aligned}$$

(The last equality uses the observation that at most one of the two intersections $* \cap \varepsilon'$ in the penultimate line is nonzero.) Thus, (V, E') is a fan. □

The following lemma further strengthens the relationship between combinatorics and topology by showing that the map $F \mapsto U_F$ is onto.

Lemma 5.23 *Let (V, E) be a fan with hypermap (D, e, n, f). Assume that (V, E) is fully surrounded. For every topological component U of $Y(V, E)$, there exists a dart $x \in D$ that leads into U.*

Proof The sets $C^0(t)$ of Lemma 5.20 depend on the initial dart x. Write $C^0(t, x)$ to make the dependence explicit.

Let $\mathbf{p} \in U$. Choose a continuous path $\varphi : [0, 1] \to \mathbb{R}^3 \setminus \{0\}$ such that $\varphi(t) \in U$ for $t < 1$ and $\varphi(1) \notin U$. Then $\mathbf{q} = \varphi(1) \in X$. If $\mathbf{q} \in C^0\{\mathbf{v}\}$ for some $\mathbf{v} \in V$, then there exists a dart x with node $\mathbf{v} = \text{node}(x)$ such that for all sufficiently small positive ϵ, there exists some $0 \le t < 1$ such that $\varphi(t) \in W^0_{\text{dart}}(x, \epsilon) \subset U_x$. Thus, x leads into U.

The other possibility is that $\mathbf{q} \in C^0\{\mathbf{v}, \mathbf{w}\}$ for some $\{\mathbf{v}, \mathbf{w}\} \in E$. In this case, there exists a unique edge $\{x, y\}$ of the hypermap such that $\mathbf{v} = \text{node}(x)$ and $\mathbf{w} = \text{node}(y)$. (That is, $x = (\mathbf{v}, \mathbf{w})$ and $y = (\mathbf{w}, \mathbf{v})$.) There is also a small neighborhood of $\mathbf{q}$ such that every point $\mathbf{q}'$ in that neighborhood takes one of the following forms.

1. $\mathbf{q}' \in C^0\{\mathbf{v}, \mathbf{w}\}$,
2. $\mathbf{q}' \in C^0(s, x) \subset U_x$ for some $0 < s < 1$,
3. $\mathbf{q}' \in C^0(s, y) \subset U_y$ for some $0 < s < 1$.

Points of the first form do not meet $Y(V, E)$. Thus, $\varphi(t) \in U_x$ or $\varphi(t) \in U_y$, in the second and third forms respectively. □

As the introduction to this section mentioned, the primary aim of the chapter is to prove that the hypermaps of certain fans are planar. The proof is a long induction. The base case of the induction consists of fans in which every face is a triangle. The following lemma gives the properties of triangles that are needed in the base case of an induction.

Lemma 5.24 (triangle attributes) *Let (V, E) be a fan with hypermap (D, e, n, f). Let $Y = Y(V, E)$. Assume that (V, E) is fully surrounded. Let F be a face of cardinality three, let $x_0 \in F$ and $x_i = f^i x_0$. Then*

1. *U_F is equal to the intersection of the three half-spaces:*

$$A_+^0(i) = \text{aff}_+^0(\{\mathbf{0}, \text{node}(x_{i+1}), \text{node}(x_{i+2})\}, \text{node}(x_i)), \quad i = 0, 1, 2.$$

2. *If a dart y leads into U_F, then $y \in F$.*

Proof *The intersection U' of the three half-spaces is a subset of U_F.* Indeed, the intersection of two half-spaces, $A_+^0(1) \cap A_+^0(2)$, is the wedge $W_{\text{dart}}^0(x_0)$. The sets $C^0(t, x_0) \subset W_{\text{dart}}^0(x_0)$ sweep out precisely the intersection of $W_{\text{dart}}^0(x_0)$ with $A_+^0(0)$, when $0 < t < 1$. The sets $C^0(t, x_0)$ belong to U_F. The claim ensues.

U_F is a subset of the intersection U'. Otherwise, let $\mathbf{p}$ be a point of U_F that does not belong to U'. Choose a continuous path $\varphi : [0, 1] \to U_F$ with $\varphi(0) \in U'$ and $\varphi(1) = \mathbf{p}$. Let $t > 0$ be the first time such that $\varphi(t) \notin U'$. Then $\mathbf{q} = \varphi(t)$ lies in the set consisting of the closed intersection of half-spaces $A_+(i)$ corresponding to $A_+^0(i)$. The point $\mathbf{q}$ also lies in one of the bounding planes. Let

$$X' = \bigcup C(i), \qquad \text{where } C(i) = C\{\text{node}(x_i), \text{node}(x_{i+1})\}.$$

Then $\mathbf{q} \in X' \subset X$. This yields an impossibility: $\mathbf{q} \in X \cap Y = \emptyset$. Thus, $U' = U_F$.

Let y be any dart that leads into U_F. Then $W_{\text{dart}}^0(y, \epsilon)$ meets U_F for all $\epsilon > 0$ sufficiently small, implying that node(y) lies in the intersection of the closed half-spaces $A_+(i)$. As previously established, this intersection is the disjoint union of U_F and X'. As node$(y) \in X$ and as X does not meet U_F, it follows that node$(y) \in X'$. The set X' is the disjoint union of the rays $C\{\text{node}(x_i)\}$ and the three blades $C^0(i)$. These blades do not meet V; hence, node$(y) =$ node(x_i) for some i. Thus, y and x_i belong to the same node. The sets $W_{\text{dart}}^0(y)$ and $W_{\text{dart}}^0(x_i)$ are disjoint for distinct darts at the same node, and this implies that $y = x_i \in F$. □

Corollary 5.25 (triangle solid angle) *Let F be a face of cardinality three in the context of Lemma 5.24. Then for $r > 0$, $U_F \cap B(\mathbf{0}, r)$ is measurable and r-radial at $\mathbf{0}$. The solid angle of U_F is given by the formula*

$$\text{sol}(U_F) = -\pi + \sum_{x \in F} \text{azim}(x).$$

Proof An intersection of half-spaces through the origin with $B(\mathbf{0}, r)$ is measurable and r-radial. The solid angle is given by Girard's formula for a spherical triangle (Lemma 3.23). □

5.3.2 conformance

The previous subsection shows that in a fully surrounded fan, the topological components of U_F have a particularly simple geometrical description as an intersection of half-spaces, when card(F) = 3. This subsection defines a class of fans (Definition 5.26) in which the faces also have a simple geometrical description.

Lemma 5.42 shows any fan that fully surrounded is conforming. We consider the definition of conforming to be useful only until Lemma 5.42 becomes available. Thereafter, properties of conforming fans may be applied to all fully surrounded fans.

Definition 5.26 (conforming) Let (V, E) be a fan with hypermap (D, e, n, f). The fan is *conforming* if the following properties hold.

1. (SURROUNDEDNESS) (V, E) is fully surrounded.
2. (BIJECTION) The map $F \mapsto U_F$ is a bijection between the faces of the hypermap and the topological components of Y.
3. (HALF-SPACE) For every face F, the topological component U_F is the intersection of the open half-spaces $\mathrm{aff}_+^0(\{\mathbf{0}, \mathrm{node}(x), \mathrm{node}(fx)\}, \mathrm{node}(f^{-1}x))$ as x runs over F.
4. (SOLID ANGLE) For every F, the intersection $B(\mathbf{0}, r) \cap U_F$ is measurable and eventually radial at $\mathbf{0}$. Moreover, the solid angle of U_F is given by the formula
$$\mathrm{sol}(U_F) = 2\pi + \sum_{x \in F} (\mathrm{azim}(x) - \pi).$$
5. (DIAGONAL) For every face F, if $x, y \in F$ are distinct with corresponding nodes $\mathrm{node}(x), \mathrm{node}(y) \in V$, then $\mathrm{node}(x)$ and $\mathrm{node}(y)$ are not parallel. Moreover, either x and y are adjacent under the face map, or
$$C^0\{\mathrm{node}(x), \mathrm{node}(y)\} \subset U_F.$$

That is, the "diagonals" of U_F are all "interior."

Conforming fans have several significant properties. They (together with the fact that every fully surrounded fan is conforming) constitute the main conclusion of this section.

Lemma 5.27 *Let (V, E) be a conforming fan. Each U_F is convex, where F is any face of* hyp(V, E).

Proof By (HALF-SPACE), U_F is the intersection of half-spaces. □

Lemma 5.28 *Let (V, E) be a conforming fan. Then* hyp(V, E) *is simple.*

Proof Let $x \in F$. By the intersection of half-spaces property, U_F is contained in the wedge $W^0_{\text{dart}}(x)$ at x. If a second dart y sits at the same node in F, then U_F is also contained in $W^0_{\text{dart}}(y)$. However, by Lemma 5.15, the wedges at a given node are disjoint. □

Lemma 5.29 *Let (V, E) be a conforming fan. Then* hyp(V, E) *is connected.*

Proof Let $[D]$ denote the set of combinatorial components of hyp(V, E). There is a well-defined, continuous (in fact, locally constant) function from Y onto $[D]$ given as follows. For $\mathbf{p} \in Y$, choose F such that $\mathbf{p} \in U_F$ and send $\mathbf{p}$ to the class of F in $[D]$. By property (BIJECTION), this map is well-defined. The map extends continuously to $C^0(\varepsilon)$ for $\varepsilon \in E$ by the following construction: for every $\mathbf{p} \in C^0(\varepsilon)$, we choose the edge $\{x, y\}$ of the hypermap associated with the edge ε and send $\mathbf{p}$ to the combinatorial component containing $\{x, y\}$. The domain

$$Y \cup \bigcup C^0(\varepsilon)$$

is connected. The continuous map from this connected set onto the discrete set is necessarily constant. As the map is onto, the set $[D]$ reduces to a singleton. □

The following lemma is the promised result on planarity.

Lemma 5.30 *Let (V, E) be a conforming fan. Then* hyp(V, E) *is planar.*

Proof The solid angle of a sphere is 4π. The set $X(V, E)$ has measure zero, so that

$$4\pi = \text{sol}(Y) = \sum_F \text{sol}(U_F) = \sum_F \Big(2\pi + \sum_{x\in F} (\text{azim}(x) - \pi)\Big). \tag{5.31}$$

In the rightmost expression, the double sum over faces and darts in a face can be replaced by a single sum over all darts. The sum of the azimuth angles of all darts at a node is 2π. Thus, the sum over all azimuth angles is $2\pi\,\#n$. Thus, the formula (5.31) becomes

$$4\pi = 2\pi\,\#f + 2\pi\,\#n - \pi\,\#D.$$

In a plain hypermap in which the edge map has no fixed points, $\#D = 2\#e$. The relation (5.31) simplifies to

$$2 + \#D = \#f + \#e + \#n.$$

This is the condition of planarity for a connected hypermap. □

5.3.3 existence

This section proves the existence of many conforming fans. The main result of this subsection (Lemma 5.42) asserts that every fully surrounded fan is conforming. The proof breaks into a series of small lemmas. The primary method to prove the existence of conforming fans is an induction on the following invariant of a fan (V, E).

Definition 5.32 ($N(V, E)$) Let

$$N(V, E) = \sum_F (k_F - 3),$$

where (V, E) is a fan, the sum runs over faces F, and k_F is the cardinality of the face F.

The following lemma gives the base case of an induction.

Lemma 5.33 *Let (V, E) be a fully surrounded fan such that $N(V, E) = 0$. Then (V, E) is conforming.*

Proof We run through the defining properties of a conforming fan (page 126). If $N(V, E) = 0$, then the hypermap is a triangulation. By Lemmas 5.21 and 5.23, every topological component of Y has the form U_F for some face F. By Lemma 5.24, U uniquely determines the face F. Thus, there is a bijection between faces of the hypermap and topological components. By Lemma 5.24, the topological component U_F is the intersection of open half-spaces, as expressed in the property (HALF-SPACE). The solid angle formula is given by Corollary 5.25. The assertion of the lemma about diagonals is trivial for a hypermap that is already a triangulation. This completes the proof in the base case $N(V, E) = 0$. □

To carry out proofs by induction, we assume the existence of a minimal counterexample and then argue by contradiction. The proofs are so long that it is necessary to break them into a series of lemmas. To organize an extended proof by contradiction, we formulate the properties of a minimal counterexample as a definition. We eventually show that minimally nonconforming fans do not exist.

Definition 5.34 (minimally nonconforming fan) A fan (V, E) is said to be *minimally nonconforming* if the following properties hold.

1. (V, E) is fully surrounded.
2. (V, E) is not conforming.
3. $N(V, E) > 0$.

4. If (V, E') is any other fully surrounded fan on the same node set V and if $N(V, E') < N(V, E)$, then (V, E') is conforming.

Remark 5.35 (reduction data) When $N(V, E) > 0$, choose a dart x that lies in a face F of the hypermap that is not a triangle. By Lemma 5.20, $C^0\{\mathbf{v}, \mathbf{w}\} \subset U_F$, where $\mathbf{v} = \text{node}(x)$, $\mathbf{w} = \text{node}(y)$, and $y = f^2x$. Form a new fan (V, E') on the same node set with $E' = E \cup \{\{\mathbf{v}, \mathbf{w}\}\}$. (See Figure 5.4 and Lemma 5.22.)

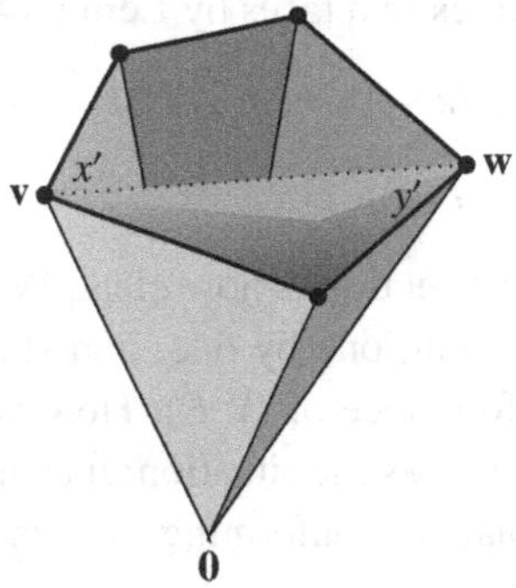

Figure 5.4 To prove results about fans by induction, we add a blade $C^0\{\mathbf{v}, \mathbf{w}\}$ between $\mathbf{v}$ and $\mathbf{w}$ to a fan to form a new fan with smaller invariant N. This partitions one of the topological components U into $U(x')$, $U(y')$, and the new blade $C^0\{\mathbf{v}, \mathbf{w}\}$. We have darts x' and y' of the new hypermap, leading into $U(x')$ and $U(y')$, respectively. The face containing the dart y' in the new hypermap is a triangle.

The following notation is used to relate the two fans (V, E) and (V, E'). Add primes to symbols denoting objects related to (V, E'). Let $x' = (\mathbf{v}, \mathbf{w})$ and $y' = (\mathbf{w}, \mathbf{v}) \in D'$, where D' is the set of darts of $\text{hyp}(V, E')$. The darts x', y' lead into topological components $U(x')$ and $U(y')$ of $Y' = Y(V, E')$ and belong to faces $F(x')$, $F(y')$ of $H' = \text{hyp}(V, E')$.

The following lemma is used to prove that if (V, E) is minimally nonconforming, then the fan (V, E') is conforming.

Lemma 5.36 *Let (V, E) be a fully surrounded fan. Assume that $N(V, E) > 0$. Let x be a dart in* hyp(V, E) *such that the face F of x is not a triangle. Let $E' = E \cup \{\{\mathbf{v}, \mathbf{w}\}\}$, $x', y', \ldots$ be the reduction data as above associated with x. Then the faces $F(x')$ and $F(y')$ are distinct, and $F(y')$ is a triangle. Moreover,* hyp(V, E) *is obtained by a double walkup along the edge $\{x', y'\}$ of* hyp(V, E') *that merges the two faces $F(x')$ and $F(y')$. Finally, $N(V, E') < N(V, E)$.*

Proof Let $H' = (D', e', n', f')$ be the hypermap of (V, E'). Write $x = (\mathbf{v}, \mathbf{u})$, $fx = (\mathbf{u}, \mathbf{w})$, $f^2x = y = (\mathbf{w}, *)$. Then it follows directly from the definition of the face map on hyp(V, E') and an inspection of the cyclic order σ that

$$(f')^3 y' = (f')^3(\mathbf{w}, \mathbf{v}) = (f')^2(\mathbf{v}, \mathbf{u}) = f'(\mathbf{u}, \mathbf{w}) = (\mathbf{w}, \mathbf{v}) = y'.$$

It follows that $F(y')$ is a triangle.

The hypermap $H = \text{hyp}(V, E)$ is obtained from $H' = \text{hyp}(V, E')$ by a double walkup transformation on the edge $\{x', y'\}$. The faces $F(x')$ and $F(y')$ are distinct by Lemma 5.24, which asserts that x' does not lead into $U(y')$. Thus, the walkup transformation merges two faces by Lemma 4.17. Then

$$N(V, E) - N(V, E') = ((k + 1) - 3) \quad - \quad ((k - 3) + (3 - 3)) = 1 > 0,$$

where k is the cardinality of $F(x')$. □

The task of the rest of the section is now clear. We run through each of the properties of a conforming fan, one by one, and show that the conformance of (V, E') implies the conformance of (V, E). However, if (V, E) is minimally nonconforming, this is an impossible situation: it cannot both conform and not conform. Hence, no minimal nonconforming fan can exist (Lemma 5.42).

Lemma 5.37 (bijection) *The property* (BIJECTION) *of conforming fans (on page 126) holds for any minimally nonconforming fan:* $F \mapsto U_F$ *is a bijection.*

Proof Let (V, E) be a minimally nonconforming fan. Choose x to obtain reduction data for (V, E). The bijection property holds for (V, E') by the minimality assumption. The proof establishes a bijection by showing that (V, E) has one more face and one more topological component than (V, E').

The two faces $F(x')$ and $F(y')$ merge into a single face F of hyp(V, E). Then

$$U = U(x') \cup U(y') \cup C^0(\varepsilon) \tag{5.38}$$

is a connected open set in Y. If $F' \neq F(x')$ and $F(y')$ is any other face in H', then $U_{F'}$ is a connected open set in Y. Moreover, the set U and sets $U_{F'}$ are pairwise disjoint and exhaust Y, so that they are precisely the topological components of Y. Some dart of F leads into U, so $U = U_F$. It follows that the number of faces is equal to the number of topological components for (V, E), so that the map $F \mapsto U_F$ is a bijection. □

Lemma 5.39 *Let (V, E) be any minimally nonconforming fan. Then it has property* (SOLID ANGLE) *of conforming fans (page 126).*

Proof Choose reduction data for (V, E). Every topological component of $Y = Y(V, E)$ except $U = U_F$ is already a topological component of Y' and the conclusion holds for components of Y'. The topological component U is a disjoint union of two components of Y' and a set $C^0(\varepsilon)$ of measure zero. Thus, U is also measurable and eventually radial. The solid angle formula is additive over the disjoint union in (5.38), so the formula holds for U. □

Lemma 5.40 *Let (V, E) be any minimally nonconforming fan. Then the fan has property* (DIAGONAL) *of conforming fans (page 126). That is, for any dart x on any face F and dart $z \in F$ that is not adjacent to x under the face map,*

$$C^0(\{\text{node}(x), \text{node}(z)\}) \subset U_F.$$

Proof By excluding trivial cases of the proof, we may assume that the dart x is used to construct reduction data E', x', y', etc. We may assume that $F(y')$ is a triangle. If $z = f^2x$, then the diagonal is precisely $C^0\{\mathbf{v}, \mathbf{w}\}$, for which the conclusion has already been established. Otherwise, z can be identified with a dart $z' \in F(x')$. Then, by minimality,

$$C^0\{\text{node}(x), \text{node}(z)\} = C^0\{\text{node}(x'), \text{node}(z')\} \subset U(x') \subset U_F.$$

□

Lemma 5.41 *Let (V, E) be a minimally nonconforming fan. Then property* (HALF-SPACE) *of conforming fans (on page 126) holds for (V, E).*

Proof Choose a dart x to give reduction data for (V, E). By the non conformance of (V, E), it is enough to consider the face F containing x. (Indeed, the other faces of hyp(V, E) can be identified with faces of hyp(V, E') and these cases are easily treated.)

The intersection U_1 of half-spaces given by property (HALF-SPACE) *lies in $U = U_F$.* Indeed, every point in $\mathbb{R}^3$ lies in the plane

$$A = \text{aff}\{\mathbf{0}, \mathbf{v}, \mathbf{w}\}$$

or in one of the two open half-spaces bounded by this plane. These half-spaces, $A^0(x')$ and $A^0(y')$, contain $U(x')$ and $U(y')$ respectively, by the minimality of (V, E) and the conformance of (V, E'). Also,

$$A^0(x') \cap U_1 \subset U(x') \subset U.$$

Similarly, $A^0(y') \cap U_1 \subset U$. We have a sequence of subsets

$$A \cap U_1 \subset A \cap W^0_{\text{dart}}(x) \cap W^0_{\text{dart}}(y) \subset C^0\{\mathbf{v}, \mathbf{w}\} \subset U.$$

Thus, $U_1 \subset U$.

For any dart z of F, the set U is a subset of the half-space with bounding plane $\{\mathbf{0}, \text{node}(z), \text{node}(fz)\}$. Indeed, if the reduction data for the dart z is used, the claim does not change. Without loss of generality, assume that $z = x$. By the minimality of (V, E), the partition (5.38) of U gives three pieces contained respectively in the three sets:

$$W^0_{\rm dart}(x'), \quad C^0\{\mathbf{v}, \mathbf{w}\}, \quad W^0_{\rm dart}(x''),$$

where $x', x'' \in D'$ correspond to the single dart x in D:

$$\mathrm{azim}(x') + \mathrm{azim}(x'') = \mathrm{azim}(x) < \pi.$$

Thus, U itself is contained in the lune

$$W^0(\{\mathbf{0}, \mathrm{node}(x)\}, \{\mathrm{node}(fx), \mathrm{node}(f^{-1}x)\}),$$

which is contained in the desired half-space. This proves the claim.

The reverse inclusion $U \subset U_1$ follows immediately from the claim. □

Lemma 5.42 (conformance) *Every fully surrounded fan is conforming.*

Proof Suppose for a contradiction that a fully surrounded fan $(V, *)$ exists that is not conforming. Among all fully surrounded nonconforming fans $(V, *)$ on the same node set, select one (V, E) that minimizes $N(V, E)$. In fact, $N(V, E) > 0$ because otherwise (V, E) is conforming by Lemma 5.33.

This is a minimally nonconforming fan. However, the preceding lemmas show that a minimally nonconforming fan actually satisfies all of the properties of a conforming fan. This contradiction gives the proof. □

5.4 Polyhedron

This section shows that a bounded polyhedron in $\mathbb{R}^3$ with nonempty interior determines a fan. The construction is elementary. Choose a point in the interior, which for convenience we take to be the origin $\mathbf{0}$. The set V_P is defined as the set of extreme points of the polyhedron. The set E_P consists of pairs of extreme points that are joined by an edge of the polyhedron. The pair (V_P, E_P) is a fan (Figure 5.5). This section describes this construction in detail.

5.4.1 background on convex sets

We begin with a review of basic terminology about affine and convex sets. The material in this subsection appears in standard textbooks on convexity [2], [48].

Definition 5.43 (affine set, affine hull, affine dimension, affinely independent, hyperplane) Recall that a set $A \subset \mathbb{R}^n$ is *affine* if for every $\mathbf{v}, \mathbf{w} \in A$ and every $t \in \mathbb{R}$,

$$t\mathbf{v} + (1-t)\mathbf{w} \in A.$$

Recall that the *affine hull* of $P \subset \mathbb{R}^n$ (denoted aff(P)) is the smallest affine set containing P. The *affine dimension* of P (written $\dim \operatorname{aff}(P)$) is $\operatorname{card}(S) - 1$, where S is a set of smallest cardinality such that

$$P \subset \operatorname{aff}(S).$$

In particular, the affine dimension of the empty set is -1. A finite set S is *affinely independent* if $\dim \operatorname{aff}(S) = \operatorname{card}(S) - 1$. A *hyperplane* in $\mathbb{R}^n$ is any set of the form

$$\{\mathbf{p} \ : \ \mathbf{u} \cdot \mathbf{p} = b\},$$

where $\mathbf{0} \neq \mathbf{u} \in \mathbb{R}^n$.

Definition 5.44 (relative interior, closure, relative boundary) Let A be the affine hull of a set $P \subset \mathbb{R}^n$. A point $\mathbf{p}$ is an *interior point* of P if some nonempty open ball $B(\mathbf{p}, r)$ is contained in P. A point $\mathbf{p}$ of P belongs to the *relative interior* of P if there is an open ball such that $B(\mathbf{p}, r) \cap A \subset P$. Let ri$(P)$ be the set of relative interior points. The *closure* of P is the set

$$\{\mathbf{p} \ : \ \forall r > 0.\ B(\mathbf{p}, r) \cap P \neq \emptyset\}.$$

The complement of ri(P) in the closure of P is the *relative boundary* of P.

Definition 5.45 (face, facet, edge, extreme point) Let P be a convex set. A *face* of P is a convex set F such that for all $\mathbf{v}, \mathbf{w} \in P$, the condition

$$\exists\, s\, t. \quad s\mathbf{v} + t\mathbf{w} \in F, \quad s > 0, \quad t > 0, \quad s + t = 1$$

implies that $\mathbf{v}, \mathbf{w} \in F$. A face F is *proper* if $F \neq \emptyset, P$. An *extreme point* is an element $\mathbf{v} \in P$ such that $\{\mathbf{v}\}$ is a face (of affine dimension zero). An *edge* is a face of P of affine dimension one. A *facet* of P is a *proper* face of affine dimension $\dim \operatorname{aff}(P) - 1$.

Remark 5.46 (convex background) We assume as background knowledge various basic facts about convex sets. For example, if P is convex and $P = \bigcup_{i=1}^r P_i$, where $r \geq 1$, then

$$\operatorname{conv}(A \cup P) = \bigcup_{i=1}^{r} \operatorname{conv}(A \cup P_i).$$

Also, $\operatorname{conv}(A \cup \operatorname{conv}(B)) = \operatorname{conv}(A \cup B)$. Furthermore, if A is convex, then $A = \operatorname{conv}(A)$. An intersection of faces of a convex set is again a face.

Remark 5.47 (affine background) Various particular facts about affine sets come up. If $U \subset \mathbb{R}^n$ is open, then $\dim \operatorname{aff}(U) = n$. In particular, we have $\dim \operatorname{aff}(\mathbb{R}^n) = n$. At the other extreme $\dim \operatorname{aff}(\{\mathbf{p}\}) = 0$. If $S_1 \subset S_2 \subset \mathbb{R}^n$, then $\dim \operatorname{aff}(S_1) \leq \dim \operatorname{aff}(S_2)$. If A is an affine set and $\mathbf{p} \notin A$, then $\dim \operatorname{aff}(A) + 1 = \dim \operatorname{aff}(A \cup \{p\})$. If A is affine of affine dimension k and if B is a hyperplane such that $A \cap B \neq \emptyset$, then $\dim \operatorname{aff}(A \cap B) \geq k - 1$. If C is a convex set, A is an affine set and $U \subset \mathbb{R}^n$ is a neighborhood of $\mathbf{p} \in C \cap A$ such that $C \cap U = A \cap U$, then $A = \operatorname{aff}(C)$.

Remark 5.48 (polysemes) The term *face* occurs in this book with two meanings: the face of a hypermap and the face of a convex set. The two contexts are sufficiently different that misunderstandings should be avoidable. Graphs, hypermaps, fans, and polyhedra all have *edges*. Fans and hypermaps have *nodes*, while polygons have *vertices*. Polyhedra have *extreme points*.

Lemma 5.49 (Krein–Milman) *Every compact convex set $P \subset \mathbb{R}^n$ is the convex hull of its set of extreme points.*

Proof See [48, Theorem 2.6.16]. □

5.4.2 background on polyhedra

The material in this subsection appears in standard textbooks on polyhedra. We follow [48].

Definition 5.50 (polyhedron) A *polyhedron* is the intersection of a finite number of closed half-spaces in $\mathbb{R}^n$.

Lemma 5.51 *An affine set in $\mathbb{R}^n$ is a polyhedron.*

Proof See [48, Cor 1.4.2]. □

Lemma 5.52 *If A is a proper affine set in $\mathbb{R}^n$, then there exists $\mathbf{u} \in \mathbb{R}^n$, with $\mathbf{u} \neq \mathbf{0}$, and $b \in \mathbb{R}$ such that $\mathbf{u} \cdot \mathbf{p} = b$ for all $\mathbf{p} \in A$. That is, every proper affine set is contained in a hyperplane.*

Proof By Lemma 5.51, the affine set A is a polyhedron and is thus contained in a closed half-space defined by an inequality $\mathbf{u} \cdot \mathbf{p} \leq c$ for some $\mathbf{u} \neq \mathbf{0}$ and c. If $\mathbf{u} \cdot \mathbf{p}$ assumes at least two distinct values, then the image of the affine set A under the map $\mathbf{p} \mapsto \mathbf{u} \cdot \mathbf{p}$ is the entire affine line $\mathbb{R}$, and the inequality is violated. Hence, $\mathbf{u} \cdot \mathbf{p}$ is constant on A. □

A polyhedron is closed and convex. A bounded polyhedron falls within the scope of the Krein–Milman theorem and is thus the convex hull of its set of extreme points.

Let $P \subset \mathbb{R}^n$ be a bounded polyhedron with affine hull A. Write

$$P = A \cap A_1^+ \cap \cdots \cap A_r^+, \tag{5.53}$$

where $A_i^+ = \{\mathbf{p} \ : \ \mathbf{u}_i \cdot \mathbf{p} \leq a_i\}$ with bounding hyperplane $A_i = \{\mathbf{p} \ : \ \mathbf{u}_i \cdot \mathbf{p} = a_i\}$ for some $\mathbf{u}_i \in \mathbb{R}^n$ and $a_i \in \mathbb{R}$. Assume that this representation is minimal in the sense that none of the factors A_i^+ may be omitted from the intersection (5.53). Let $F_i = A_i \cap P$.

Lemma 5.54 *Let $P \subset \mathbb{R}^n$ be a bounded polyhedron. Then*

1. *The facets of P are F_i, $i = 1, \ldots, r$.*
2. *The relative boundary of P is $F_1 \cup \cdots \cup F_r$.*
3. *Every proper face is the intersection of the facets that contain it.*
4. *Every face of a face of P is a face of P.*
5. *A point is an extreme point of a face F if and only if it is an extreme point of P that is contained in F.*
6. *If F and F' are two faces of P with meeting relative interiors, then $F = F'$.*

Proof See [48, Thm 3.2.1] for the first three conclusions. See [48, Th 2.6.5] for the proof of the fourth conclusion.

An extreme point singleton of F is a face of P (of dimension zero) by the fourth conclusion. Hence, every extreme point of F is an extreme point of P. Conversely, a face of P in F is *a fortiori* a face of F. This gives the fifth conclusion. See [48, Cor 2.6.7] for the final conclusion. □

Corollary 5.55 *A face of a polyhedron is itself a polyhedron.*

Proof By Lemma 5.54, each facet is defined by a system of linear inequalities. (See Lemma 5.51.) A proper face is an intersection of finitely many facets, and is therefore given by the conjunction of the inequalities defining the various facets. □

Lemma 5.56 *If P is a bounded polyhedron of positive affine dimension with facets $F_1, \ldots, F_r$ and if $\varnothing \neq A \subset P$, then*

$$P = \bigcup_{i=1}^{r} \operatorname{conv}(A \cup F_i).$$

Proof Choose $\mathbf{u} \in A$. Then

$$\bigcup_{i=1}^{r} \operatorname{conv}(\{\mathbf{u}\} \cup F_i) \subset \bigcup_{i=1}^{r} \operatorname{conv}(A \cup F_i) \subset P.$$

Hence, it is enough to show that any $\mathbf{v} \in P$, with $\mathbf{v} \neq \mathbf{u}$, lies in one of the sets on the left. By the boundedness of P, the ray $\mathrm{aff}_+\{\mathbf{u}, \{\mathbf{v}\}\}$ meets P in an interval $\mathrm{conv}\{\mathbf{u}, \mathbf{w}\}$, where $\mathbf{w}$ lies on the relative boundary of P. By Lemma 5.54, in fact, $\mathbf{w}$ lies in a facet F_i. This gives

$$\mathbf{v} \in \mathrm{conv}\{\mathbf{u}, \mathbf{w}\} \subset \mathrm{conv}(\{\mathbf{u}\} \cup F_i).$$

□

Lemma 5.57 *Let $S \subset \mathbb{R}^n$ be a finite set. Then $P = \mathrm{conv}(S)$ is a polyhedron. Assume moreover that S is an affinely independent set with cardinality at least two. Then S is the set of extreme points of P. Furthermore, F is a facet of P if and only if $F = \mathrm{conv}(S \setminus \{\mathbf{u}\})$ for some $\mathbf{u} \in S$.*

Proof The proof is left to the reader. □

Lemma 5.58 *Let P be a bounded polyhedron with $\mathbf{0}$ as an interior point. Let F and F' be proper faces of P; let $\mathbf{p} \in F$ and $\mathbf{p}' \in F'$; and let t and t' be positive scalars such that $t\mathbf{p} = t'\mathbf{p}'$. Then $t = t'$.*

Proof The faces F, F' are subsets of facets of P. We may assume without loss of generality that F and F' are themselves facets. Without loss of generality, assume for a contradiction that $t > 1$ and $t' = 1$. Then $t\mathbf{p} = \mathbf{p}' \in P$. By Lemma 5.54, $F = A \cap P$, where $A = \{\mathbf{q} : \mathbf{q} \cdot \mathbf{u} = a\}$ for some $\mathbf{u}$ and a. Also,

$$C\{\mathbf{p}\} \cap P = \mathrm{conv}\{\mathbf{0}, \mathbf{p}\} = \{\mathbf{q} : 0 \leq \mathbf{q} \cdot \mathbf{u} \leq a\}.$$

In particular, $t\mathbf{p} \notin P$ for $t > 1$. This contradiction gives $t = t'$. □

The following lemma is similar.

Lemma 5.59 *Let P be a bounded polyhedron with facets F and F'. Let $\mathbf{p} \in F$, $\mathbf{p}' \in F'$, $\mathbf{p}_0 \in P$, and $\mathbf{p}_0 \notin F \cup F'$; and let t and t' be positive scalars such that*

$$(1 - t)\,\mathbf{p}_0 + t\,\mathbf{p} = (1 - t')\,\mathbf{p}_0 + t'\mathbf{p}'.$$

Then $t = t'$.

5.4.3 fan and polyhedron

We finally deliver our promise of showing that a fan can be associated with a bounded polyhedron (Figure 5.5). Here is the definition of the set V_P of nodes and set E_P of edges.

Definition 5.60 (V_P, E_P) Let P be a bounded polyhedron. Let V_P be the set of extreme points of P. Let E_P be the set of pairs $\{\mathbf{v}, \mathbf{w}\}$ of extreme points such that $\mathrm{conv}\{\mathbf{v}, \mathbf{w}\}$ is an edge of P.

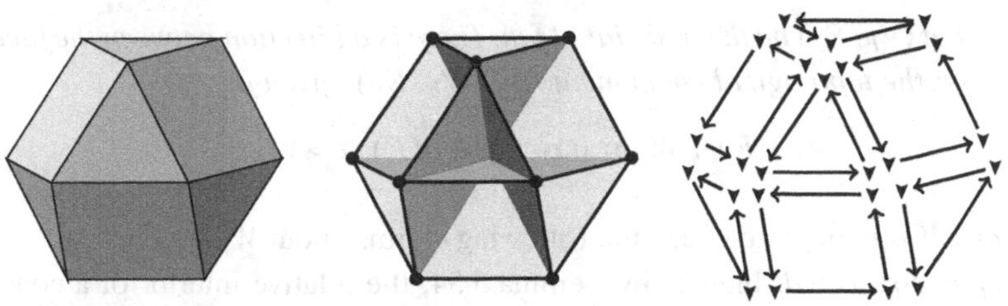

Figure 5.5 A bounded polyhedron P, its fan (V_P, E_P), and the face permutation of the front half of its hypermap hyp(V_P, E_P). The set of facets of P is in bijection with the set of topological components of $Y(V_P, E_P)$ and with the set of faces of the hypermap. The set of edges of a single facet F of P is in bijection with (the set of darts in) the corresponding face of the hypermap.

Lemma 5.61 (fan of a polyhedron) *Let P be a bounded polyhedron in $\mathbb{R}^3$ with the interior point $\mathbf{0}$. Then (V_P, E_P) is a fan.*

Proof The properties of a fan can be checked one by one. By the Krein–Milman lemma, the set V_P of extreme points is nonempty. By Lemma 5.54, the polyhedron has finitely many faces, so that V_P is finite. Since $\mathbf{0}$ is an interior point, it does not meet any proper face. This implies that $\mathbf{0} \notin V_P$ and that for all $\varepsilon \in E_P$, $\mathbf{0} \notin \text{conv}(\varepsilon)$.

Suppose for a contradiction that $\{\mathbf{v}, \mathbf{w}\} \in E_P$ and that $\mathbf{v}$ and $\mathbf{w}$ are parallel. As $\mathbf{0} \notin \text{conv}(\varepsilon)$, some relation has the form $s\,\mathbf{v} = t\,\mathbf{w}$ for some $s, t > 0$. By Lemma 5.58, $s = t$ and $\mathbf{v} = \mathbf{w}$, which is contrary to the definition of edge as a face of dimension one.

Finally, we check the intersection property $C(\varepsilon) \cap C(\varepsilon') = C(\varepsilon \cap \varepsilon')$. By Lemma 5.58,

$$C(\varepsilon) \cap C(\varepsilon') = \{\mathbf{0}\} \cup \{t\,\mathbf{p} \ : \ \mathbf{p} \in \text{conv}(\varepsilon) \cap \text{conv}(\varepsilon') \text{ and } t \geq 0\}.$$

The sets conv(ε) and conv(ε') are both faces of P. The intersection is again a face of P and hence the convex hull of its set of extreme points; that is, the convex hull of $\varepsilon \cap \varepsilon'$. Thus,

$$C(\varepsilon) \cap C(\varepsilon') = \{\mathbf{0}\} \cup \{t\,\mathbf{p} \ : \ \mathbf{p} \in \text{conv}(\varepsilon \cap \varepsilon')\} = C(\varepsilon \cap \varepsilon').$$

Thus, all the defining properties of a fan are satisfied. □

We can relate the combinatorial properties of polyhedron to the topological properties of the fan.

Lemma 5.62 *Let P be a bounded polyhedron in $\mathbb{R}^3$ with $\mathbf{0}$ as an interior* *point. Let (V_P, E_P) be the associated fan. There is a bijection between the facets of P and the topological components of $Y(V_P, E_P)$, given by*

$$F \mapsto W_F = \{t\mathbf{p} \ :\ \mathbf{p} \in \mathrm{ri}(F),\ t > 0\}.$$

Proof It is enough to check the following claims about W_F.

W_F is connected. Indeed, by Lemma 5.54, the relative interior of a convex polyhedron is the intersection of an affine set with open half-spaces, which is the intersection of convex sets, and is therefore convex. The set $\mathrm{ri}(F)$ is convex and is therefore connected. The positive half-line $I = \{t \ :\ t > 0\}$ is also connected. The continuous image of the connected set $\mathrm{ri}(F) \times I$ of these two sets is W_F. Hence, W_F is connected.

The set W_F is open. Indeed, this is a standard ϵ-argument. Let A be the affine hull of F. For any $\mathbf{p} \in \mathrm{ri}(F)$, select $r > 0$ such that $B(\mathbf{p}, r) \cap A \subset \mathrm{ri}(F)$. Select $r' > 0$ and $0 < \epsilon < 1$ such that for all $\mathbf{q} \in B(\mathbf{p}, r')$, there exists t such that $|t| < \epsilon$ and $(1 + t)\mathbf{q} \in A$. After shrinking r', if $\mathbf{q} \in B(\mathbf{p}, r')$, then $(1 + t)\mathbf{q} \in B(\mathbf{p}, r) \cap A \subset \mathrm{ri}(F)$. That is, $B(\mathbf{p}, r') \subset W_F$. Hence, W_F is open.

The sets W_F are pairwise disjoint, and the map $F \mapsto W_F$ is one-to-one. Indeed, select any two facets F, F' for which $W_F \cap W_{F'} \neq \emptyset$. That is, there exist $\mathbf{p} \in \mathrm{ri}(F)$, $\mathbf{p}' \in \mathrm{ri}(F')$, and $t, t' > 0$ such that $t\mathbf{p} = t'\mathbf{p}'$. By Lemma 5.58, $t = t'$ and $\mathbf{p} = \mathbf{p}' \in \mathrm{ri}(F) \cap \mathrm{ri}(F')$. By the final statement of Lemma 5.54, this implies that $F = F'$.

The union of the sets W_F is $Y(V_P, E_P)$. Indeed, select any $\mathbf{p} \in Y(V_P, E_P)$. As $\mathbf{0}$ lies in the interior of the bounded polyhedron, we may rescale $\mathbf{p}$ by a positive scalar t so that $t\mathbf{p}$ lies in the boundary of P, and hence (by Lemma 5.54) in a facet F. If $t\mathbf{p} \in \mathrm{ri}(F)$, then $\mathbf{p} \in W_F$, as desired. Otherwise, $t\mathbf{p}$ lies in the relative boundary of F. The facets of a three dimensional polyhedron have dimension two, and the facets forming the relative boundary of F have dimension one. These faces are edges of P. Thus, $t\mathbf{p}$ lies in an edge of P, so that $\mathbf{p} \in X(V_P, E_P)$, which is contrary to the assumption that $\mathbf{p} \in Y(V_P, E_P)$. The claim ensues.

It follows that the sets W_F are the topological components of $Y(V_P, E_P)$. □

The following lemma shows that the fan (V_P, E_P) is completely surrounded. Hence, by Lemma 5.42, all of the properties of conforming fans from the previous section apply to this fan.

Lemma 5.63 *Let P be a bounded polyhedron with $\mathbf{0}$ as an interior point. Then* $\mathrm{azim}(x) < \pi$ *for every dart x in the hypermap* $\mathrm{hyp}(V_P, E_P)$.

Proof The result is a consequence of the lemma that follows. □

Lemma 5.64 *Let P be a bounded polyhedron with interior point $\mathbf{0}$. Let $\mathbf{v}$ be an extreme point of P. Let A be an open half space whose bounding plane contains $\mathbf{0}$ and $\mathbf{v}$. Then there exists an extreme point $\mathbf{w}$ of P, such that $\mathbf{w} \in A$ and* $\mathrm{conv}\{\mathbf{v}, \mathbf{w}\}$ *is an edge of P.*

Proof Use the definition of extreme point, to choose a half space

$$\{\mathbf{p} \,:\, \mathbf{u}_1 \cdot \mathbf{p} \le b_1\} \supset P,$$

where equality holds for a point $\mathbf{p} \in P$ if and only if $\mathbf{p} = \mathbf{v}$. Since $\mathbf{0}$ is an interior point of P, we may assume that $b_1 > 0$.

Let V_P be the set of extreme points of P. Select b_1' with $0 < b_1' < b_1$ such that

$$\{\mathbf{p} \,:\, \mathbf{u}_1 \cdot \mathbf{p} \ge b_1'\} \cap V_P = \{\mathbf{v}\}.$$

Let P' be the polyhedron $P \cap \{\mathbf{p} \,:\, \mathbf{u}_1 \cdot \mathbf{p} = b_1'\}$.

We claim that there exists an extreme point $\mathbf{w}'$ of P' in A. Indeed, near $\mathbf{0}$ we can find a point

$$\mathbf{p} \in A \cap P \cap \{\mathbf{p} \,:\, \mathbf{u}_1 \cdot \mathbf{p} < b_1'\}.$$

Then $\mathrm{conv}\{\mathbf{p}, \mathbf{v}\} \cap P'$ is a nonempty subset of A. In particular, P' meets A. Since P' is the convex hull of its extreme points, some extreme point $\mathbf{w}'$ of P' is an element of A.

Use the definition of extreme point, to choose a half space

$$\{\mathbf{p} \,:\, \mathbf{u}_2 \cdot \mathbf{p} \le b_2\} \supset P',$$

where equality holds for a point $\mathbf{p} \in P'$ if and only if $\mathbf{p} = \mathbf{w}'$. By adjusting the parameters $(\mathbf{u}_2, b_2)$ by a multiple of $(\mathbf{u}_1, b_1')$, we may assume without loss of generality that $\mathbf{u}_2 \cdot \mathbf{v} = b_2$.

If $\mathbf{p} \in P$, there exists $\mathbf{p}' \in P'$ and $t \ge 0$ such that $\mathbf{p} = \mathbf{v} + t(\mathbf{p}' - \mathbf{v})$. Then

$$\begin{aligned}\mathbf{u}_2 \cdot \mathbf{p} &= \mathbf{u}_2 \cdot (\mathbf{v} + t(\mathbf{p}' - \mathbf{v}))\\ &= b_2 + t(\mathbf{u}_2 \cdot \mathbf{p}' - b_2) \le b_2,\end{aligned}$$

where equality holds if and only if $\mathbf{p} = \mathbf{v} + t(\mathbf{w}' - \mathbf{v})$. It follows that

$$\{\mathbf{p} \,:\, \mathbf{u}_2 \cdot \mathbf{p} = b_2\} \cap P$$

is an edge of P containing $\mathbf{v}$ and $\mathbf{w}'$. Let $\mathbf{w} \ne \mathbf{v}$ be an extreme point of this edge. This extreme point satisfies the conclusions of the lemma. □

We can relate the combinatorial properties of the polyhedron to the combinatorial properties of the hypermap.

Lemma 5.65 *Let $P \subset \mathbb{R}^3$ be a bounded polyhedron with interior point $\mathbf{0}$. The facets of P are in bijection with the faces of* $\mathrm{hyp}(V_P, E_P)$*, under the correspondence*

$$F \leftrightarrow F' \text{ if and only if } W_F = U_{F'}.$$

Proof By the (BIJECTION) property of conforming fans, the faces of the hypermap are in bijection with the set $[Y]$ of topological components of $Y(V_P, E_P)$. The facets of P are also in bijection with $[Y]$, by Lemma 5.62. □

Lemma 5.66 *Let $P \subset \mathbb{R}^3$ be a bounded polyhedron with interior point $\mathbf{0}$. Under the bijection $F \leftrightarrow F'$ of Lemma 5.65, there are bijections among the following three sets:*

1. *the set of edges of the facet F;*
2. *the face F' of* $\mathrm{hyp}(V_P, E_P)$*;*
3. *a subset E_1 of edges of the fan (V_P, E_P):*

$$E_1 = \{\varepsilon \in E_P \ :\ C^0(\varepsilon) \text{ meets the closure of } U_{F'}\}.$$

Proof It is sufficient to give bijections of the first two sets onto E_1. Let $\mathrm{hyp}(V_P, E_P) = (D, e, n, f)$. Let F be a facet of P and let F' be the corresponding face of $\mathrm{hyp}(V_p, E_P)$. Write $\bar{U}$ for the closure of $W_F = U_{F'}$. As the closure of a convex set, $\bar{U}$ is convex. In fact, whenever $\mathbf{v}, \mathbf{w} \in \bar{U}$, then $C\{\mathbf{v}, \mathbf{w}\} \subset \bar{U}$.

By the construction of E_P, there is an injective map from the set of edges of F to E_P that sends each edge of P to its set $\varepsilon = \{\mathbf{v}, \mathbf{w}\}$ of extreme points. *We claim that this injection sends (1) into E_1.* Indeed, the extreme points $\mathbf{v}, \mathbf{w}$ belong to $F \subset \bar{U}$ and by convexity and rescaling, $C^0\{\mathbf{v}, \mathbf{w}\} \subset \bar{U}$ as well.

We claim that this map sends (1) onto E_1. Indeed, let $\varepsilon' \in E_1$ and let $\mathbf{p} \in C^0(\varepsilon') \cap \bar{U}$. Let $\mathbf{q} : \mathbb{N} \to W_F$ be a sequence converging to $\mathbf{p}$. By rescaling the sequence, we may assume that $\mathbf{q}' : \mathbb{N} \to \mathrm{ri}(F)$ converges to $\mathbf{p}' \in F \cap C^0(\varepsilon') \cap \bar{U}$. This limit lies in $X(V_P, E_P)$, and thus in the boundary of F, which is a union of edges. We obtain $\mathbf{p}' \in \mathrm{conv}\, \varepsilon \cap C^0(\varepsilon')$, where ε is the set of extreme points of an edge of F. By the (INTERSECTION) property of fans, $\varepsilon = \varepsilon'$. This shows that the map sends (1) onto E_1.

Next, we construct a bijection from F' to E_1. There is a map from F' to E_P given by $x = (\mathbf{v}, \mathbf{w}) \mapsto \{\mathbf{v}, \mathbf{w}\}$.

This map is injective. Otherwise, $y = ex \in F'$ for some $x \in F'$. By the simplicity of conforming fans (Lemma 5.28), the two darts y and fx in the same face and the same node are equal: $fx = ex = e^{-1}x = nfx$, giving a fixed point fx under the node map n. By the definition of the node map on $\mathrm{hyp}(V_P, E_P)$, this forces edge set $E(\mathbf{w})$ to contain at most one element, which is contrary to the fact that every azimuth angle of the dart is less than π in a conforming fan.

The image is contained in E_1. Indeed, if $x = (\mathbf{v}, \mathbf{w}) \in F'$, then x at node $\mathbf{v}$ leads into $U_{F'}$ and fx at node $\mathbf{w}$ leads into $U_{F'}$, which gives $\mathbf{v}, \mathbf{w} \in \bar{U}$ and by convexity and rescaling, $C^0\{\mathbf{v}, \mathbf{w}\} \subset \bar{U}$.

Finally, we claim that the map from (2) to E_1 *is onto*. Let $\{\mathbf{v}, \mathbf{w}\} \in E_1$ and write $x = (\mathbf{v}, \mathbf{w})$ and $y = (\mathbf{w}, \mathbf{v})$. Fix $\mathbf{q} \in C^0\{\mathbf{v}, \mathbf{w}\} \cap \bar{U}$. The proof of Lemma 5.23 constructs a neighborhood of $\mathbf{q}$ that is a subset of $U_x \cup U_y \cup C^0\{\mathbf{v}, \mathbf{w}\}$. The topological component $U_{F'}$ meets that neighborhood, forcing it to equal a topological component U_x or U_y. Say $U_x = U_{F'}$. By the property (BIJECTION) of conforming fans (Definition 5.26), $x \in F'$. It follows that $\{\mathbf{v}, \mathbf{w}\} \in E_1$ is the image of $x \in F'$. □

PART THREE

THE KEPLER CONJECTURE

6

Packing

Summary. *This chapter comprises much of the core material of the book. At last we take up the topic of dense sphere packings. Associated with a sphere packing V in* $\mathbb{R}^3$ *are various subsidiary decompositions of space. This chapter focuses on three such decompositions: the Voronoi decomposition into polyhedra, the Rogers decomposition into simplices, and the Marchal decomposition into cells. Each of these decompositions leads to a bound on the density of sphere packings. The bounds in the first two cases are not sharp. The third decomposition leads to a sharp bound* $\pi/\sqrt{18}$ *on the density of sphere packing in three dimensions. The final sections of this chapter undertake a detailed study of the properties of the Marchal cell decomposition.*

6.1 The Primitive State of Our Subject Revealed

6.1.1 definition

Informally, a *packing* is an arrangement of congruent balls in Euclidean three space that are nonoverlapping in the sense that the interiors of the balls are pairwise disjoint. By convention, we take the radius of the congruent balls to be 1. Let V be the set of centers of the balls in a packing. The choice of unit radius for the balls implies that any two points in V have distance at least 2 from each other. Formally, the packing is identified with the set of centers V.

A packing in which no further balls can be added is said to be *saturated* (Figure 6.1).

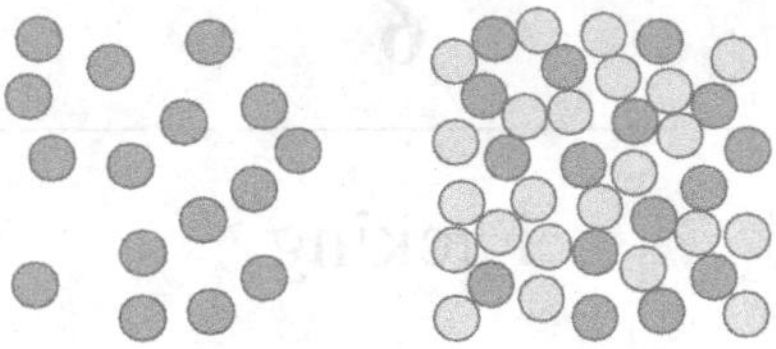

Figure 6.1 A random packing of disks in $\mathbb{R}^2$ and a saturated extension.

Definition 6.1 (saturated, packing) A *packing* $V \subset \mathbb{R}^3$ is a set such that

$$\forall \mathbf{u}, \mathbf{v} \in V.\ \|\mathbf{u} - \mathbf{v}\| < 2 \Rightarrow (\mathbf{u} = \mathbf{v}).$$

A set V is *saturated* if for every $\mathbf{p} \in \mathbb{R}^3$ there exists some $\mathbf{u} \in V$ such that $\|\mathbf{u} - \mathbf{p}\| < 2$.

Let $B(\mathbf{p}, r)$ denote the open ball in Euclidean three-space at center $\mathbf{p}$ and radius r. The open ball is measurable with measure $4\pi r^3/3$. Set $V(\mathbf{p}, r) = V \cap B(\mathbf{p}, r)$.

Lemma 6.2 *Let V be a packing and let $\mathbf{p} \in \mathbb{R}^3$. Then the set $V(\mathbf{p}, r)$ is finite.*

Proof Let $\mathbf{p} = (p_1, p_2, p_3)$. The floor function gives the map

$$(v_1, v_2, v_3) \mapsto (\lfloor 2(v_1 - p_1) \rfloor, \lfloor 2(v_2 - p_2) \rfloor, \lfloor 2(v_3 - p_3) \rfloor).$$

It is a one-to-one map from $V(\mathbf{p}, r)$ into the set $\mathbb{Z}^3 \cap B(\mathbf{0}, 2r+1)$. By Lemma 3.28 the range of this one-to-one map is finite. Hence, the domain $V(\mathbf{p}, r)$ of the map is also finite.[1] □

6.1.2 Voronoi cell

Geometric decompositions of space give a way to estimate the density of sphere packings. A popular decomposition of space is the Voronoi cell decomposition (Figure 6.2).

Definition 6.3 (Voronoi cell, Ω) Let $V \subset \mathbb{R}^3$ and $\mathbf{v} \in V$. The *Voronoi cell* $\Omega(V, \mathbf{v})$ is the set of points at least as close to $\mathbf{v}$ as to any other point in V.

Lemma 6.4 (Voronoi partition) *If V is a saturated packing, then*

$$\mathbb{R}^3 = \bigcup \{\Omega(V, \mathbf{v}) \ :\ \mathbf{v} \in V\}. \tag{6.5}$$

[1] An alternative proof uses the open cover of the compact ball $\bar{B}(\mathbf{p}, r)$ by the sets $\bar{B}(\mathbf{p}, r) \setminus V$ and $B(\mathbf{v}, 1)$ for $\mathbf{v} \in V$. By compactness, the cover is necessarily finite.

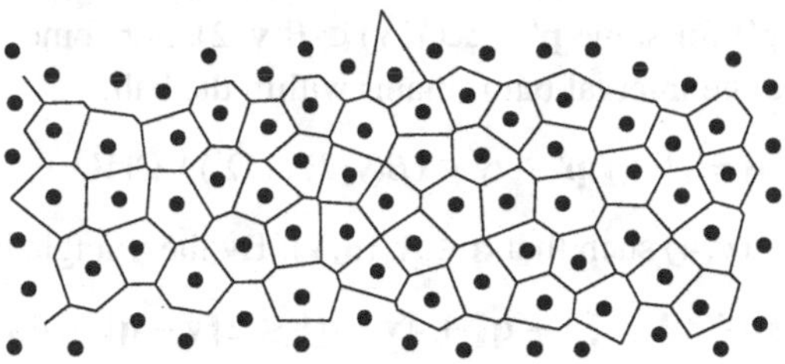

Figure 6.2 Voronoi cells of a packing in $\mathbb{R}^2$.

Proof If V is a saturated packing, then every point $\mathbf{p}$ has distance less than 2 from some point of V. The set $V(\mathbf{p}, 2)$ is finite by Lemma 6.2. Hence, $\mathbf{p}$ is at least as close to some $\mathbf{v} \in V$ as it is to any other $\mathbf{w} \in V$. This means that $\mathbf{p} \in \Omega(V, \mathbf{v})$. □

We use half-spaces to separate one Voronoi cell from another.

Definition 6.6 (half-space)

$$A(\mathbf{u}, \mathbf{v}) = \{\mathbf{p} \in \mathbb{R}^3 \ : \ 2(\mathbf{v} - \mathbf{u}) \cdot \mathbf{p} = \|\mathbf{v}\|^2 - \|\mathbf{u}\|^2\},$$
$$A_+(\mathbf{u}, \mathbf{v}) = \{\mathbf{p} \in \mathbb{R}^3 \ : \ 2(\mathbf{v} - \mathbf{u}) \cdot \mathbf{p} \leq \|\mathbf{v}\|^2 - \|\mathbf{u}\|^2\},$$

when $\mathbf{u}, \mathbf{v} \in \mathbb{R}^3$. The plane $A(\mathbf{u}, \mathbf{v})$ is the *bisector* of $\{\mathbf{u}, \mathbf{v}\}$ and $A_+(\mathbf{u}, \mathbf{v})$ is the *half-space* of points at least as close to $\mathbf{u}$ as to $\mathbf{v}$.

Each Voronoi cell is a bounded polyhedron.

Lemma 6.7 (Voronoi polyhedron) *Let $V \subset \mathbb{R}^3$ be a saturated packing. Then $\Omega(V, \mathbf{v}) \subset B(\mathbf{v}, 2)$. Also, $\Omega(V, \mathbf{v})$ is a polyhedron defined by the intersection of the finitely many half-spaces $A_+(\mathbf{v}, \mathbf{u})$ for $\mathbf{u} \in V(\mathbf{v}, 4) \setminus \{\mathbf{v}\}$.*

Proof The Voronoi cell $\Omega(V, \mathbf{v})$ is the intersection of the half-spaces $A_+(\mathbf{v}, \mathbf{u})$ as $\mathbf{u}$ runs over $V \setminus \{\mathbf{v}\}$.

Let $\mathbf{p} \notin B(\mathbf{v}, 2)$. By saturation, there exists $\mathbf{u} \in V$ such that $\|\mathbf{p} - \mathbf{u}\| < 2$. Then

$$\|\mathbf{p} - \mathbf{u}\| < 2 \leq \|\mathbf{p} - \mathbf{v}\|.$$

Hence, $\mathbf{p} \notin \Omega(V, \mathbf{v})$. This proves the first conclusion.

Let Ω' be the intersection of the half-spaces $A_+(\mathbf{v}, \mathbf{u})$ as $\mathbf{u}$ runs over $V(\mathbf{v}, 4)$. Clearly, $\Omega(V, \mathbf{v}) \subset \Omega'$. Assume for a contradiction that $\mathbf{p} \in \Omega' \setminus \Omega(V, \mathbf{v})$. The intersection of the ray $\mathrm{aff}_+\{\mathbf{v}, \{\mathbf{p}\}\}$ with $\Omega(V, \mathbf{v})$ is a closed and bounded convex subset of the line. By general principles of convex sets, this intersection is an interval conv$\{\mathbf{v}, \mathbf{p}'\}$ for some $\mathbf{p}' \in \Omega(V, \mathbf{v}) \subset B(\mathbf{v}, 2)$. For some small $t > 0$, the point lies beyond the interval but remains within the ball:

$$\mathbf{q} = (1 + t)\mathbf{p}' - t\mathbf{v} \in (B(\mathbf{v}, 2) \cap \Omega') \setminus \Omega(V, \mathbf{v}).$$

Choose $\mathbf{u} \in V \setminus V(\mathbf{v}, 4)$ such that $\mathbf{q} \in A_+(\mathbf{u}, \mathbf{v})$. By the triangle inequality,

$$\|\mathbf{u} - \mathbf{v}\| \leq \|\mathbf{u} - \mathbf{q}\| + \|\mathbf{v} - \mathbf{q}\| \leq 2\|\mathbf{v} - \mathbf{q}\| < 4.$$

This contradicts the assumption $\mathbf{u} \notin V(\mathbf{v}, 4)$.

The number of half-spaces $A_+(\mathbf{v}, \mathbf{u})$ for $\mathbf{u} \in V(\mathbf{v}, 4)$ is finite by Lemma 6.2. A set defined by the intersection of a finite number of closed half-spaces is a polyhedron. □

Lemma 6.8 (Voronoi compact) *Let V be a saturated packing. For every $\mathbf{v} \in V$, the Voronoi cell $\Omega(V, \mathbf{v})$ is compact, convex, and measurable.*

Proof By the previous lemma, it is a bounded polyhedron. Every bounded polyhedron is compact, convex, and measurable. □

6.1.3 reduction to a finite packing

We finally state the main result of this book, the Kepler conjecture. The proof fills most of this book. This section describes the outline of the proof and gives references to the sources of the details of the proof.

Theorem* 6.9 (Kepler's conjecture on dense packings) *No packing of congruent balls in Euclidean three space has density greater than that of the face-centered cubic (FCC) packing.*

Remark 6.10 This density is $\pi/\sqrt{18} \approx 0.74$. There are other packings, such as the HCP or the FCC packing with finitely many balls removed, that attain this same density.

The Kepler conjecture is a statement about space-filling packings. A space-filling packing is specified by a countable number of real coordinates – three for the position of each of countably many balls. The first task in resolving the conjecture is to reduce the problem to one involving only a finite number of balls. This is accomplished by Lemma 6.13.

The relevant concepts are *negligibility* and *FCC-compatibility*, given as follows. FCC-compatibility means that the Voronoi cells on average have volume at least that of those in the FCC-packing. Negligibility means that the error term is insignificant.

Definition 6.11 (negligible, FCC-compatible) A function $G : V \to \mathbb{R}$ on a set $V \subset \mathbb{R}^3$ is *negligible* if there is a constant c_1 such that for all $r \geq 1$,

$$\sum_{\mathbf{v} \in V(\mathbf{0},r)} G(\mathbf{v}) \leq c_1 r^2.$$

A function $G : V \to \mathbb{R}$ is *FCC-compatible* if for all $\mathbf{v} \in V$,

$$4\sqrt{2} \leq \mathrm{vol}(\Omega(V, \mathbf{v})) + G(\mathbf{v}).$$

Remark 6.12 The value $\mathrm{vol}(\Omega(V, \mathbf{v})) + G(\mathbf{v})$ may be interpreted as an *adjusted* volume of the Voronoi cell. The constant $4\sqrt{2}$ that appears in the definition of FCC-compatibility is the volume of the Voronoi cell in the HCP and FCC packings. (See Chapter 1.) The corrected volume is at least the volume of these Voronoi cells when the correction term G is FCC-compatible.

The density $\delta(V, \mathbf{p}, r)$ of a packing V within a bounded region of space is defined as a ratio. The numerator is volume of $B(V, \mathbf{p}, r)$, defined as the intersection with $B(\mathbf{p}, r)$ of the union of all balls $B(\mathbf{v}, 1)$ in the packing V. The denominator is the volume of $B(\mathbf{p}, r)$.

Lemma 6.13 (reduction to finite dimensions) *If there exists a negligible FCC-compatible function $G : V \to \mathbb{R}$ for a saturated packing V, then there exists a constant $c = c(V)$ such that for all $r \geq 1$,*

$$\delta(V, \mathbf{0}, r) \leq \pi \big/ \sqrt{18} + c/r.$$

Proof The volume of $B(V, \mathbf{0}, r)$ is at most the product of the volume $4\pi/3$ of each ball with the number of centers in $B(\mathbf{0}, r + 1)$. Hence,

$$\mathrm{vol}\, B(V, \mathbf{0}, r) \leq \mathrm{card}(V(\mathbf{0}, r + 1))\, 4\pi/3. \tag{6.14}$$

Each truncated Voronoi cell is contained in a ball of radius 2 that is concentric with the unit ball in that cell. The volume of the large ball $B(\mathbf{0}, r + 3)$ is at least the combined volume of all truncated Voronoi cells centered in $B(\mathbf{0}, r+1)$. This observation, combined with FCC-compatibility and negligibility, gives

$$\begin{aligned} 4\sqrt{2}\, \mathrm{card}(V(\mathbf{0}, r + 1)) &\leq \sum_{\mathbf{v} \in V(\mathbf{0},r+1)} (G(\mathbf{v}) + \mathrm{vol}(\Omega(V, \mathbf{v}))) \\ &\leq c_1(r + 1)^2 + \mathrm{vol}\, B(\mathbf{0}, r + 3) \\ &\leq c_1(r + 1)^2 + (1 + 3/r)^3 \mathrm{vol}\, B(\mathbf{0}, r). \end{aligned} \tag{6.15}$$

Recall that $\delta(V, \mathbf{0}, r) = \mathrm{vol}\, B(V, \mathbf{0}, r)/\mathrm{vol}\, B(\mathbf{0}, r)$. Divide Inequality 6.14 through by vol $B(\mathbf{0}, r)$. Use Inequality 6.15 to eliminate $\text{card}(V(\mathbf{0}, r+1))$ from the resulting inequality. This gives

$$\delta(V, \mathbf{0}, r) \le \frac{\pi}{\sqrt{18}}(1 + 3/r)^3 + c_1 \frac{(r+1)^2}{r^3 4\sqrt{2}}.$$

The result follows for an appropriately chosen constant c (depending on c_1). □

Remark 6.16 (Kepler conjecture in precise terms) The precise meaning of the *sphere packing problem* or the *Kepler conjecture* is to prove the bound bound $\delta(V, \mathbf{0}, r) \le \pi/\sqrt{18} + c/r$ for every saturated packing V. The error term c/r comes from the boundary effects of a bounded container holding the balls. The error tends to zero as the radius r of the container tends to infinity. Thus, by the preceding lemma, the existence of a negligible FCC-compatible function provides the solution to the packing problem. The strategy is to define a negligible function and then to solve an optimization problem in finitely many variables to establish that the function is also FCC-compatible.

6.2 Rogers Simplex

Rogers gave a bound on the density of sphere packings in Euclidean space of arbitrary dimension [36]. His bound states that the density of a packing in n-dimensions cannot exceed the ratio of the volume of $A \cap T$ to the volume of T, where T is a regular tetrahedron of side length 2 and A is the set of $n+1$ balls of unit radius placed at the extreme points of T. In two dimensions, the Rogers's bound is sharp and gives a solution to the sphere packing problem. In three dimensions the bound is approximately 0.7797, which differs significantly from the optimal value $\pi/\sqrt{18} \approx 0.74$. Rogers's bound is the unattainable density that would result if regular tetrahedra could tile space.[2]

To prove his bound, Rogers gives a partition of Euclidean space into simplices with extreme points in a packing V. This section develops the basis properties of Rogers simplices. The next section modifies the simplices to obtain a sharp bound on the density of packings.

[2] Aristotle erroneously believed that regular tetrahedra tile space: "It is agreed that there are only three plane figures which can fill a space, the triangle, the square, and the hexagon, and only two solids, the pyramid and the cube." [1].

6.2.1 faces

The Rogers partition is a refinement of the Voronoi cell decomposition. In preparation for this decomposition, this subsection goes into greater detail about the structure of the faces of a Voronoi cell. We parameterize various faces of the Voronoi cell by lists of points in a saturated packing V (Figure 6.3).

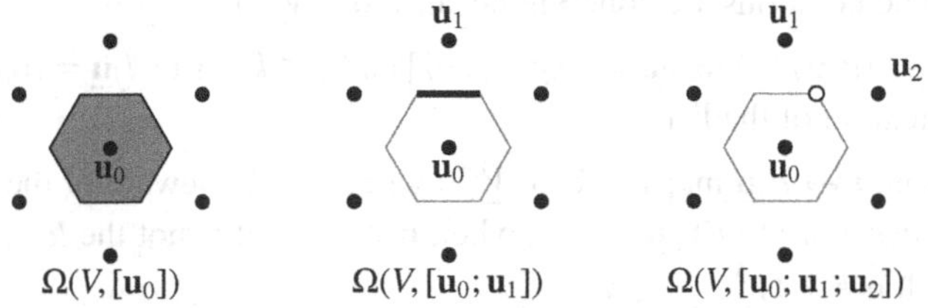

Figure 6.3 We use lists $[\mathbf{u}_0; \cdots; \mathbf{u}_k]$ of points in a packing to select a Voronoi cell (shaded), one of its edges (thick segment), and one of its extreme points (white dot). For simplicity, we illustrate the two-dimensional analogue.

Definition 6.17 (Ω reprise) Let V be a saturated packing. The notation $\Omega(V, *)$ can be *overloaded* to denote intersections of Voronoi cells, when the second argument is a set or list of points. If $W \subset V$, then the intersection of the family of Voronoi cells is $\Omega(V, W)$:

$$\Omega(V, W) = \bigcap \{\Omega(V, \mathbf{u}) \ : \ \mathbf{u} \in W\}.$$

Define Ω on lists to be the same as its value on point sets:

$$\Omega(V, [\mathbf{u}_0; \ldots; \mathbf{u}_k]) = \Omega(V, \{\mathbf{u}_0; \ldots; \mathbf{u}_k\}).$$

An intersection of Voronoi cells can be written in many equivalent forms:

$$\Omega(V, \mathbf{v}) \cap \Omega(V, \mathbf{u}) = \Omega(V, \{\mathbf{u}, \mathbf{v}\}) = \Omega(V, \mathbf{v}) \cap A_+(\mathbf{u}, \mathbf{v}) = \Omega(V, \mathbf{v}) \cap A(\mathbf{u}, \mathbf{v}) = \cdots.$$

Definition 6.18 ($\underline{V}$) Let V be a saturated packing. When $k = 0, 1, 2, 3$, let $\underline{V}(k)$ be the set of lists $\underline{\mathbf{u}} = [\mathbf{u}_0; \ldots; \mathbf{u}_k]$ of length $k + 1$ with $\mathbf{u}_i \in V$ such that

$$\dim \text{aff}(\Omega(V, [\mathbf{u}_0; \ldots; \mathbf{u}_j])) = 3 - j \tag{6.19}$$

for all $0 < j \leq k$. (Recall that $\dim \text{aff}(X)$ is the affine dimension of X from Definition 5.43.) Set $\underline{V}(k) = \varnothing$ for $k > 3$.

In particular, V can be identified with $\underline{V}(0)$ under the natural bijection $\mathbf{v} \mapsto [\mathbf{v}]$, and $\underline{V}(1)$ is the set of lists $[\mathbf{u}; \mathbf{v}]$ of distinct elements such that the Voronoi cells at $\mathbf{u}$ and $\mathbf{v}$ have a common facet (Lemma 6.23).

Notation 6.20 (underscore) The underscores follow a special syntax. In $\underline{V}(k)$, the underscore is a function

$$\underline{\ \ } : \{V : V \text{ saturated packing}\} \times \mathbb{N} \to *.$$

The syntax is somewhat different in $\underline{\mathbf{u}}$. Here the underscore is not a function, but part of its name, following a general typographic convention to mark lists of points. The notations are coherent because $\underline{\mathbf{u}} \in \underline{V}(k)$.

Notation 6.21 ($d_j\underline{\mathbf{u}}$) When $\underline{\mathbf{u}} = [\mathbf{u}_0; \ldots; \mathbf{u}_k]$ and $j \le k$, write $d_j\underline{\mathbf{u}} = [\mathbf{u}_0; \ldots; \mathbf{u}_j]$ for the truncation of the list.

Truncation $\underline{\mathbf{u}} \mapsto d_j\underline{\mathbf{u}}$ maps $\underline{V}(k)$ to $\underline{V}(j)$ when $j \le k$. Beware of the index: k is the *codimension* of $\Omega(V, \underline{\mathbf{u}})$ in $\mathbb{R}^3$, when $\underline{\mathbf{u}} \in \underline{V}(k)$; it is not the *length* of the list $\underline{\mathbf{u}}$ (which is $k + 1$).[3]

Lemma 6.22 (Voronoi face) *Let $V \subset \mathbb{R}^3$ be a saturated packing. Let $\underline{\mathbf{u}} = [\mathbf{u}_0; \ldots; \mathbf{u}_k] \in \underline{V}(k)$. Then $\Omega(V, \underline{\mathbf{u}})$ is a face of $\Omega(V, \mathbf{u}_0)$.*

Proof This follows directly from the definition of face on page 133. The set $\Omega(V, \underline{\mathbf{u}})$ is an intersection of the convex sets $\Omega(V, \mathbf{u}_i)$ and is therefore convex. Also, $\Omega(V, \underline{\mathbf{u}})$ is the intersection of $\Omega(V, \mathbf{u}_0)$ with the planes $A(\mathbf{u}_0, \mathbf{u}_i)$, where $i > 0$. Let $\mathbf{p}, \mathbf{q} \in \Omega(V, \mathbf{u}_0)$ and assume

$$\mathbf{p}' = s\mathbf{p} + t\mathbf{q} \in \Omega(V, \underline{\mathbf{u}}), \quad \text{for some } s > 0, \quad t > 0, \quad s + t = 1.$$

Then $\mathbf{p}' \in A(\mathbf{u}_0, \mathbf{u}_i)$. Each plane $A(\mathbf{u}_0, \mathbf{u}_i)$ is a face of the corresponding half-space $A_+(\mathbf{u}_0, \mathbf{u}_i)$ containing $\mathbf{p}$ and $\mathbf{q}$. By the definition of face, $\mathbf{p}, \mathbf{q}$ must also lie in $A(\mathbf{u}_0, \mathbf{u}_i)$. It follows that $\mathbf{p}, \mathbf{q}$ also lie in $\Omega(V, \underline{\mathbf{u}})$. By the definition of face, $\Omega(V, \underline{\mathbf{u}})$ is a face of $\Omega(V, \mathbf{u}_0)$. □

Lemma 6.23 (facets) *Let $V \subset \mathbb{R}^3$ be a saturated packing. Let $\underline{\mathbf{u}} \in \underline{V}(k)$ for some $k < 3$. Then F is a facet of $\Omega(V, \underline{\mathbf{u}})$ if and only if there exists $\underline{\mathbf{v}} \in \underline{V}(k + 1)$ such that $F = \Omega(V, \underline{\mathbf{v}})$ and $d_k\underline{\mathbf{v}} = \underline{\mathbf{u}}$.*

Proof Use Lemma 6.7 to write the polyhedron $\Omega(V, \underline{\mathbf{u}})$ in the form of Equation 5.53:

$$\Omega(V, \underline{\mathbf{u}}) = A \cap A_\pm(\mathbf{v}_1, \mathbf{u}_0) \cap \cdots \cap A_\pm(\mathbf{v}_r, \mathbf{u}_0),$$

where A is the affine hull of $\Omega(V, \underline{\mathbf{u}})$, $\mathbf{v}_i \in V$, where $A_-(\mathbf{v}, \mathbf{w}) = A_+(\mathbf{w}, \mathbf{v})$ with the signs $\pm$ chosen as needed, and r is as small as possible. By Lemma 5.54, if F is any facet of $\Omega(V, \underline{\mathbf{u}})$, then there exists an $i \le r$ such that

$$F = \Omega(V, \underline{\mathbf{u}}) \cap A(\mathbf{v}_i, \mathbf{u}_0) = \Omega(V, \underline{\mathbf{v}}),$$

[3] By convention aa k-simplex is presented as a $k + 1$-tuple. Because of this shift by one, the notation $d_j\underline{\mathbf{u}}$ also differs by the same shift.

where $\underline{\mathbf{v}} = [\mathbf{u}_0; \cdots; \mathbf{u}_k; \mathbf{v}_i]$ is the list that appends $\mathbf{v}_i$ to $\underline{\mathbf{u}}$. Also,

$$\dim \operatorname{aff}(\Omega(V, \underline{\mathbf{v}})) = \dim \operatorname{aff}(F) = \dim \operatorname{aff}(\Omega(V, \underline{\mathbf{u}})) - 1 = 3 - k - 1,$$

because F is a facet. It follows that $\underline{\mathbf{v}} \in \underline{V}(k + 1)$. This proves the implication in the forward direction.

To prove the converse, let $\underline{\mathbf{v}} \in \underline{V}(k + 1)$, where $d_k\underline{\mathbf{v}} = \underline{\mathbf{u}}$. Elementary verifications show that $\Omega(V, \underline{\mathbf{v}}) \subset \Omega(V, \underline{\mathbf{u}})$ and that this set is nonempty if $k < 3$. By Lemma 6.22 and Lemma 5.54, $\Omega(V, \underline{\mathbf{v}})$ is a face of $\Omega(V, \underline{\mathbf{u}})$. By the definition of $\underline{V}(*)$,

$$\dim \operatorname{aff}(\Omega(V, \underline{\mathbf{v}})) = 3 - (k + 1) = \dim \operatorname{aff}(\Omega(V, \underline{\mathbf{u}})) - 1.$$

It follows that $\Omega(V, \underline{\mathbf{v}})$ is a facet of $\Omega(V, \underline{\mathbf{u}})$. □

6.2.2 partitioning space

Each Rogers simplex is given as the convex hull of its set of extreme points. The extreme points $\omega(d_i\underline{\mathbf{u}})$ are defined by recursion.

Definition 6.24 (ω) Let V be a saturated packing and let $\underline{\mathbf{u}} = [\mathbf{u}_0; \ldots] \in \underline{V}(k)$ for some k. Define points $\omega_j = \omega_j(V, \underline{\mathbf{u}}) \in \mathbb{R}^3$ by recursion over $j \le k$ (Figure 6.4).

$$\begin{aligned} \omega_0 &= \mathbf{u}_0, \\ \omega_{j+1} &= \text{the closest point to } \omega_j \text{ on } \Omega(V, d_{j+1}\underline{\mathbf{u}}). \end{aligned}$$

Set $\omega(V, \underline{\mathbf{u}}) = \omega_k(V, \underline{\mathbf{u}})$, when $\underline{\mathbf{u}} \in \underline{V}(k)$. The set V is generally fixed and is dropped from the notation.

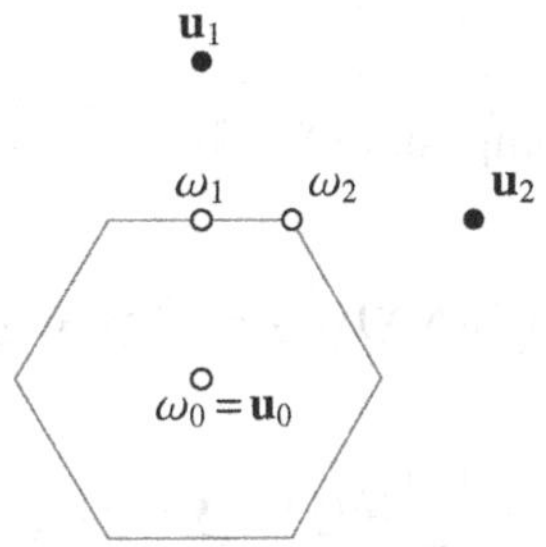

Figure 6.4 The points ω_j are constructed in a nested sequence of faces of a Voronoi cell. For simplicity, we illustrate the two-dimensional analogue. The hexagon is a Voronoi cell about $\mathbf{u}_0$ and $\omega_j = \omega_j(V, [\mathbf{u}_0; \mathbf{u}_1; \mathbf{u}_2])$.

The point $\omega(\underline{\mathbf{u}})$ exists when $\underline{\mathbf{u}} \in V(k)$. Indeed, the set $\Omega(V, \underline{\mathbf{u}})$ is nonempty, convex, and compact. Thus, by convex analysis, the closest point $\omega(\underline{\mathbf{u}})$ exists uniquely.

The point ω_k depends on $\underline{\mathbf{u}}$ through its projection to $d_k\underline{\mathbf{u}}$ so that

$$\omega_k(\underline{\mathbf{u}}) = \omega_k(d_k\underline{\mathbf{u}}) = \omega(d_k\underline{\mathbf{u}}).$$

Definition 6.25 (R, Rogers simplex) Let $V \subset \mathbb{R}^3$ be a saturated packing. For $\underline{\mathbf{u}} \in \underline{V}(k)$, let

$$R(\underline{\mathbf{u}}) = \mathrm{conv}\{\omega(d_0\underline{\mathbf{u}}), \omega(d_1\underline{\mathbf{u}}), \ldots, \omega(d_k\underline{\mathbf{u}})\}.$$

The set $R(\underline{\mathbf{u}})$ is called the Rogers simplex of $\underline{\mathbf{u}}$.

Each Voronoi cell can be partitioned into Rogers simplices (Figure 6.5).

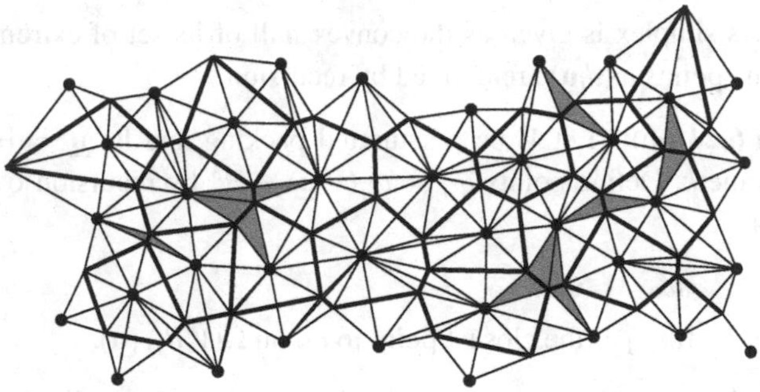

Figure 6.5 The Rogers partition of a packing. For simplicity, we illustrate the two-dimensional analogue. Heavy edges are facets of Voronoi cells. The Rogers simplices that are not right triangles are shaded.

Lemma 6.26 (Rogers decomposition) *For any saturated packing $V \subset \mathbb{R}^3$, and any $\mathbf{u}_0 \in V$,*

$$\Omega(V, \mathbf{u}_0) = \bigcup\{R(\underline{\mathbf{v}}) \ :\ \underline{\mathbf{v}} \in \underline{V}(3),\ d_0\underline{\mathbf{v}} = [\mathbf{u}_0]\}. \tag{6.27}$$

Consequently,

$$\mathbb{R}^3 = \bigcup\{R(\underline{\mathbf{v}}) \ :\ \underline{\mathbf{v}} \in \underline{V}(3)\}.$$

Proof The proof uses standard facts about convex sets and polyhedra from Section 5.4.

By the covering of $\mathbb{R}^3$ by Voronoi cells by (6.5), it is enough to show that each Voronoi cell is covered by Rogers simplices.

Let $\underline{\mathbf{u}} \in \underline{V}(j)$ for $j < 3$. Consider the following set:

$$N = \left\{ k \in \mathbb{N} \ : \ j \le k \le 3, \ \ \Omega(V, \underline{\mathbf{u}}) = \bigcup_{\underline{\mathbf{v}} \in \underline{V}(k),\ d_j \underline{\mathbf{v}} = \underline{\mathbf{u}}} \operatorname{conv}(O_k \cup \Omega(V, \underline{\mathbf{v}})) \right\},$$

where $O_k = \{\omega(d_j\underline{\mathbf{v}}), \ldots, \omega(d_{k-1}\underline{\mathbf{v}})\}$.

We claim $N = \{j, \ldots, 3\}$. Indeed, to see that $j \in N$, we note that

$$\Omega(V, \underline{\mathbf{u}}) = \operatorname{conv}(\Omega(V, \underline{\mathbf{u}})),$$

which holds by the convexity of the polyhedron $\Omega(V, \underline{\mathbf{u}})$. We assume that $k \in N$ and consider the membership condition of N for $k + 1$. We may assume that $k + 1 \le 3$. Then

$$\begin{aligned}
& \bigcup_{\underline{\mathbf{v}} \in \underline{V}(k+1),\ d_j\underline{\mathbf{v}} = \underline{\mathbf{u}}} \operatorname{conv}(O_{k+1} \cup \Omega(V, \underline{\mathbf{v}})) \\
&= \bigcup_{\underline{\mathbf{v}} \in \underline{V}(k+1),\ d_j\underline{\mathbf{v}} = \underline{\mathbf{u}}} \operatorname{conv}(O_k \cup \operatorname{conv}(\{\omega(d_k\underline{\mathbf{v}})\} \cup \Omega(V, \underline{\mathbf{v}}))) \\
&= \bigcup_{\underline{\mathbf{w}} \in \underline{V}(k),\ d_j\underline{\mathbf{w}} = \underline{\mathbf{u}},} \ \bigcup_{\underline{\mathbf{v}} \in \underline{V}(k+1),\ d_k\underline{\mathbf{v}} = \underline{\mathbf{w}}} \operatorname{conv}(O_k \cup \operatorname{conv}(\{\omega(d_k\underline{\mathbf{v}})\} \cup \Omega(V, \underline{\mathbf{v}}))) \\
&= \bigcup_{\underline{\mathbf{w}} \in \underline{V}(k),\ d_j\underline{\mathbf{w}} = \underline{\mathbf{u}}} \operatorname{conv}(O_k \cup \Omega(V, \underline{\mathbf{w}})) \\
&= \Omega(V, \underline{\mathbf{u}}).
\end{aligned}$$

The induction hypothesis is used in the last step. This proves $k + 1 \in N$, and induction gives $N = \{j, \ldots, 3\}$.

Consider the extreme case $j = 0$ and $k = 3$. The set $\Omega(V, \underline{\mathbf{v}})$ reduces to $\{\omega(\underline{\mathbf{v}})\}$ and the convex hull becomes

$$\operatorname{conv}(O_k \cup \Omega(V, \underline{\mathbf{v}})) = R(\underline{\mathbf{v}})$$

when $\underline{\mathbf{v}} \in \underline{V}(3)$. This gives

$$\Omega(V, \mathbf{u}_0) = \bigcup \{R(\underline{\mathbf{v}}) \ : \ \underline{\mathbf{v}} \in \underline{V}(3),\ d_0\underline{\mathbf{v}} = [\mathbf{u}_0]\}. \tag{6.28}$$

This proves the lemma. □

Although the Rogers simplex $R(\underline{\mathbf{u}})$ need not determine the parameter $\underline{\mathbf{u}}$ (Figure 6.6), the intersection of two different Rogers simplices is a null set.

Lemma 6.29 (Rogers disjoint) *Let V be a saturated packing and let $\underline{\mathbf{u}}, \underline{\mathbf{v}} \in \underline{V}(3)$ be lists such that $R(\underline{\mathbf{u}}) \neq R(\underline{\mathbf{v}})$. Then the intersection*

$$R(\underline{\mathbf{u}}) \cap R(\underline{\mathbf{v}})$$

is contained in a plane (and hence has measure zero).

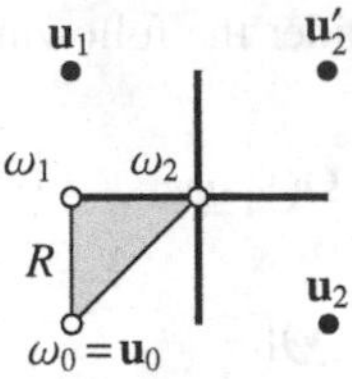

Figure 6.6 Two different parameters $\underline{\mathbf{u}}, \underline{\mathbf{v}} \in \underline{V}(k)$ can determine the same Rogers simplex $R(\underline{\mathbf{u}}) = R(\underline{\mathbf{v}})$. In this example, $\underline{\mathbf{u}} = [\mathbf{u}_0; \mathbf{u}_1; \mathbf{u}_2]$ and $\underline{\mathbf{v}} = [\mathbf{u}_0; \mathbf{u}_1; \mathbf{u}_2']$.

This result and the previous lemma show that the simplices $R(\underline{\mathbf{u}})$ partition Euclidean three-space.

Proof We may assume that the affine dimension of $R(\underline{\mathbf{u}})$ is three, for otherwise $R(\underline{\mathbf{u}})$ is contained in a plane. Similarly, we may assume that the affine dimension of $R(\underline{\mathbf{v}})$ is three.

Let $\underline{\mathbf{u}} = [\mathbf{u}_0; \ldots]$ and $\underline{\mathbf{v}} = [\mathbf{v}_0; \ldots]$. Let k be the first index such that

$$\Omega(V, [\mathbf{u}_0; \ldots; \mathbf{u}_k]) \neq \Omega(V, [\mathbf{v}_0; \ldots; \mathbf{v}_k]).$$

Such an index k exists. Indeed, the definition of points $\omega(d_i\underline{\mathbf{u}})$ depends on $\underline{\mathbf{u}}$ only through the sets $\Omega(V, d_j\underline{\mathbf{u}})$. Hence, $R(\mathbf{u}) \neq R(\mathbf{v})$ implies that the two sequences $\Omega(V, *)$ must differ at some index. We have $\omega_i(\underline{\mathbf{u}}) = \omega_i(\underline{\mathbf{v}})$, for $i < k$.

Select $\mathbf{w} \in R(\underline{\mathbf{u}}) \cap R(\underline{\mathbf{v}})$. Write

$$\mathbf{w} = \sum_{j=0}^{3} s_j \omega_j(\underline{\mathbf{u}}) = \sum_{j=0}^{3} t_j \omega_j(\underline{\mathbf{v}}), \text{ where } \sum_{j=0}^{3} s_j = \sum_{j=0}^{3} t_j = 1.$$

Set $\sigma_i = \sum_{j=i}^{3} s_j$.

We claim that $s_i = t_i$, and $\sigma_{i+1} = \sum_{j=i+1}^{3} t_j$, for $i = 0, \ldots, k-1$. Indeed, the proof is an induction on i. Assume that the claim holds for all indices less than i so that

$$\sum_{j=i}^{3} s_j \omega_j(\underline{\mathbf{u}}) = \sum_{j=i}^{3} t_j \omega_j(\underline{\mathbf{v}}).$$

We apply Lemma 5.59 to the points

$$\mathbf{p}_0 = \omega_i(\underline{\mathbf{u}}) = \omega_i(\underline{\mathbf{v}}), \quad \mathbf{p} = \sum_{j=i+1}^{3} \frac{s_j}{\sigma_i} \omega_j(\underline{\mathbf{u}}), \quad \mathbf{p}' = \sum_{j=i+1}^{3} \frac{t_j}{\sigma_i} \omega_j(\underline{\mathbf{v}})$$

in the polyhedron $\Omega(V, [\mathbf{u}_0; \ldots; \mathbf{u}_i])$ to obtain the induction step $s_i = t_i$.

Let

$$\Omega' = \Omega(V, [\mathbf{u}_0; \ldots; \mathbf{u}_k]) \cap \Omega(V, [\mathbf{v}_0; \ldots; \mathbf{v}_k]) = \Omega(V, [\mathbf{u}_0; \ldots; \mathbf{u}_k; \mathbf{v}_k]).$$

The claim implies that

$$\frac{1}{\sigma_k} \sum_{j=k}^{3} s_j \omega_j(\underline{\mathbf{u}}) = \frac{1}{\sigma_k} \sum_{j=k}^{3} t_j \omega_j(\underline{\mathbf{v}}) \in \Omega'.$$

It follows that the intersection $R(\mathbf{u}) \cap R(\mathbf{v})$ lies in the convex hull C of

$$\{\omega([\mathbf{u}_0]), \ldots, \omega([\mathbf{u}_0; \ldots; \mathbf{u}_{k-1}])\}$$

and Ω'. The set Ω' lies in a facet of $\Omega(V, d_k \underline{\mathbf{u}})$. Hence, the affine dimension of Ω' is at most $3 - k - 1 = 2 - k$. In general, if a set A has affine dimension r, then the affine dimension of $\mathrm{conv}(\{\mathbf{p}\} \cup A)$ is at most $r + 1$. It follows that the affine dimension of C is at most $k + (2 - k) = 2$. The intersection is thus contained in a plane. □

6.2.3 circumcenter

The extreme points of a Rogers simplex are closely related to the circumcenter of various subsets of V. This subsection develops the connection between Rogers simplices and circumcenters.

Definition 6.30 (circumcenter, circumradius) Let $S \subset \mathbb{R}^n$. A point $\mathbf{p}$ is a *circumcenter* of S if it is an element in the affine hull of S that is equidistant from every $\mathbf{v} \in S$. If S has circumcenter $\mathbf{p}$, then the common distance $\|\mathbf{p} - \mathbf{v}\|$ for all $\mathbf{v} \in S$ is the *circumradius* of S.

The circumcenter comes as a solution to a system of linear equations. We pause to review a standard result from the theory of linear algebra, asserting the existence of a solution to a system of equations. Recall that a finite set S is *affinely independent* if $\dim \mathrm{aff}(S) = \mathrm{card}(S) - 1$.

Lemma 6.31 (linear systems) *Let $S = \{\mathbf{v}_0, \ldots, \mathbf{v}_m\} \subset \mathbb{R}^n$ be an affinely independent set of cardinality $m + 1$. Then every system of equations*

$$\mathbf{p} \cdot (\mathbf{v}_i - \mathbf{v}_0) = b_i - b_0, \quad \textit{for } i = 1, \ldots, m$$

has a unique solution $\mathbf{p}$ that lies in the affine hull of S.

Proof This is a standard result from linear algebra. We sketch a proof for the sake of completeness.

Let $\mathbf{w}_i = \mathbf{v}_i - \mathbf{v}_0$ and replace $b_i - b_0$ with b_i. The lemma reduces to the following claim. Let $S' = \{\mathbf{w}_1, \ldots, \mathbf{w}_m\}$ be a *linearly independent* set of cardinality m. Then every system of equations

$$\mathbf{p} \cdot \mathbf{w}_i = b_i, \quad \text{for } i = 1, \ldots, m$$

has a unique solution in $\mathbf{p}$ that lies in the linear span of S'.

A solution is unique. Indeed, the difference $\mathbf{p} = \mathbf{p}' - \mathbf{p}'' = \sum s_i \mathbf{w}_i$ of two solutions $\mathbf{p}', \mathbf{p}''$ satisfies

$$\|\mathbf{p}\|^2 = \mathbf{p} \cdot \mathbf{p} = \sum s_i \mathbf{w}_i \cdot (\mathbf{p}' - \mathbf{p}'') = \sum s_i(b_i - b_i) = 0.$$

It follows that $\mathbf{p} = \mathbf{0}$ and $\mathbf{p}' = \mathbf{p}''$. This proves uniqueness.

Let W be the linear span of $\mathbf{w}_1, \ldots, \mathbf{w}_m$. The image of the map $W \to \mathbb{R}^m$, $\mathbf{p} \mapsto (\mathbf{p} \cdot \mathbf{w}_1, \ldots, \mathbf{p} \cdot \mathbf{w}_m)$ is a linear space and is therefore an affine set.

A solution exists; that is, the image is all of $\mathbb{R}^m$. Otherwise, by Lemma 5.52 some equation must hold; that is, there exists $\mathbf{u} \neq \mathbf{0}$ such that $\mathbf{u} \cdot \mathbf{q} = b$ for every point $\mathbf{q}$ in the image. As $\mathbf{0}$ lies in the image, $b = 0$. Write $\mathbf{p} = \sum u_i \mathbf{w}_i \in W$ and let $\mathbf{q} \in \mathbb{R}^m$ be the image of $\mathbf{p} \in W$. Then

$$\|\mathbf{p}\|^2 = \mathbf{p} \cdot \mathbf{p} = \sum u_i(\mathbf{p} \cdot \mathbf{w}_i) = \mathbf{u} \cdot \mathbf{q} = 0.$$

Thus $\mathbf{p} = \mathbf{0}$ so that $\mathbf{u} = \mathbf{0}$. We have reached a contradiction. □

Lemma 6.32 (circumcenter exists) *Let $S \subset \mathbb{R}^n$ be a nonempty affinely independent set. Then there exists a unique circumcenter of S.*

Proof A point $\mathbf{p}$ is a circumcenter if and only if it is a point in the affine hull of S that satisfies the system of equations:

$$\|\mathbf{p} - \mathbf{v}_i\|^2 = \|\mathbf{p} - \mathbf{v}_0\|^2, \qquad i = i, \ldots, m.$$

Equivalently,

$$\mathbf{p} \cdot (\mathbf{v}_i - \mathbf{v}_0) = b_i - b_0, \qquad i = 1, \ldots, m,$$

where $b_i = \|\mathbf{v}_i\|^2/2$. By Lemma 6.31, this system of equations has a unique solution. □

The following lemma describes the structure of the affine hull of a face of a Voronoi cell. It describes the affine hull as an intersection of half-spaces and shows that it meets $\mathrm{aff}(S)$ orthogonally at the circumcenter of S.

Lemma 6.33 *Let V be a saturated packing and let $k \leq 3$. Let $S = \{\mathbf{u}_0, \ldots, \mathbf{u}_k\}$, where $\underline{\mathbf{u}} = [\mathbf{u}_0, \ldots, \mathbf{u}_k] \in \underline{V}(k)$. Then*

1. $\dim \mathrm{aff}(S) = k$. *(In particular,* $\mathrm{card}\{\mathbf{u}_0, \ldots, \mathbf{u}_k\} = k + 1$, *and S is affinely independent.)*

2. $\operatorname{aff} \Omega(V, \underline{\mathbf{u}}) = \cap_{i=1}^{k} A(\mathbf{u}_0, \mathbf{u}_i)$.
3. $\operatorname{aff} \Omega(V, \underline{\mathbf{u}}) \cap \operatorname{aff}(S) = \{\mathbf{q}\}$, *where* $\mathbf{q}$ *is the circumcenter of* S.
4. $(\operatorname{aff} \Omega(V, \underline{\mathbf{u}}) - \mathbf{q}) \perp (\operatorname{aff}(S) - \mathbf{q})$, *where* $X - \mathbf{q}$ *denotes the translate of a set* X *by* $-\mathbf{q}$, *and* $(\perp)$ *is the orthogonality relation.*

Proof The proof is by induction on k.

The lemma holds when $k = 0$. Indeed, $\Omega(V, \mathbf{u}_0)$ contains an open ball centered at $\mathbf{u}_0$, so its affine hull is $\mathbb{R}^3$. This is the first conclusion. The other conclusions reduce to trivial facts: $\dim \operatorname{aff} \mathbb{R}^3 = 3$, $\dim \operatorname{aff}\{\mathbf{u}_0\} = 0$, $\mathbb{R}^3 \cap \{\mathbf{u}_0\} = \{\mathbf{u}_0\}$, and $\mathbb{R}^3 \perp \{\mathbf{0}\}$.

Assume the induction hypothesis for k. We may assume that $k < 3$ because otherwise there is nothing further to prove. Let $\underline{\mathbf{u}} \in \underline{V}(k+1)$. Let $\underline{\mathbf{v}} = d_k\underline{\mathbf{u}} \in \underline{V}(k)$. Let $\mathbf{q}_k$ be the circumcenter of (the point set of) $\underline{\mathbf{v}}$. Write $A_j = \cap_{i=1}^{j} A(\mathbf{u}_0, \mathbf{u}_i)$; $B_j = \operatorname{aff}(\Omega(V, d_j\underline{\mathbf{u}}))$; $C_j = \operatorname{aff}\{\mathbf{u}_0, \ldots, \mathbf{u}_j\}$; $S_j = \{\mathbf{u}_0, \ldots, \mathbf{u}_j\}$. By the induction hypothesis $A_k = B_k$.

We claim $\dim \operatorname{aff} S_{k+1} = k+1$. Otherwise, by general background facts about affine sets, $\mathbf{u}_{k+1} \in C_k$. Write $\mathbf{u}_{k+1} - \mathbf{q}_k = \sum_{i \le k} t_i(\mathbf{u}_i - \mathbf{q}_k)$. If $\mathbf{p} \in A_k$, then by the orthogonality induction hypothesis:

$$(\mathbf{u}_{k+1} - \mathbf{q}_k) \cdot (\mathbf{p} - \mathbf{q}_k) = \sum t_i(\mathbf{u}_i - \mathbf{q}_k) \cdot (\mathbf{p} - \mathbf{q}_k) = 0,$$

and consequently

$$\|\mathbf{u}_{k+1} - \mathbf{p}\|^2 - \|\mathbf{u}_0 - \mathbf{p}\|^2 = \|\mathbf{u}_{k+1} - \mathbf{q}_k\|^2 - \|\mathbf{u}_0 - \mathbf{q}_k\|^2.$$

Thus, if A_k meets $A(\mathbf{u}_0, \mathbf{u}_{k+1})$ at some point $\mathbf{p}$, then both sides of this equation vanish and $A_k \subset A(\mathbf{u}_0, \mathbf{u}_{k+1})$. This is contrary to $0 \le \dim \operatorname{aff}(A_{k+1}) = \dim \operatorname{aff}(A_k) - 1$, which holds because $\underline{\mathbf{u}} \in \underline{V}(k+1)$ with $k < 3$.

We claim that $B_{k+1} = A_{k+1}$. Indeed, by definition, $B_{k+1} \subset A_{k+1} \subset A_k$. Also,

$$\dim \operatorname{aff} B_{k+1} = 3 - (k+1) \le \dim \operatorname{aff} A_{k+1} \le \dim \operatorname{aff} A_k = 3 - k.$$

Hence, by general background on affine sets, if $A_{k+1} \ne A_k$, then $B_{k+1} = A_{k+1}$. Suppose for a contradiction that $A_k = A_{k+1}$. Then $\Omega(V, \underline{\mathbf{v}}) \subset \Omega(V, \underline{\mathbf{u}}) = \Omega(V, \underline{\mathbf{v}}) \cap A(\mathbf{u}_0, \mathbf{u}_{k+1}) \subset \Omega(V, \underline{\mathbf{v}})$, so that $B_k = B_{k+1}$. This contradicts the defining conditions of $\underline{V}(k+1)$.

We claim that $A_{k+1} \cap C_{k+1} = \{\mathbf{q}_{k+1}\}$. Indeed, by the definition of A_{k+1}, any point in this affine set is equidistant from every point of S_{k+1}. By the definition of C_{k+1}, the intersection lies in the affine hull of S_{k+1}. This uniquely characterizes the circumcenter.

Finally, $(A_{k+1} - \mathbf{q}_{k+1}) \perp (C_{k+1} - \mathbf{q}_{k+1})$. Indeed, if $\mathbf{p} \in A_{k+1}$, then

$$\begin{aligned} 0 &= \|\mathbf{p} - \mathbf{u}_i\|^2 - \|\mathbf{p} - \mathbf{u}_0\|^2 \\ &= \|(\mathbf{p} - \mathbf{q}_{k+1}) - (\mathbf{u}_i - \mathbf{q}_{k+1})\|^2 - \|(\mathbf{p} - \mathbf{q}_{k+1}) - (\mathbf{u}_0 - \mathbf{q}_{k+1})\|^2 \\ &= -2(\mathbf{p} - \mathbf{q}_{k+1}) \cdot (\mathbf{u}_i - \mathbf{u}_0). \end{aligned}$$

Since the linear span of the points $\mathbf{u}_i - \mathbf{u}_0$ is all of $C_{k+1} - \mathbf{q}_{k+1}$, the final claim and the proof by induction ensue. □

Definition 6.34 (h) If $\underline{\mathbf{u}} = [\mathbf{u}_0; \mathbf{u}_0; \ldots; \mathbf{u}_k]$ is a list of points in $\mathbb{R}^n$, then let $h(\underline{\mathbf{u}})$ be the circumradius of its point set $\{\mathbf{u}_0, \ldots, \mathbf{u}_k\}$.

Remark 6.35 The constant $r = \sqrt{2}$ is the smallest real number r such that there exist four cocircular points in the plane with pairwise distances at least 2 and with circumradius r (Figure 6.7). The four points are the vertices of a square of side length 2. Eight two-dimensional Rogers simplices meet at the circumcenter of the square, but when $r < \sqrt{2}$, only six Rogers simplices meet at the circumcenter. In general, at $r = \sqrt{2}$, certain degeneracies start to appear in n-dimensions that cannot occur for a smaller radius. To avoid degeneracies, many lemmas in this section assume that the circumradius is less than $\sqrt{2}$.

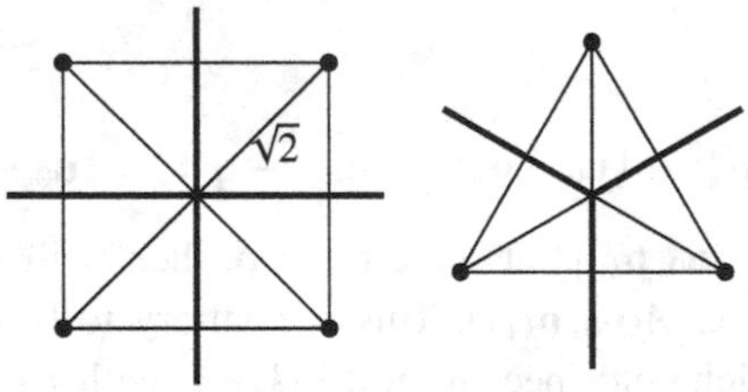

Figure 6.7 There are eight two-dimensional Rogers simplices of diameter $\sqrt{2}$ that meet at the center of a square. Whenever the diameter of the Rogers simplex is less than $\sqrt{2}$, the simplex is one of exactly six that meet at an extreme point of a Voronoi cell. Heavy edges are facets of Voronoi cells.

Lemma 6.36 (nondegeneracy) *Let $V \subset \mathbb{R}^3$ be a saturated packing. Let $S \subset V$ be an affinely independent set with circumcenter* $\mathbf{p}$. *Assume that the circumradius of S is less than $\sqrt{2}$. Then $\|\mathbf{v} - \mathbf{p}\| > \|\mathbf{u} - \mathbf{p}\|$ for all $\mathbf{u} \in S$ and all $\mathbf{v} \in V \setminus S$.*

Proof Assume for a contradiction that there is a point $\mathbf{w} \in V \setminus S$ satisfying

$$\|\mathbf{w} - \mathbf{p}\| \leq \|\mathbf{u} - \mathbf{p}\|, \quad \text{for all } \mathbf{u} \in S. \tag{6.37}$$

The angles $\mathrm{arc}_V(\mathbf{p}, \{\mathbf{v}, \mathbf{u}\})$ are obtuse for distinct elements $\mathbf{v}, \mathbf{u}$ of $S \cup \{\mathbf{w}\}$ because of the law of cosines and

$$\|\mathbf{p} - \mathbf{u}\| < \sqrt{2}, \quad \|\mathbf{p} - \mathbf{v}\| < \sqrt{2}, \quad \|\mathbf{u} - \mathbf{v}\| \geq 2.$$

Let $S = \{\mathbf{u}_0, \ldots, \mathbf{u}_k\}$. A case-by-case argument follows for each $k \in \{0, 1, 2, 3\}$.

0. The case $k = 0$ is trivial.
1. In the case $k = 1$, the points $\mathbf{p}, \mathbf{u}_0, \mathbf{u}_1$ are collinear and cannot give two obtuse angles.
2. In this case, let $\mathbf{w}'$ be the projection of $\mathbf{w}$ to the plane containing $\mathbf{p}, \mathbf{u}_0, \mathbf{u}_1, \mathbf{u}_2$. Under orthogonal projection, the angles remain obtuse:

   $$(\mathbf{u}_i - \mathbf{p}) \cdot (\mathbf{w} - \mathbf{p}) = (\mathbf{u}_i - \mathbf{p}) \cdot (\mathbf{w}' - \mathbf{p}) < 0.$$

   The four points $\mathbf{w}', \mathbf{u}_0, \mathbf{u}_1$, and $\mathbf{u}_2$ can be arranged cyclically around $\mathbf{p}$, according to the polar cycle, each forming an obtuse angle with the next. A circle around $\mathbf{p}$ cannot give four obtuse angles because the sum is 2π.
3. In this case, assume that $\mathbf{u}_0, \ldots, \mathbf{u}_3$ are labeled according to the azimuth cycle around the line $\mathrm{aff}\{\mathbf{p}, \mathbf{w}\}$. Consider the dihedral angle

   $$\gamma = \gamma_i = \mathrm{dih}(\{\mathbf{p}, \mathbf{w}\}, \{\mathbf{u}_i, \mathbf{u}_{i+1}\})$$

   of the simplex $\{\mathbf{p}, \mathbf{w}, \mathbf{u}_i, \mathbf{u}_{i+1}\}$ along the edge $\{\mathbf{p}, \mathbf{w}\}$. By the spherical law of cosines, the angle γ of the spherical triangle with sphere center $\mathbf{p}$ is given in terms of the edges as

   $$\cos c - \cos a \cos b = \sin a \sin b \cos \gamma.$$

   The angles a, b, c are obtuse, so that both terms on the left-hand side are negative. Thus, $\gamma > \pi/2$. The angle $\mathrm{azim}(\mathbf{p}, \mathbf{w}, \mathbf{u}_i, \mathbf{u}_{i+1})$ is then also greater than $\pi/2$ by Lemma 2.80. This is impossible, as the sum of the four azimuth angles γ is 2π by Lemma 2.94.

□

With nondegeneracy established, we can now give further details about the extreme points of a Rogers simplex and their relationship to the circumcenter of a subset S of the packing V.

Lemma 6.38 (Rogers simplex and circumcenter) *Let V be a saturated packing. Let $\underline{\mathbf{u}} = [\mathbf{u}_0; \ldots; \mathbf{u}_k] \in \underline{V}(k)$ for some $k \leq 3$, and let $S = \{\mathbf{u}_0, \ldots, \mathbf{u}_k\}$ be the point set of $\underline{\mathbf{u}}$. Assume that $h(\underline{\mathbf{u}}) < \sqrt{2}$. Then*

1. *$\omega(\underline{\mathbf{u}})$ is the circumcenter of S.*
2. $\omega(\underline{\mathbf{u}}) \in \mathrm{conv}(S)$.

3. *The set* $\{\omega(d_j\underline{\mathbf{u}}) : j \le k\}$ *has affine dimension* k.
4. *The sequence* $h(d_j\underline{\mathbf{u}})$ *is strictly increasing in* j.

Proof The four conclusions of the lemma are proved separately.

1. *We claim that* $\omega(\underline{\mathbf{u}})$ *is the circumcenter of* S. Indeed, by definition, if $\underline{\mathbf{u}} \in \underline{V}(k)$, then

$$\dim \operatorname{aff} \Omega(V, [\mathbf{u}_0; \ldots; \mathbf{u}_k]) = 3 - k.$$

The case $k = 0$ of the claim is trivially satisfied. Assume by induction the result holds for natural numbers up to k.

Now consider the case $k + 1$. Let $\underline{\mathbf{u}} \in \underline{V}(k + 1)$ and let S_{k+1} be the point set of $\underline{\mathbf{u}}$. By the induction hypothesis, $\omega(d_k\underline{\mathbf{u}})$ is the circumcenter of the point set of $d_k\underline{\mathbf{u}}$. Let $\mathbf{p}$ be the point in $A = \operatorname{aff}(\Omega(V, \underline{\mathbf{u}}))$ closest to $\omega(d_k\underline{\mathbf{u}})$. By Lemma 6.33, the circumcenter of S_{k+1} is the point of intersection of orthogonal affine sets $\operatorname{aff}(S_{k+1})$ and A. Thus, the circumcenter equals the unique point $\mathbf{p}$ of A closest to the point $\omega(d_k\underline{\mathbf{u}})$ in $\operatorname{aff}(S_{k+1})$. By Lemma 6.36, $\mathbf{p} \in \Omega(V, \underline{\mathbf{u}})$. Thus, $\mathbf{p} = \omega(\underline{\mathbf{u}})$. The claim ensues.
2. *We claim* $\omega(\underline{\mathbf{u}}) \in \operatorname{conv}(S)$. Otherwise, select $\mathbf{v} \in S$ such that $\operatorname{aff}(S')$ separates $\omega(\underline{\mathbf{u}})$ from $\mathbf{v}$, where $S' = S \setminus \{\mathbf{v}\}$. Let $\mathbf{p}'$ (resp. $\mathbf{p} = \omega(\underline{\mathbf{u}})$) be the circumcenter of S' (resp. S). When $\mathbf{u} \in S'$, the law of cosines gives

$$\|\mathbf{u} - \mathbf{p}\|^2 = \|\mathbf{u} - \mathbf{p}'\|^2 + \|\mathbf{p}' - \mathbf{p}\|^2$$
$$\|\mathbf{v} - \mathbf{p}\|^2 \ge \|\mathbf{v} - \mathbf{p}'\|^2 + \|\mathbf{p}' - \mathbf{p}\|^2.$$

This gives $\|\mathbf{v} - \mathbf{p}'\| \le \|\mathbf{u} - \mathbf{p}'\|$, which is contrary to Lemma 6.36.
3. *The set* $\{\omega(d_j\underline{\mathbf{u}}) : j \le k\}$ *has affine dimension* k. Lemma 6.33 implies that the vectors $\omega(d_{i+1}\underline{\mathbf{u}}) - \omega(d_i\underline{\mathbf{u}})$ are mutually orthogonal. Thus, the claim about affine dimension easily follows if we show that these vectors are nonzero. Otherwise, the circumcenter $\omega(d_i\underline{\mathbf{u}})$ of $S_i = \{\mathbf{u}_0, \ldots, \mathbf{u}_i\}$ has an equally close point $\mathbf{u}_{i+1} \in V \setminus S_i$, which is impossible by Lemma 6.36.
4. *The sequence* $h(d_j\underline{\mathbf{u}})$ *is strictly increasing in* j. Indeed, by the Pythagorean theorem,

$$\|\omega(d_j\underline{\mathbf{u}}) - \omega(d_0\underline{\mathbf{u}})\|^2 = \sum_{i=1}^{j} \|\omega(d_i\underline{\mathbf{u}}) - \omega(d_{i-1}\underline{\mathbf{u}})\|^2. \tag{6.39}$$

So the result follows from the previous claim.

□

Lemma 6.40 *Let* V *be a saturated packing. Let* $\underline{\mathbf{u}} = [\mathbf{u}_0; \ldots] \in \underline{V}(k)$ *for some* k. *Then* $h(\underline{\mathbf{u}}) \le \|\omega(\underline{\mathbf{u}}) - \mathbf{u}_0\|$. *Moreover, if* $h(\underline{\mathbf{u}}) < \sqrt{2}$, *then* $h(\underline{\mathbf{u}}) = \|\omega(\underline{\mathbf{u}}) - \mathbf{u}_0\|$.

Proof By construction, the point $\omega(\underline{\mathbf{u}})$ belongs to $\Omega(V, \underline{\mathbf{u}})$ and is therefore equidistant to the points in $S = \{\mathbf{u}_0, \ldots, \mathbf{u}_k\}$. The orthogonal projection of $\omega(\underline{\mathbf{u}})$ to aff(S) is the circumcenter of S. The orthogonal projection cannot increase distances, and the inequality ensues. If $h(\underline{\mathbf{u}}) < \sqrt{2}$, then $\omega(\underline{\mathbf{u}})$ is already the circumcenter by Lemma 6.38, so that equality holds. □

Lemma 6.41 *Let V be a saturated packing. Let $\underline{\mathbf{u}}, \underline{\mathbf{v}} \in \underline{V}(3)$. Suppose that $R(\underline{\mathbf{u}}) = R(\underline{\mathbf{v}})$ and that* dim aff $R(\underline{\mathbf{u}}) = 3$. *Then $\omega_i(\underline{\mathbf{u}}) = \omega_i(\underline{\mathbf{v}})$, for $i = 0, 1, 2, 3$.*

Proof Let $R = R(\underline{\mathbf{u}}) = R(\underline{\mathbf{v}})$. The set

$$W = \{\omega_0(\underline{\mathbf{u}}), \cdots, \omega_3(\underline{\mathbf{u}})\}$$

is characterized as the set of extreme points of the simplex R. It is the same for both $\underline{\mathbf{u}}$ and $\underline{\mathbf{v}}$. Since $R \subset \Omega(V, \mathbf{u}_0) \cap \Omega(V, \mathbf{v}_0)$, and the Rogers simplex R has full dimension, we must have $\mathbf{u}_0 = \mathbf{v}_0$. Inductively, we may determine $\omega_i = \omega_i(\underline{\mathbf{u}}) = \omega_i(\underline{\mathbf{v}})$ as follows. The point ω_0 is $\mathbf{u}_0 = \mathbf{v}_0$. The point ω_{i+1} is the closest point of conv$(W \setminus \{\omega_0, \ldots, \omega_i\})$ to ω_i. Note that conv$(W \setminus \{\omega_0, \ldots, \omega_i\})$ is a subset containing ω_{i+1} of the set $\Omega(V, d_{i+1}\underline{\mathbf{u}})$ that is used to define $\omega_{i+1}(\underline{\mathbf{u}})$ (Definition 6.24). This description of the points ω_i is independent of $\underline{\mathbf{u}} \in \underline{V}(3)$ such that $R = R(\underline{\mathbf{u}})$. □

Lemma 6.42 *Let V be a saturated packing. Let $\underline{\mathbf{u}} = [\mathbf{u}_0; \ldots] \in \underline{V}(3)$ be such that* dim aff $R(\underline{\mathbf{u}}) = 3$. *Select $k \leq 3$ such that $h(d_k\underline{\mathbf{u}}) < \sqrt{2}$. Suppose that $R(\underline{\mathbf{u}}) = R(\underline{\mathbf{v}})$, for some $\underline{\mathbf{v}} \in \underline{V}(3)$. Then*

$$d_k\underline{\mathbf{u}} = d_k\underline{\mathbf{v}}.$$

Proof Write ω_i for $\omega_i(\underline{\mathbf{u}})$. By Lemma 6.41, these points are determined by $R(\underline{\mathbf{u}})$. By Lemma 6.38, $h(d_i\underline{\mathbf{u}}) < \sqrt{2}$, and ω_i is the circumcenter of $\{\mathbf{u}_0, \ldots, \mathbf{u}_i\}$, for all $i \leq k$.

Since $R(\underline{\mathbf{u}}) = \text{conv}\{\omega_0, \ldots, \omega_3\}$ has affine dimension 3, the points $\omega_0, \ldots, \omega_k$ are affinely independent. These circumcenters are constructed as points in the affine hull of $\{\mathbf{u}_0, \ldots, \mathbf{u}_k\}$. Hence $\mathbf{u}_0, \ldots, \mathbf{u}_k$ are also affinely independent.

By Lemma 6.36, we have the following recursive description of the points $\mathbf{u}_i$ in terms of ω_i, for $i \leq k$. The point $\mathbf{u}_0$ is ω_0. The point $\mathbf{u}_{i+1}$ is the unique $\mathbf{v} \in V \setminus \{\mathbf{u}_0, \ldots, \mathbf{u}_i\}$ such that

$$\|\mathbf{v} - \omega_{i+1}\| = \|\mathbf{u}_0 - \omega_{i+1}\|.$$

This description of $[\mathbf{u}_0; \ldots; \mathbf{u}_k] = d_k\underline{\mathbf{u}}$ depends on $\mathbf{u}$ only through $R(\underline{\mathbf{u}})$. □

6.2.4 Delaunay simplex

The Delaunay decomposition of space into simplices is dual to the Voronoi cell. It is presented as a collection of k-simplices with vertices in V, for $k = 1, 2, 3$. The Delaunay 1-simplices are defined as the edges between two points in a packing V whose their Voronoi cells have a common facet.[4] A 2-simplex is given with vertices at three points in V if their Voronoi cells have a common edge. A 3-simplex is given for every four points in V whose Voronoi cells have a common extreme point. A Delaunay 3-simplex is the convex hull of four points in the packing V.

Under a nondegeneracy condition (on the circumradius of the set of points), we may construct a Delaunay simplex as a union of Rogers simplices. To this end, we examine the set of all Rogers simplices around a common extreme point. The convex hull of a nondegenerate set $S \subset V$ of four points consists of $4!$ Rogers simplices, each facet of the convex hull consists of $3!$ pieces, and so forth (Lemma 6.48). In brief, the Rogers simplices give every nondegenerate Delaunay simplex an identical simplicial structure.

Recall that $\mathrm{Sym}(k + 1)$ is the *group* of all permutations on the set $\{0, \ldots, k\}$. Let $\underline{\mathbf{u}} = [\mathbf{u}_0, \ldots, \mathbf{u}_k]$ be a list of length $k + 1$. For any *permutation* $\rho \in \mathrm{Sym}(k + 1)$, let $\rho_*(\underline{\mathbf{u}})$ be the *rearrangement* given by

$$\rho_*(\underline{\mathbf{u}})_i = \mathbf{u}_{\rho^{-1}i},$$

where $\mathbf{u}_i$ denotes the ith element of a list $\underline{\mathbf{u}}$.

The following lemma shows that rearrangements have the same extreme point of a Rogers simplex.

Lemma 6.43 (extreme point rearrangement) *Let V be a saturated packing. Let $\underline{\mathbf{u}} \in \underline{V}(k)$. Assume that $h(\underline{\mathbf{u}}) < \sqrt{2}$. Let $\underline{\mathbf{v}}$ be any rearrangement of $\underline{\mathbf{u}}$ under a permutation. Then $\underline{\mathbf{v}} \in \underline{V}(k)$ and $\omega(\underline{\mathbf{u}}) = \omega(\underline{\mathbf{v}})$.*

Proof Let $\underline{\mathbf{v}} = [\mathbf{v}_0; \ldots ; \mathbf{v}_k]$. Let $S_j = \{\mathbf{v}_0, \ldots, \mathbf{v}_j\}$, $\Omega_j = \Omega(V, d_j\underline{\mathbf{v}})$, $A_j = \cap_{i=1}^{j} A(\mathbf{v}_0, \mathbf{v}_i)$, and $a_j = \dim \mathrm{aff}(A_j)$, for $0 \le j \le k$. By convention, set $A_0 = \mathbb{R}^3$, so that $a_0 = \dim \mathrm{aff}(A_0) = 3$. Also, set $a_{-1} = 4$ by convention.

The set S_k is the point set of $\underline{\mathbf{u}}$, which is affinely independent by Lemma 6.33. The set S_j is also affinely independent. Let $\mathbf{p}_j$ be the circumcenter of S_j. The circumradius of S_j is at most the circumradius of S_k, which by assumption is less than $\sqrt{2}$.

We claim that $\dim \mathrm{aff}\, \Omega_j = a_j$, *when* $0 \le j \le k$. By Lemma 6.36, if $\mathbf{p} = \mathbf{p}_j$,

[4] The Delaunay decomposition may be degenerate if the points of V are not in general position. This book confines itself to the nondegenerate situation.

then

$$\|\mathbf{v} - \mathbf{p}\| > \|\mathbf{u} - \mathbf{p}\| \text{ for all } \mathbf{u} \in S_j \text{ and for all } \mathbf{v} \in V \setminus S_j. \tag{6.44}$$

Select a small neighborhood U of $\mathbf{p}_j$ such that (6.44) holds for all $\mathbf{p} \in U_j$. By the definition of Voronoi cell, $\Omega_j \cap U = A_j \cap U$. By background facts on affine sets $\dim \operatorname{aff} \Omega_j = \dim \operatorname{aff} A_j = a_j$. This gives the claim.

To prove the lemma, we prove the following claim by simultaneous induction on j. For all $0 \le j \le k$ we have

$$a_j \ge a_{j-1} - 1 \ge 3 - j.$$
$$a_j = 3 - j \text{ if and only if } a_i = 3 - i \text{ for all } 0 \le i \le j.$$

The base case $j = 0$ is trivial. Assume the induction hypothesis for j.

We have $A_{j+1} = A_j \cap A(\mathbf{v}_0, \mathbf{v}_{j+1})$. The intersection contains $\mathbf{p}_{j+1}$ and is therefore nonempty. By general background facts on the intersection of an affine set with a hyperplane, $a_{j+1} \ge a_j - 1$. By the induction hypothesis, $a_j - 1 \ge 3 - (j+1)$. If $a_{j+1} = 3 - (j+1)$, then $a_j = 3 - j$ and by the induction hypothesis $a_i = 3 - i$ for all $0 \le i \le j$. This completes the proof of the claim by induction.

We have $a_k = \dim \operatorname{aff} A_k = \dim \operatorname{aff} \Omega_k$. However, $\Omega_k = \Omega(V, \underline{\mathbf{u}})$, and since $\underline{\mathbf{u}} \in \underline{V}(k)$, it follows that $3 - k = \dim \operatorname{aff} \Omega(V, \underline{\mathbf{u}}) = a_k$. By the established claims, $a_i = 3 - i$ for all $0 \le i \le k$. This proves $\underline{\mathbf{v}} \in \underline{V}(k)$.

Finally, $\omega(\underline{\mathbf{u}}) = \omega(\underline{\mathbf{v}})$ because both equal the circumcenter of the point set S_k. □

The next lemma shows that the map from permutations to Rogers simplices is one-to-one.

Lemma 6.45 (permutations one-to-one) *Let V be a saturated packing and let $\underline{\mathbf{u}} \in \underline{V}(k)$. Assume that $h(\underline{\mathbf{u}}) < \sqrt{2}$. Let $\rho \in \operatorname{Sym}(k+1)$ such that $R(\underline{\mathbf{u}}) = R(\rho_* \underline{\mathbf{u}})$. Then $\rho = I$.*

Proof We assume that $\rho \ne I$, write $\underline{\mathbf{v}} = \rho_* \underline{\mathbf{u}}$, and prove that $R(\underline{\mathbf{u}}) \ne R(\underline{\mathbf{v}})$. By Lemma 6.38, the sets $\{\omega(d_j \underline{\mathbf{u}}) : j \le k\}$ and $\{\omega(d_j \underline{\mathbf{v}}) : j \le k\}$ are each affinely independent of cardinality $k + 1$. By Lemma 5.57, these are sets of extreme points of $R(\underline{\mathbf{u}})$ and $R(\underline{\mathbf{v}})$, respectively. Thus, it is enough to show that the sets of extreme points are unequal.

Let j be the largest index such that $d_j \underline{\mathbf{u}} = d_j \underline{\mathbf{v}}$. The assumption $\rho \ne I$ implies that $j < k$. Let $\mathbf{p}$ be the circumcenter of $\{\mathbf{u}_0, \ldots, \mathbf{u}_{j+1}\}$. By Lemma 6.36,

$$\|\mathbf{u}_0 - \mathbf{p}\| = \|\mathbf{u}_{j+1} - \mathbf{p}\| < \|\mathbf{v}_{j+1} - \mathbf{p}\|.$$

Thus, $\omega(d_{j+1} \underline{\mathbf{u}}) \ne \omega(d_{j+1} \underline{\mathbf{v}})$. The result ensues. □

To prepare for Lemma 6.48, we need a preliminary lemma that does some index shuffling for us. It gives an explicit representatives of the cosets of Sym($k+1$) in Sym($k+2$).

Definition 6.46 Let $\underline{\mathbf{u}}$ be any list. For each i, let $\underline{\mathbf{u}}^i = [\mathbf{u}_0; \ldots; \hat{\mathbf{u}}_i; \ldots]$ be the list that drops the ith entry.

Lemma 6.47 (coset representatives) *There is a bijection between the set*

$$\{(i, \sigma) \ : \ 0 \le i \le k+1, \quad \sigma \in \mathrm{Sym}(k+1)\}$$

and Sym($k+2$) *such that for any list* $\underline{\mathbf{u}}$ *of length* $k+2$

$$(\rho_* \underline{\mathbf{u}})_j = \begin{cases} (\sigma_*(\underline{\mathbf{u}}^i))_j & 0 \le j \le k \\ \underline{\mathbf{u}}_i & j = k+1. \end{cases}$$

Proof The bijection sends (i, σ) to the permutation ρ, where

$$\rho^{-1} j = \begin{cases} \sigma^{-1} j, & \sigma^{-1} j < i \\ (\sigma^{-1} j) + 1 & \sigma^{-1} j \ge i \\ i & j = k+1. \end{cases}$$

This has the required properties. □

This lemma shows that each (nondegenerate) Delaunay simplex can be partitioned as a union of Rogers simplices, indexed by the permutation group (Figure 6.8).

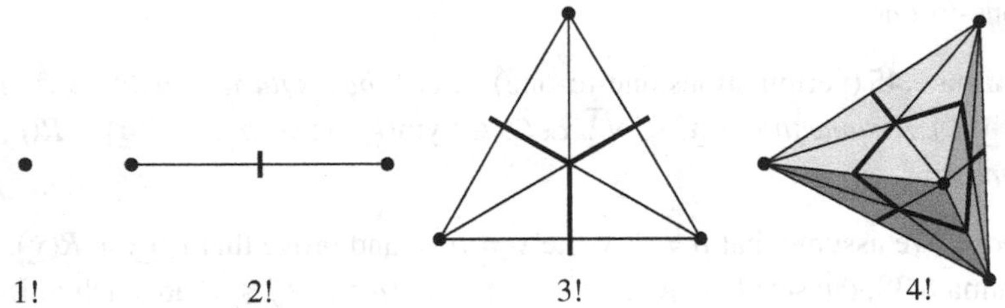

Figure 6.8 In dimension k, the union of $(k+1)!$ nondegenerate Rogers simplices is again a simplex.

Lemma 6.48 (Delaunay simplex) *Let V be a saturated packing and let* $\underline{\mathbf{u}} = [\mathbf{u}_0; \ldots; \mathbf{u}_k] \in \underline{V}(k)$. *Assume that* $h(\underline{\mathbf{u}}) < \sqrt{2}$. *Then*

$$\mathrm{conv}\{\mathbf{u}_0, \ldots, \mathbf{u}_k\} = \bigcup \{R(\rho_* \underline{\mathbf{u}}) \ : \ \rho \in \mathrm{Sym}(k+1)\}.$$

Proof The proof is by induction on k. The base case of the induction $k = 0$ reduces to the trivial assertion: $\mathrm{conv}\{\mathbf{u}_0\} = \mathrm{conv}\{\mathbf{u}_0\}$.

We claim $\underline{\mathbf{u}}^i \in \underline{V}(k)$, *when* $\underline{\mathbf{u}} = [\mathbf{u}_0; \ldots; \mathbf{u}_{k+1}] \in \underline{V}(k+1)$. Indeed, some permutation $\rho \in \mathrm{Sym}(k+2)$ carries $\underline{\mathbf{u}}$ to $\underline{\mathbf{v}} = [\mathbf{u}_0; \ldots; \hat{\mathbf{u}}_i; \ldots; \mathbf{u}_{k+1}; \mathbf{u}_i]$. By Lemma 6.43, $\underline{\mathbf{v}} \in \underline{V}(k+1)$, so that $\underline{\mathbf{u}}^i = d_k \underline{\mathbf{v}} \in \underline{V}(k)$.

By the induction hypothesis

$$\mathrm{conv}(S \setminus \{\mathbf{u}_i\}) = \bigcup \{R(\sigma_* \underline{\mathbf{u}}^i) \ :\ \sigma \in \mathrm{Sym}(k+1)\}, \tag{6.49}$$

where $S = \{\mathbf{u}_0, \ldots, \mathbf{u}_{k+1}\}$. By Lemma 5.57, the facets of the polyhedron $\mathrm{conv}(S)$ are the sets $\mathrm{conv}(S \setminus \{\mathbf{u}_i\})$. Lemma 5.56 gives the partition

$$\mathrm{conv}(S) = \bigcup_{i=0}^{k+1} \mathrm{conv}(\{\omega(\underline{\mathbf{u}})\} \cup \mathrm{conv}(S \setminus \{\mathbf{u}_i\})). \tag{6.50}$$

Substitute the formula (6.49) into (6.50) and use the bijection of Lemma 6.47 to replace the double union by a single union over $\rho \in \mathrm{Sym}(k+2)$. Background facts in affine geometry then simplify the expression to the desired formula. The proof by induction ensues. □

In summary of this section, by construction, the Rogers simplices $R(\underline{\mathbf{u}})$ are compatible with the Voronoi decomposition of space. Under mild restrictions on the circumradius, they can also by Lemma 6.48 be reassembled into simplices (the Delaunay simplices) with extreme points at the centers of the packing.

6.3 Cells

6.3.1 definition

Marchal [31] has proposed an approach to sphere packings that gives some improvements to the original proof in [22]. He gives a partition of space into cells that is a variant of Rogers's partition into the simplices $R(\underline{\mathbf{u}})$. The main part of the construction is the decomposition obtained by truncating Voronoi cells with a ball of radius $\sqrt{2}$. In a few carefully chosen situations, the simplices $R(\underline{\mathbf{u}})$ are assembled into larger convex cells (Delaunay cells), as suggested by Lemma 6.48.

Definition 6.51 (Marchal cells) Let V be a saturated packing. Let

$$\underline{\mathbf{u}} = [\mathbf{u}_0; \ldots; \mathbf{u}_3] \in \underline{V}(3).$$

Define $\xi(\underline{\mathbf{u}})$ as follows. If $\sqrt{2} \le h(d_2\underline{\mathbf{u}})$, then let $\xi(\underline{\mathbf{u}}) = \omega(d_2\underline{\mathbf{u}})$. If $h(d_2\underline{\mathbf{u}}) < \sqrt{2} \le h(\underline{\mathbf{u}})$, define $\xi(\underline{\mathbf{u}})$ to be the unique point in

$$\mathrm{conv}\{\omega(d_2\underline{\mathbf{u}}), \omega(\underline{\mathbf{u}})\}$$

at distance $\sqrt{2}$ from $\mathbf{u}_0$. A set $\mathrm{cell}(\underline{\mathbf{u}}, i) \subset \mathbb{R}^3$ is associated with $\underline{\mathbf{u}}$ and $i = 0, 1, 2, 3, 4$.

0. The 0-cell of $\underline{\mathbf{u}}$ is defined to be empty unless $\sqrt{2} \le h(\underline{\mathbf{u}})$. If this inequality holds, then the 0-cell is

$$\mathrm{cell}(\underline{\mathbf{u}}, 0) = R(\underline{\mathbf{u}}) \setminus B(\mathbf{u}_0, \sqrt{2}).$$

1. The 1-cell of $\underline{\mathbf{u}}$ is defined to be empty unless $\sqrt{2} \le h(\underline{\mathbf{u}})$. If this inequality holds, then the 1-cell is

$$\mathrm{cell}(\underline{\mathbf{u}}, 1) = (R(\underline{\mathbf{u}}) \cap \bar{B}(\mathbf{u}_0, \sqrt{2})) \setminus \mathrm{rcone}^0(\mathbf{u}_0, \mathbf{u}_1, a), \quad a = h(d_1\underline{\mathbf{u}})/\sqrt{2}.$$

(The term $\mathrm{rcone}^0(\mathbf{u}_0, \mathbf{u}_1, a)$ is empty when $a > 1$.)

2. The 2-cell of $\underline{\mathbf{u}}$ is defined to be empty unless $h(d_1\underline{\mathbf{u}}) < \sqrt{2} \le h(\underline{\mathbf{u}})$. If this inequality holds, then the 2-cell is (with a as above)

$$\mathrm{cell}(\underline{\mathbf{u}}, 2) = \mathrm{rcone}(\mathbf{u}_0, \mathbf{u}_1, a) \cap \mathrm{rcone}(\mathbf{u}_1, \mathbf{u}_0, a) \cap \mathrm{aff}_+(\{\mathbf{u}_0, \mathbf{u}_1\}, \{\xi(\underline{\mathbf{u}}), \omega(\underline{\mathbf{u}})\}).$$

3. The 3-cell of $\underline{\mathbf{u}}$ is defined to be empty unless $h(d_2\underline{\mathbf{u}}) < \sqrt{2} \le h(\underline{\mathbf{u}})$. If this inequality holds, then $\xi(\underline{\mathbf{u}}) \in \mathrm{conv}\{\omega(d_2\underline{\mathbf{u}}), \omega(\underline{\mathbf{u}})\}$ and the 3-cell is

$$\mathrm{cell}(\underline{\mathbf{u}}, 3) = \mathrm{conv}\{\mathbf{u}_0, \mathbf{u}_1, \mathbf{u}_2, \xi(\underline{\mathbf{u}})\}.$$

4. The 4-cell of $\underline{\mathbf{u}}$ is defined to be empty unless $h(\underline{\mathbf{u}}) < \sqrt{2}$. If this inequality holds, the 4-cell is

$$\mathrm{cell}(\underline{\mathbf{u}}, 4) = \mathrm{conv}\{\mathbf{u}_0, \mathbf{u}_1, \mathbf{u}_2, \mathbf{u}_3\}.$$

The 0- and 1-cells are subsets of a Rogers simplex R (Figure 6.9). Yet, the 2-, 3-, and 4-cells lie in a union of simplices. The index i in $\mathrm{cell}(\underline{\mathbf{u}}, i)$ indicates the number of points of V that are extreme points of the cell (Figures 6.9 and 6.10).

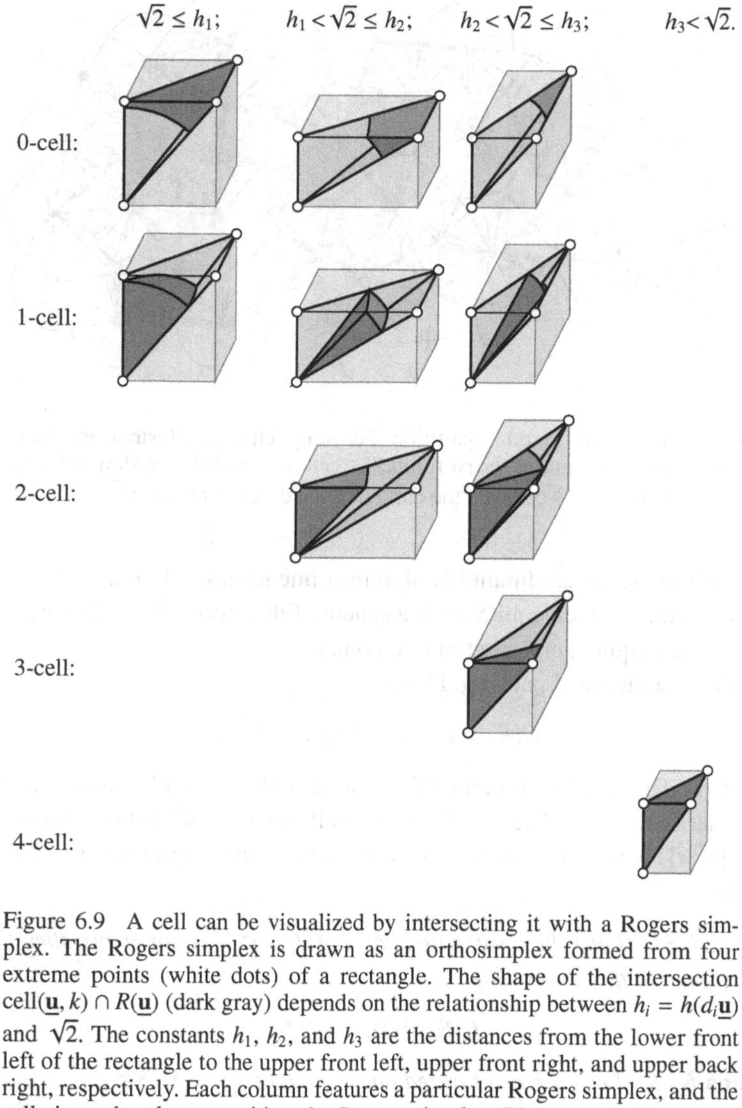

Figure 6.9 A cell can be visualized by intersecting it with a Rogers simplex. The Rogers simplex is drawn as an orthosimplex formed from four extreme points (white dots) of a rectangle. The shape of the intersection cell($\underline{\mathbf{u}}, k$) $\cap$ $R(\underline{\mathbf{u}})$ (dark gray) depends on the relationship between $h_i = h(d_i\underline{\mathbf{u}})$ and $\sqrt{2}$. The constants h_1, h_2, and h_3 are the distances from the lower front left of the rectangle to the upper front left, upper front right, and upper back right, respectively. Each column features a particular Rogers simplex, and the cells in each column partition the Rogers simplex. The empty cells are not illustrated.

6.3.2 informal discussion

A *cell*, short for Marchal cell, can be described in an alternative intuitive way. If $S \subset \mathbb{R}^3$, let

$$\text{equi}(S, r) = \{\mathbf{p} \;:\; \|\mathbf{p} - \mathbf{v}\| = r \text{ for all } \mathbf{v} \in S\}.$$

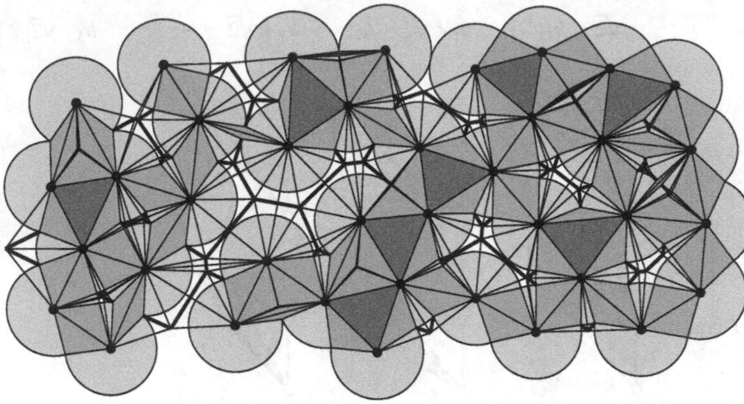

Figure 6.10 The Marchal partition. For simplicity, we illustrate the two-dimensional analogue of the partition. The cells are shaded according to level: 0, 1, 2, and 3. The 3-cells are the darkest. Each disk has radius $\sqrt{2}$.

If S is a finite set of cardinality k, of affine dimension $k-1$, and with circumradius less than r, then $\mathrm{equi}(S, r)$ is a sphere of dimension $3-k$. In particular, if $k = 3$, then $\mathrm{equi}(S, r)$ is a set of two points.

Let V be a saturated packing. Define

$$C(S) = \mathrm{conv}(S \cup \mathrm{equi}(S, \sqrt{2}))$$

for $S \subset V$. The set $C(S)$ is empty if the circumradius of S is greater than $\sqrt{2}$.

The set $C(\emptyset)$ is $\mathbb{R}^3$. The set $C(\{\mathbf{w}\})$ is a ball of radius $\sqrt{2}$ with center $\mathbf{w}$. The set $C(\{\mathbf{v}, \mathbf{w}\})$ is a double cone, $C(\{\mathbf{u}, \mathbf{v}, \mathbf{w}\})$ a bipyramid, and $C(\{\mathbf{t}, \mathbf{u}, \mathbf{v}, \mathbf{w}\})$ is a simplex.

Lemma 6.52 *Let V be a saturated packing. If $S \subset V$ is not empty, then $C(S)$ is contained in the union of sets*

$$C(S \setminus \{\mathbf{v}\}), \quad \mathbf{v} \in S.$$

Lemma 6.53 *Let V be a saturated packing and let $S \subset V$. Set*

$$C'(S) = C(S) \setminus \bigcup_{S \subset S' \subset V} C(S'),$$

where $S' \subset V$ runs over subsets of cardinality $k = \mathrm{card}(S) + 1$ that contain S. Then $C'(S)$ equals a union of k-cells, up to a null set.

In other words, up to a null set, the union of 0-cells is the set of points outside the balls $C(\{\mathbf{v}\})$, for $\mathbf{v} \in V$. The union of the 1-cells is the set of points inside the balls $C(\{\mathbf{v}\})$ but outside the double cones $C(\{\mathbf{u}, \mathbf{w}\})$, and so forth.

It is possible to base a construction of cells on this lemma, dispensing entirely with Voronoi cells and Rogers simplices. It is quick and intuitive. We have followed a longer path that gives more detail about the structure of cells.

At first, the definition of cells seems unmotivated. Some history might help. The 1998 proof of the Kepler conjecture partitioned space into a hybrid of truncated Voronoi cells and Delaunay-like simplices (Figure 6.11). In vague terms, the Delaunay simplices are tuned for detail. The Voronoi cells are coarsely tuned, suitable for rough hewing. Delaunay simplices articulate the foreground, while Voronoi cells fill the background. The solution to the problem lies in the right balance between foreground and background. Too many Delaunay simplices and the details overwhelm. Too many Voronoi cells and the estimates become too weak. The central geometrical insights of the original proof are expressed as rules that delineate foreground against background, Delaunay against Voronoi.

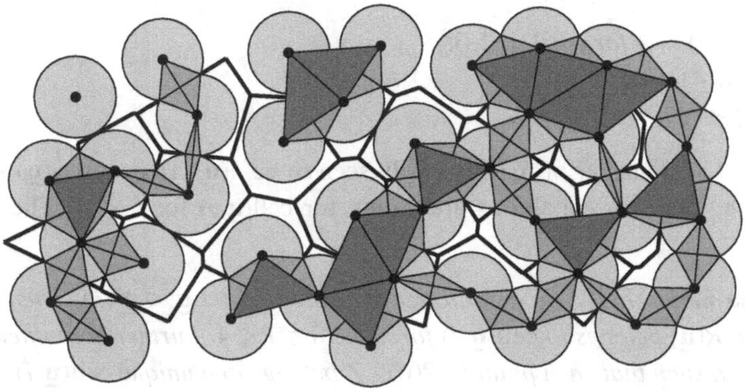

Figure 6.11 The hybrid partition of space that was used to prove the Kepler conjecture in 1998. For simplicity, we illustrate the two-dimensional analogue of that partition.

Cells give a hybrid decomposition. A 4-cell is a Delaunay simplex. The 0- and 1-cells are parts of a Voronoi cell. The 2- and 3-cells are gradations between the two.

Examples show the shortcomings of a nonhybrid approach. Recall that the density of the face centered cubic packing is $\pi/\sqrt{18} \approx 0.74048$. Numerical evidence shows that an approach based entirely on Delaunay simplices should give a bound of about 0.740873, a failure that comes tantalizingly close [18]. The dodecahedral theorem, which asserts that the Voronoi cell of smallest volume in the regular dodecahedron, gives the bound of about 0.755 [23]. Thus, the pure Voronoi cell strategy fails as well. The pure approaches can be modified in ways that are conjectured to produce sharp bounds. These modifications are complex and daunting.

A common practice that started with L. Fejes Tóth is to truncate Voronoi cells by intersecting them with a ball concentric with the cell. Different authors use different radii for the truncating sphere: $7/\sqrt{27} \approx 1.347$ [12], $\sqrt{2}$, 1.385, and 1.255 [22], $\sqrt{2}$ [31], and $\sqrt{3}\tan\pi/5 \approx 1.258$ [23]. A larger radius retains more information and complexity than a smaller radius. The 0-cells are the refuse that lie outside the ball of truncation and are inconsequential to the proof.

6.3.3 cell partition

Lemma 6.54 *Let V be a saturated packing. Let* $\underline{\mathbf{u}} \in \underline{V}(3)$*. The following are equivalent.*

1. $\mathrm{cell}(\underline{\mathbf{u}}, i) = \emptyset$ *for* $i = 0, 1, 2, 3$.
2. $\mathrm{cell}(\underline{\mathbf{u}}, 4) \neq \emptyset$.
3. $h(\underline{\mathbf{u}}) < \sqrt{2}$.

Proof The diameter of $R(\underline{\mathbf{u}})$ is easily seen to be $h(\underline{\mathbf{u}})$. Hence, if $h(\underline{\mathbf{u}}) < \sqrt{2}$ all of the defining conditions are empty for $\mathrm{cell}(\underline{\mathbf{u}}, i)$ for $i < 4$. The result ensues. □

Lemma 6.55 *Let V be a saturated packing and let* $\underline{\mathbf{u}} \in \underline{V}(3)$*. Then every point in* $R(\underline{\mathbf{u}})$ *belongs to* $\mathrm{cell}(\underline{\mathbf{u}}, i)$ *for some* $0 \le i \le 4$*. Furthermore, there is a null set Z such that each point in* $R(\underline{\mathbf{u}}) \setminus Z$ *belongs to a unique* $\mathrm{cell}(\underline{\mathbf{u}}, i)$.

Proof Explicitly, the null set is the union of $R(\underline{\mathbf{u}}) \setminus R^0(\underline{\mathbf{u}})$ (which lies in a finite union of planes), the sphere of radius $\sqrt{2}$ at $\mathbf{u}_0$, the difference $\mathrm{rcone}(\mathbf{u}_0, \mathbf{u}_1, a) \setminus \mathrm{rcone}^0(\mathbf{u}_0, \mathbf{u}_1, a)$, and the plane $\mathrm{aff}\{\mathbf{u}_0, \mathbf{u}_1, \xi(\underline{\mathbf{u}})\}$. Let $\mathbf{p} \in R(\underline{\mathbf{u}})$. To make the cases disjoint, each of the following cases assumes that the conditions of preceding cases fail. It is convenient to reorder the cases to make the 4-cell appears first.

4. If $h(\underline{\mathbf{u}}) < \sqrt{2}$, then $\mathbf{p} \in \mathrm{cell}(\underline{\mathbf{u}}, 4)$.
0. If $\|\mathbf{p} - \mathbf{u}_0\| \ge \sqrt{2}$, then $\mathbf{p} \in \mathrm{cell}(\underline{\mathbf{u}}, 0)$.
1. If $\mathbf{p} \notin \mathrm{rcone}^0(\mathbf{u}_0, \mathbf{u}_1, h(d_1\underline{\mathbf{u}})/\sqrt{2})$, then $\mathbf{p} \in \mathrm{cell}(\underline{\mathbf{u}}, 1)$.
2. If $\mathbf{p} \in \mathrm{aff}_+(\{\mathbf{u}_0, \mathbf{u}_1\}, \{\xi(\underline{\mathbf{u}}), \omega(\underline{\mathbf{u}})\})$, then $\mathbf{p} \in \mathrm{cell}(\underline{\mathbf{u}}, 2)$.
3. If $\mathbf{p} \in \mathrm{aff}_+(\{\mathbf{u}_0, \mathbf{u}_1\}, \{\mathbf{u}_2, \xi(\underline{\mathbf{u}})\})$, then $\mathbf{p} \in \mathrm{cell}(\underline{\mathbf{u}}, 3)$.

When the corresponding strict inequalities are used, we obtain uniqueness for $R(\underline{\mathbf{u}}) \setminus Z$. □

Definition 6.56 (i-rearrangement) Let $\underline{\mathbf{u}} = [\mathbf{u}_0; \ldots; \mathbf{u}_k], \underline{\mathbf{v}} = [\mathbf{v}_0; \ldots; \mathbf{v}_k]$ be two lists of the same length. One is an *i-rearrangement* of the other if $\rho_* \underline{\mathbf{u}} = \underline{\mathbf{v}}$ for some $\rho \in \mathrm{Sym}(k+1)$ such that $\rho(j) = j$ when $j \geq i$.

In particular, if $\underline{\mathbf{u}}, \underline{\mathbf{v}}$ are 0- or 1-rearrangements of one another, then $\underline{\mathbf{u}} = \underline{\mathbf{v}}$. The constraint $\rho(j) = j$ is vacuous when $j > k$.

Lemma 6.57 *Let V be a saturated packing, let $\underline{\mathbf{u}} \in \underline{V}(3)$, and let $i \in \{2, 3, 4\}$. Assume that $h(d_{i-1}\underline{\mathbf{u}}) < \sqrt{2}$. Let $\underline{\mathbf{v}}$ be an i-rearrangement of $\underline{\mathbf{u}}$. Then $\underline{\mathbf{v}} \in \underline{V}(3)$ and $\omega_j(\underline{\mathbf{u}}) = \omega_j(\underline{\mathbf{v}})$, for $j = i-1, \ldots, 3$.*

Proof Let $S_j = \{\mathbf{u}_0, \ldots, \mathbf{u}_j\}$, for $j \geq i-1$, where $\underline{\mathbf{u}} = [\mathbf{u}_0; \ldots]$. Since $\underline{\mathbf{v}} = [\mathbf{v}_0; \ldots]$ is an i-rearrangement of $\underline{\mathbf{u}}$, we have $S_j = \{\mathbf{v}_0, \ldots, \mathbf{v}_j\}$ and $\Omega(V, d_j\underline{\mathbf{u}}) = \Omega(V, d_j\underline{\mathbf{v}})$, for all $j \geq i-1$. By Lemma 6.43, $\omega_{i-1}(\underline{\mathbf{u}}) = \omega_{i-1}(\underline{\mathbf{v}})$. By the recursive definition of the points ω_j, we then have $\omega_j(\underline{\mathbf{u}}) = \omega_j(\underline{\mathbf{v}})$, for $j = i-1, \ldots, 3$.

We show that $\underline{\mathbf{v}} \in \underline{V}(3)$ by checking the defining condition

$$\dim \mathrm{aff}\, \Omega(V, d_j\underline{\mathbf{v}}) = 3 - j, \text{ for } 0 < j \leq 3.$$

When $j \leq i-1$, the conclusion $d_{i-1}\underline{\mathbf{v}} \in \underline{V}(i-1)$ of Lemma 6.43 implies the condition. When $i - 1 < j$, the identity $\Omega(V, d_J\underline{\mathbf{v}}) = \Omega(V, d_j\underline{\mathbf{u}})$ and $\underline{\mathbf{u}} \in \underline{V}(3)$ imply the condition. The result ensues. □

Lemma 6.58 *Let V be a saturated packing, let $\underline{\mathbf{u}}, \underline{\mathbf{v}} \in \underline{V}(3)$, and let $i \in \{0, 1, 2, 3, 4\}$. If $\underline{\mathbf{u}}$ is an i-rearrangement of $\underline{\mathbf{v}}$, then* $\mathrm{cell}(\underline{\mathbf{u}}, i) = \mathrm{cell}(\underline{\mathbf{v}}, i)$.

Proof The statement follows from the definition of cells. □

Lemma 6.59 *Let V be a saturated packing, let $\underline{\mathbf{u}} \in \underline{V}(3)$, and $k \in \{0, 1, 2, 3, 4\}$. Assume that* $\mathrm{cell}(\underline{\mathbf{u}}, k)$ *is not empty. Then each k-rearrangement $\underline{\mathbf{v}}$ of $\underline{\mathbf{u}}$ lies in $\underline{V}(3)$. Moreover,* $\mathrm{cell}(\underline{\mathbf{u}}, k)$ *is contained in the union of $R(\underline{\mathbf{v}})$, as $\underline{\mathbf{v}}$ runs over all k-rearrangements of $\underline{\mathbf{u}}$.*

Proof If $k = 0$ or $k = 1$, then by definition, $\mathrm{cell}(\underline{\mathbf{u}}, k) \subset R(\underline{\mathbf{u}})$.

Assume that $2 \leq k \leq 4$. The nonemptiness hypothesis implies $h(d_{k-1}\underline{\mathbf{u}}) < \sqrt{2}$. Lemma 6.57 implies that $\underline{\mathbf{v}} \in \underline{V}(3)$.

The definition of the cells can be used to show directly that

$$\mathrm{cell}(\underline{\mathbf{u}}, k) \subset \mathrm{conv}\{\mathbf{u}_0, \mathbf{u}_1, \ldots, \mathbf{u}_{k-1}, \omega_k(\underline{\mathbf{u}}), \ldots, \omega_3(\underline{\mathbf{u}})\}.$$

The definition of k-rearrangement, Lemma 6.48, and Lemma 6.57 partition this convex hull according to k-rearrangements of $\underline{\mathbf{u}}$. The result ensues. □

Lemma 6.60 *Let V be a saturated packing, let $\underline{\mathbf{u}}, \underline{\mathbf{v}} \in \underline{V}(3)$, and let $k \in \{0, 1, 2, 3, 4\}$. Suppose that $R(\underline{\mathbf{u}}) = R(\underline{\mathbf{v}})$, that $R(\underline{\mathbf{u}})$ has affine dimension three, and that* $\text{cell}(\underline{\mathbf{u}}, k)$ *is not empty. Then* $\text{cell}(\underline{\mathbf{u}}, k) = \text{cell}(\underline{\mathbf{v}}, k)$.

Proof We break the proof into cases, according to k. Assume $k = 4$. By the definition of the cell, the nonemptiness condition implies that $h(\underline{\mathbf{u}}) < \sqrt{2}$. By Lemma 6.40 and Lemma 6.42, $\underline{\mathbf{u}} = \underline{\mathbf{v}}$.

Assume that $k = 3$. The nonemptiness condition gives $h(d_2\underline{\mathbf{u}}) < \sqrt{2} \le h(\underline{\mathbf{u}})$. By Lemma 6.40 and Lemma 6.42, $d_2\underline{\mathbf{u}} = d_2\underline{\mathbf{v}}$. The point $\omega(\underline{\mathbf{u}})$ is determined by $R(\mathbf{u})$ by Lemma 6.41. The point $\xi(\underline{\mathbf{u}})$ is determined by $\omega(d_2\underline{\mathbf{u}})$ and $\omega(\underline{\mathbf{u}})$. Finally, $\text{cell}(\underline{\mathbf{u}}, 3)$ is determined by $d_2\underline{\mathbf{u}}$ and $\xi(\underline{\mathbf{u}})$. The conclusion $\text{cell}(\underline{\mathbf{u}}, 3) = \text{cell}(\underline{\mathbf{v}}, 3)$ ensues.

The other cases are similar. Assume that $k = 2$ and that $\text{cell}(\underline{\mathbf{u}}, 2)$ is not empty. If $h(d_2\underline{\mathbf{u}}) < \sqrt{2}$, then $d_2\underline{\mathbf{u}} = d_2\underline{\mathbf{v}}$, and the cell is determined by $d_2\underline{\mathbf{u}}$ and $\xi(\underline{\mathbf{u}})$, as in the case $k = 3$. If $h(d_1\underline{\mathbf{u}}) < \sqrt{2} \le h(d_2\underline{\mathbf{u}})$, then $d_1\underline{\mathbf{u}} = d_1\underline{\mathbf{v}}$, and the cell is determined by $d_1\underline{\mathbf{u}}$ and the points $\omega_j(\underline{\mathbf{u}})$, which in turn depend only on $R(\underline{\mathbf{u}})$, by Lemma 6.41.

The cases $k = 0$ and $k = 1$ are similar, but even more trivial, and are left to the reader. □

The following lemma and Lemma 6.55 show that the cells partition $\mathbb{R}^3$.

Lemma 6.61 *Let V be a saturated packing, let $\underline{\mathbf{u}}, \underline{\mathbf{v}} \in \underline{V}(3)$, and let $k, k' \in \{0, 1, 2, 3, 4\}$. Suppose that* $\text{cell}(\underline{\mathbf{u}}, k) \cap \text{cell}(\underline{\mathbf{v}}, k')$ *has positive measure. Then $k = k'$ and* $\text{cell}(\underline{\mathbf{u}}, k) = \text{cell}(\underline{\mathbf{v}}, k)$.

Proof Select $\underline{\mathbf{w}} \in \underline{V}(3)$ such that

$$R(\underline{\mathbf{w}}) \cap \text{cell}(\underline{\mathbf{u}}, k) \cap \text{cell}(\underline{\mathbf{v}}, k')$$

has positive measure. In particular, $R(\underline{\mathbf{w}})$ has affine dimension three. By Lemmas 6.59 and 6.58 and 6.29, we may replace $\underline{\mathbf{u}}$ with a k-rearrangement, and $\underline{\mathbf{v}}$ with a k'-rearrangement to assume without loss of generality that $R(\underline{\mathbf{w}}) = R(\underline{\mathbf{u}}) = R(\underline{\mathbf{v}})$.

Lemma 6.60 implies that $\text{cell}(\underline{\mathbf{u}}, k) = \text{cell}(\underline{\mathbf{w}}, k)$ and $\text{cell}(\underline{\mathbf{v}}, k') = \text{cell}(\underline{\mathbf{w}}, k')$. Since $R(\mathbf{w}) \cap \text{cell}(\underline{\mathbf{w}}, k') \cap \text{cell}(\underline{\mathbf{w}}, k)$ has positive measure, Lemma 6.55 implies that $k = k'$. The result ensues. □

6.3.4 edges of cells

Definition 6.62 Let V be a saturated packing and let $\underline{\mathbf{u}} = [\mathbf{u}_0; \ldots] \in \underline{V}(3)$. Let $X = \text{cell}(\underline{\mathbf{u}}, k)$. When $X \ne \emptyset$, define $V(X) = \{\mathbf{u}_0, \ldots, \mathbf{u}_{k-1}\}$. In particular, if X is a 0-cell $V(X) = \emptyset$.

Lemma 6.63 *Let V be a saturated packing and let $\underline{\mathbf{u}} = [\mathbf{u}_0; \ldots] \in \underline{V}(3)$. Let $X = \text{cell}(\underline{\mathbf{u}}, k)$. If $X \neq \emptyset$, then $V(X) = V \cap X$. In particular, the set $V(X)$ is well-defined.*

Proof If $i \leq k-1$, then $\mathbf{u}_i \in V$ and $X = \text{cell}(\underline{\mathbf{v}}, k)$, for some k-rearrangement $\underline{\mathbf{v}} = [\mathbf{u}_i; \ldots]$ of $\underline{\mathbf{u}}$. By the definition of $\text{cell}(\underline{\mathbf{v}}, k)$, we find that $\mathbf{v}_0 = \mathbf{u}_i$ belongs to $\text{cell}(\underline{\mathbf{v}}, k)$, when $k \geq 1$. This implies that $V(X) \subset V \cap X$.

Conversely, let $\mathbf{v} \in V \cap \text{cell}(\underline{\mathbf{u}}, k)$. It can be checked from definitions that $\text{cell}(\underline{\mathbf{u}}, 0) \subset \Omega(V, \mathbf{u}_0)$ and

$$\text{cell}(\underline{\mathbf{u}}, k) \subset \Omega(V, \mathbf{u}_0) \cup \cdots \cup \Omega(V, \mathbf{u}_{k-1}), \quad \text{when } k \geq 1.$$

This implies $\mathbf{v} \in V \cap \Omega(V, \mathbf{u}_i)$ for some $i \leq k$. This forces $\mathbf{v} = \mathbf{u}_i$. Hence $V(X) = V \cap X$. □

Lemma 6.64 *Let V be a saturated packing and let $\underline{\mathbf{u}} \in \underline{V}(3)$. Then $X = \text{cell}(\underline{\mathbf{u}}, k)$ is measurable and eventually radial at each $\mathbf{v} \in V$. Furthermore, the cell X is bounded away from every $v \in V \setminus V(X)$, so that the solid angle of X is zero, except at $\mathbf{v} \in V(X)$.*

Proof The first claim of the lemma follows from the fact that $R(\underline{\mathbf{u}})$ is a simplex, and $R(\underline{\mathbf{u}}) \cap V = \{\mathbf{u}_0\}$. Each cell is compact, and is bounded away from every point not in the cell. Lemma 6.63 implies the second claim of the lemma. □

Lemma 6.65 *Let V be a saturated packing. For every $\mathbf{v} \in V$,*

$$\sum_{X\,:\,\mathbf{v} \in V(X)} \text{sol}(X, \mathbf{v}) = 4\pi,$$

where the sum runs over all cells X such that $\mathbf{v} \in V(X)$.

Proof Indeed, the cells partition $\mathbb{R}^3$ and $\text{sol}(B(\mathbf{v}, \epsilon)) = 4\pi$. □

Definition 6.66 (tsol) Define the *total solid angle* of a cell X to be

$$\text{tsol}(X) = \sum_{\mathbf{v} \in V(X)} \text{sol}(X, \mathbf{v}).$$

Definition 6.67 (edge) Let $E(X)$ be the set of *extremal edges* of the k-cell X in a saturated packing V. More precisely, let

$$E(X) = \{\{\mathbf{u}_i, \mathbf{u}_j\} \;:\; \mathbf{u}_i \neq \mathbf{u}_j \in V(X)\}.$$

In particular, $E(X)$ is empty for 0 and 1-cells and contains $\binom{k}{2}$ pairs when $2 \leq k \leq 4$.

Definition 6.68 (dih) Let V be a saturated packing. Let X be a k-cell, where $2 \le k \le 4$. Let $\varepsilon \in E(X)$. We define the dihedral angle $\mathrm{dih}(X, \varepsilon)$ of X along ε as follows. Explicitly, if X is a null set, then set $\mathrm{dih}(X, \varepsilon) = 0$. Otherwise, choose $\underline{\mathbf{u}} = [\mathbf{u}_0; \mathbf{u}_1; \mathbf{u}_2; \mathbf{u}_3] \in \underline{V}(3)$ such that $X = \mathrm{cell}(\underline{\mathbf{u}}, k)$ and $\varepsilon = \{\mathbf{u}_0, \mathbf{u}_1\}$. Set $\mathrm{dih}(X, \varepsilon) = \mathrm{dih}_V(\{\mathbf{u}_0, \mathbf{u}_1\}, \{\mathbf{v}, \mathbf{w}\})$, where

$$\{\mathbf{v}, \mathbf{w}\} = \begin{cases} \{\xi(\underline{\mathbf{u}}), \omega(\underline{\mathbf{u}})\} & k = 2 \\ \{\mathbf{u}_2, \xi(\underline{\mathbf{u}})\} & k = 3 \\ \{\mathbf{u}_2, \mathbf{u}_3\} & k = 4. \end{cases}$$

This is independent of the choice $\underline{\mathbf{u}}$ defining X.

Each edge $\varepsilon = \{\mathbf{u}, \mathbf{v}\} \in E(X)$ has a half-length $h(\varepsilon) = \|\mathbf{u} - \mathbf{v}\|/2$. This definition of h is compatible with the previous definition of the circumradius of lists in the sense that $h([\mathbf{u}; \mathbf{v}]) = h(\varepsilon)$.

Lemma 6.69 *Let V be a saturated packing. Assume that $\mathbf{u}_0, \mathbf{u}_1 \in V$ satisfy $\|\mathbf{u}_0 - \mathbf{u}_1\| < 2\sqrt{2}$. Set $\varepsilon = \{\mathbf{u}_0, \mathbf{u}_1\}$. Then*

$$\sum_{X\,:\,\varepsilon \in E(X)} \mathrm{dih}(X, \varepsilon) = 2\pi.$$

The sum runs over cells X such that $\varepsilon \in E(X)$.

Proof Consider the set $C = B(\mathbf{u}_0, r) \cap \mathrm{rcone}^0(\mathbf{u}_0, \mathbf{u}_1, a)$, where r and a are small positive real numbers. From the definition of k-cells, it follows that we can choose r and a sufficiently small so that if X is a k cell that meets C in a set of positive measure, then $k \ge 2$ and there exists $\underline{\mathbf{u}} \in \underline{V}(3)$ such that $X = \mathrm{cell}(\underline{\mathbf{u}}, k)$ and $d_1\underline{\mathbf{u}} = [\mathbf{u}_0; \mathbf{u}_1]$. Moreover,

$$C \cap X = C \cap A, \quad A = \mathrm{aff}_+(\{\mathbf{u}_0, \mathbf{u}_1\}, \{\mathbf{v}, \mathbf{w}\}),$$

where A is the lune of Definition 3.14 and $\mathbf{v}, \mathbf{w}$ are chosen as in Definition 6.68. By Lemma 3.27 and Definition 6.68, the volume of this intersection is

$$\mathrm{vol}(C \cap A) = \mathrm{vol}(C)\,\mathrm{dih}_V(\{\mathbf{u}_0, \mathbf{u}_1\}, \{\mathbf{v}, \mathbf{w}\})/(2\pi) = \mathrm{vol}(C)\,\mathrm{dih}(X, \varepsilon)/(2\pi).$$

The set of cells meeting C in a set of positive measure gives a partition of C into finitely many measurable sets. This gives

$$\mathrm{vol}(C) = \sum_{X\,:\,\varepsilon \in E(X)} \mathrm{vol}(C \cap X) = \mathrm{vol}(C) \sum_{X\,:\,\varepsilon \in E(X)} \mathrm{dih}(X, \varepsilon)/(2\pi).$$

The calculation of volumes in Chapter 3 gives $\mathrm{vol}(C) > 0$. The conclusion follows by canceling $\mathrm{vol}(C)$ from both sides of the equation. □

6.3.5 A conjecture

This section shows how the existence of a FCC-compatible negligible function is a consequence of an explicit inequality related to the distances $h(\underline{\mathbf{u}})$, where $\underline{\mathbf{u}} \in \underline{V}(1)$.

Definition 6.70 (sol_0, τ_0, m_1, m_2, h_+, M) Define the following constants and functions:

$$\mathrm{sol}_0 = 3\arccos(1/3) - \pi \tag{6.71}$$
$$\tau_0 = 4\pi - 20\,\mathrm{sol}_0 \tag{6.72}$$
$$m_1 = \mathrm{sol}_0\, 2\sqrt{2}/\tau_0 \approx 1.012 \tag{6.73}$$
$$m_2 = (6\,\mathrm{sol}_0 - \pi)\sqrt{2}/(6\tau_0) \approx 0.0254 \tag{6.74}$$
$$h_+ = 1.3254 \text{ (exact rational value).} \tag{6.75}$$

Let $M : \mathbb{R} \to \mathbb{R}$ be the following piecewise polynomial function (Figure 6.12):

$$M(h) = \begin{cases} \dfrac{\sqrt{2}-h}{\sqrt{2}-1}\,\dfrac{h_+ - h}{h_+ - 1}\,\dfrac{17h - 9h^2 - 3}{5} & h \le \sqrt{2} \\ 0 & h > \sqrt{2}. \end{cases} \tag{6.76}$$

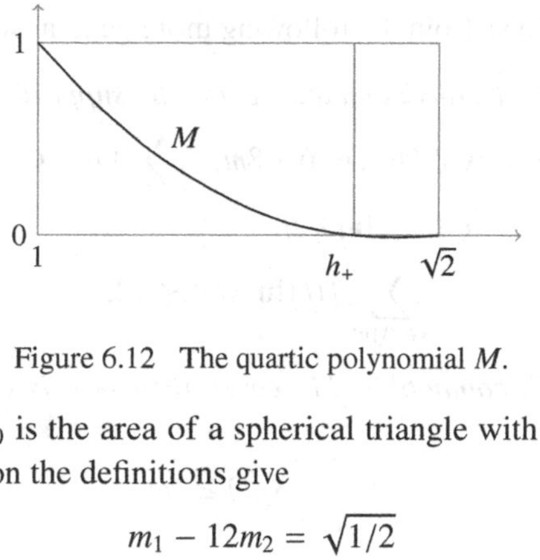

Figure 6.12 The quartic polynomial M.

The constant sol_0 is the area of a spherical triangle with sides $\pi/3$. Simple calculations based on the definitions give

$$m_1 - 12m_2 = \sqrt{1/2} \tag{6.77}$$

and

$$M(1) = 1, \quad M(h_+) = 0, \quad M(\sqrt{2}) = 0. \tag{6.78}$$

Definition 6.79 (γ) For any cell X of a saturated packing, define the functional $\gamma(X, *)$ on $\{f : \mathbb{R} \to \mathbb{R}\}$ by

$$\gamma(X, f) = \mathrm{vol}(X) - \left(\frac{2m_1}{\pi}\right)\mathrm{tsol}(X) + \left(\frac{8m_2}{\pi}\right) \sum_{\varepsilon \in E(X)} \mathrm{dih}(X, \varepsilon) f(h(\varepsilon)). \tag{6.80}$$

Theorem* 6.81 *Let V be any saturated packing and let X be any cell of V. Then*

$$\gamma(X, M) \geq 0, \tag{6.82}$$

where M is the function defined in (6.76).

Remark 6.83 We do not use this inequality, and its proof is omitted. The only published proof [31] is not satisfactory because it plots sample level curves of the function and reaches conclusions based on the visual appearance of these level curves.

Conjecture 6.84 (Marchal) *For any packing V and any $\mathbf{u} \in V$,*

$$\sum_{\mathbf{v} \in V \setminus \{\mathbf{u}\}} M(h([\mathbf{u}; \mathbf{v}])) \leq 12.$$

This book proves a variant of the conjecture.

Theorem 6.85 *The conjecture* (6.84) *and inequality* (6.82) *imply that for every saturated packing V, there exists a negligible FCC-compatible function $G : V \to \mathbb{R}$.*

The theorem follows from the following more general statement.

Lemma 6.86 *Let f be any bounded, compactly supported function. Set*

$$G(\mathbf{u}_0, f) = -\mathrm{vol}(\Omega(V, \mathbf{u}_0)) + 8m_1 - \sum 8m_2 f(h([\mathbf{u}_0; \mathbf{u}])),$$

with sum running over $\mathbf{u} \in V \setminus \{\mathbf{u}_0\}$. If

$$\sum_{\mathbf{v} \in V \setminus \{\mathbf{u}\}} f(h([\mathbf{u}; \mathbf{v}])) \leq 12,$$

then $G(*, f)$ is FCC-compatible. Moreover, if there exists a constant c_0 such that for all $r \geq 1$

$$\sum_{X \subset B(\mathbf{0}, r)} \gamma(X, f) \geq c_0 r^2,$$

then $G(*, f)$ is negligible.

Theorem 6.85 is the special case $f = M$. Inequality 6.82 implies that we can take $c_0 = 0$ in the lemma.

Proof The function $G(*, f)$ is FCC-compatible (page 149) directly by equation (6.77) and the assumption of the lemma:

$$\begin{aligned} 4\sqrt{2} &= 8m_1 - 8(12m_2) \\ &\leq 8m_1 - 8m_2 \sum_{\mathbf{u}\in V\setminus\{\mathbf{u}_0\}} f(h([\mathbf{u}_0;\mathbf{u}])) \\ &= \mathrm{vol}(\Omega(V,\mathbf{u}_0)) + G(\mathbf{u}_0, f). \end{aligned}$$

The issue is to prove that it is negligible. More explicitly, we show that there is a constant c such that for all $r \geq 1$:

$$\begin{aligned} -\sum G(\mathbf{u}, f) &= \sum \mathrm{vol}(\Omega(V,\mathbf{u})) - \sum 8m_1 + \sum \sum_{\mathbf{v}\in V\setminus\{\mathbf{u}\}} 8m_2 f(h([\mathbf{u};\mathbf{v}])) \\ &\geq \sum_{X\subset B(\mathbf{0},r)} \gamma(X, f) + cr^2, \end{aligned} \tag{6.87}$$

where all unmarked sums run over $\mathbf{u} \in V(\mathbf{0}, r)$. The lemma follows from this inequality, the assumption of the lemma, and the definition of negligible (Definition 6.11).

Lemmas 3.30 and 6.2 show that the number of points of V near the boundary of $B(\mathbf{0}, r)$ is at most cr^2, for some c.

The function $\gamma(X, f)$ is defined as a sum of three terms (6.80). The sum of $\gamma(X, f)$ over all cells in a large ball $B(\mathbf{0}, r)$ is a sum of the contributions $T_1(r) + T_2(r) + T_3(r)$ from the three separate terms defining γ. The sum of $-G$ in equation (6.87) is a sum of three corresponding terms $T_i'(r)$. It is enough to work term by term, producing constants c_i such that

$$T_i'(r) \geq T_i(r) + c_i r^2, \quad i = 1, 2, 3.$$

The sum of the volumes of the Voronoi cells $\mathbf{u} \in B(\mathbf{0}, r)$ is not exactly the volume of $B(\mathbf{0}, r)$ because of the contribution at the boundary of $B(\mathbf{0}, r)$ of Voronoi cells that are only partly contained in $B(\mathbf{0}, r)$. Similarly, the sum of the various k-cells for $X \subset B(\mathbf{0}, r)$ is not exactly the volume of $B(\mathbf{0}, r)$ because of contributions from the boundary. The boundary contributions have order r^2. See Section 3.3 for order r^2 calculations. Thus,

$$T_1' = \sum_{\mathbf{u}\in V(\mathbf{0},r)} \mathrm{vol}(\Omega(V,\mathbf{u})) \geq \sum_{X\subset B(\mathbf{0},r)} \mathrm{vol}(X) + c_1 r^2 = T_1 + c_1 r^2.$$

The estimates on the other terms are similar. The solid angles around each point sum to 4π. In Landau big O notation, this gives

$$\begin{aligned}\sum_{X\subset B(\mathbf{0},r)} \operatorname{tsol}(X) &= \sum_{X\subset B(\mathbf{0},r)} \sum_{\mathbf{u}\in V(X)} \operatorname{sol}(X,\mathbf{u}) \\ &= \sum_{\mathbf{u}\in V(\mathbf{0},r)} \sum_{X\,:\,\mathbf{u}\in V(X)} \operatorname{sol}(X,\mathbf{u}) + O(r^2) \\ &= \sum_{\mathbf{u}\in V(\mathbf{0},r)} 4\pi + O(r^2).\end{aligned}$$

Hence

$$T_2' = -\sum_{V(\mathbf{0},r)} 8m_1 = -\sum_{X\subset B(\mathbf{0},r)} \left(\frac{2m_1}{\pi}\right) \operatorname{tsol}(X) + O(r^2) = T_2 + O(r^2).$$

Similarly, the dihedral angles around each edge sum to 2π. A factor of two enters the following calculation because there are two ordered pairs for each unordered pair $\varepsilon = \{\mathbf{u}_0, \mathbf{u}_1\}$:

$$\begin{aligned}&\sum_{X\subset B(\mathbf{0},r)} \sum_{\varepsilon\in E(X)} \operatorname{dih}(X,\varepsilon) f(h(\varepsilon)) \\ &= \sum_{\varepsilon\subset B(\mathbf{0},r)} \sum_{X\,:\,\varepsilon\in E(X)} \operatorname{dih}(X,\varepsilon) f(h(\varepsilon)) + O(r^2) \\ &= \sum_{\varepsilon\subset B(\mathbf{0},r)} 2\pi f(h(\varepsilon)) + O(r^2) \\ &= \sum_{\mathbf{u}_0\in V(\mathbf{0},r)} \sum_{\mathbf{u}_1\in V(\mathbf{0},r)} \pi f(h(\mathbf{u}_0,\mathbf{u}_1)) + O(r^2).\end{aligned}$$

Finally,

$$\begin{aligned}T_3' &= \sum \sum 8m_2 f(h(\underline{\mathbf{u}})) \\ &\geq \left(\frac{8m_2}{\pi}\right) \sum_{X\subset B(\mathbf{0},r)} \sum_{\varepsilon\in E(X)} \operatorname{dih}(X,\varepsilon) f(h(\varepsilon)) + O(r^2) \\ &= T_3 + O(r^2).\end{aligned}$$

□

6.4 Clusters

This section introduces a variant of Conjecture 6.84. In this variant, a piecewise linear function L replaces the piecewise polynomial function M. More crucially, the support of the function L is contained in $[2, 2.52]$. By contrast, the function M is positive on a large interval: $[2, 2.6508)$. This difference in the support of the function creates a large difference in the difficulty of the conjectures.

The conjecture formulated in this section also implies the existence of FCC-compatible negligible functions. To prove this existence result, it is helpful to group cells together into new aggregates, called *clusters*. This section makes a detailed study of clusters in order to produce a negligible function. The aim of this section is to prove a variant (Theorem 6.97) of Theorem 6.85 that uses the function L rather than M.

Recall that $M(h_+) = 0$, where $h_+ = 1.3254$.

Definition 6.88 (L, h_0, h_-) Set

$$h_0 = 1.26.$$

Let $L : \mathbb{R} \to \mathbb{R}$ be the piecewise linear function

$$L(h) = \begin{cases} \dfrac{h_0 - h}{h_0 - 1}, & h \le h_0 \\ 0, & h \ge h_0. \end{cases}$$

It follows from the definition that

$$L(1) = 1 \text{ and } L(h_0) = 0.$$

Let $h_- \approx 1.23175$ be the unique root of the quartic polynomial $M(h) - L(h)$ that lies in the interval $[1.231, 1.232]$.

The inequality $L(h) \ge M(h)$ holds except when $h \in [h_-, h_+]$.

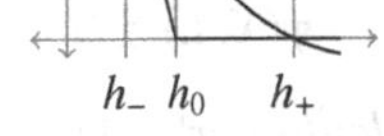

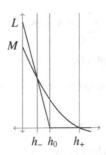

Figure 6.13 Detail of the quartic M and piecewise linear function L on the domain $[1.2, 1.35]$.

Definition 6.89 (critical edge, EC, wt) A *critical edge* ε of a saturated packing V is an unordered pair that appears as an element of $E(X)$ for some k-cell

X of the packing V such that $h(\varepsilon) \in [h_-, h_+]$. Let $\mathrm{EC}(X)$ be the set of critical edges that belong to $E(X)$. If X is any cell such that $\mathrm{EC}(X)$ is nonempty, let the *weight* $\mathrm{wt}(X)$ of X be $1/\mathrm{card}(\mathrm{EC}(X))$.

Definition 6.90 (β_0, β) Set

$$\beta_0(h) = 0.005(1 - (h - h_0)^2/(h_+ - h_0)^2).$$

(See Figure 6.14.) If X is a 4-cell with exactly two critical edges and if those edges are opposite, then set

$$\beta(\varepsilon, X) = \beta_0(h(\varepsilon)) - \beta_0(h(\varepsilon')), \text{ where } \mathrm{EC}(X) = \{\varepsilon, \varepsilon'\}.$$

Otherwise, for all other edges in all other cells, set $\beta(\varepsilon, X) = 0$.

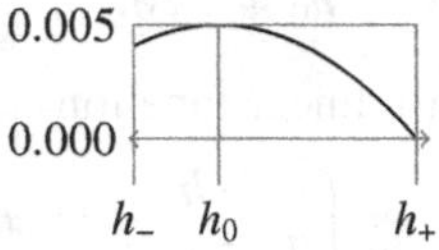

Figure 6.14 The functions β_0.

Definition 6.91 (cell cluster, Γ) Let V be a saturated packing. Let $\varepsilon \in \mathrm{EC}(X)$ be a critical edge of a k-cell X of V for some $2 \leq k \leq 4$. A *cell cluster* is the set

$$\mathrm{CL}(\varepsilon) = \{X \ : \ \varepsilon \in \mathrm{EC}(X)\}$$

of all cells around ε. Define

$$\Gamma(\varepsilon) = \sum_{X \in \mathrm{CL}(\varepsilon)} \gamma(X, L)\mathrm{wt}(X) + \beta(\varepsilon, X).$$

The following weak form of Theorem 6.81 is sufficient for our needs.

Lemma 6.92 *Let V be any saturated packing and let X be any cell of V such that* $\mathrm{EC}(X)$ *is empty. Then*

$$\gamma(X, L) \geq 0.$$

Proof This is a *computer calculation*[5] [21]. □

Theorem 6.93 (cell cluster inequality) *Let* $\mathrm{CL}(\varepsilon)$ *be any cell cluster of a critical edge* ε *in a saturated packing V. Then* $\Gamma(\varepsilon) \geq 0$.

[5] `[TSKAJXY]`

Proof sketch The proof of this cell cluster inequality is a *computer calculation*[6] [21], which is the most delicate computer estimate in the book. It reduces the cell cluster inequality to hundreds of nonlinear inequalities in at most six variables. In degenerate cells with a face of area zero, Euler's formula (Lemma 2.73) should be used to calculate solid angles, because the standard dihedral angle formula for solid angles can lead to the evaluation of $\arctan_2$ at the branch point $(0, 0)$, which is numerical unstable and is best avoided. □

Example 6.94 *We construct an example of a cell cluster in the form of an octahedron, with four* 4*-cells joined along a common critical edge* ε. *Assume that all of the edges of the octahedron have length* 2, *except for one of length* y *for some edge that does not meet* ε. *If* $y \in [2h_-, 2h_+]$, *then one of the four simplices has weight* $1/2$ *and the other three simplices have weight* 1. *The parameter* y *determines the cell cluster up to isometry. We plot the function* $f(y) = \Gamma(\varepsilon)$ *as a function of* y. *We also plot the function*

$$g(y) = \sum_{X \in \mathrm{CL}(\varepsilon)} \gamma(X, L)\mathrm{wt}(X).$$

As the plot shows, the function g *is not positive. This shows that without the small correction term* β, *the cell cluster inequality is false. Numerical evidence suggests that the global minimum of* $\Gamma(\varepsilon)$ *occurs when the cell cluster has the form of an octahedron with parameter* $y = 2h_+$ *and value* $f(2h_+) \approx 0.0013$.

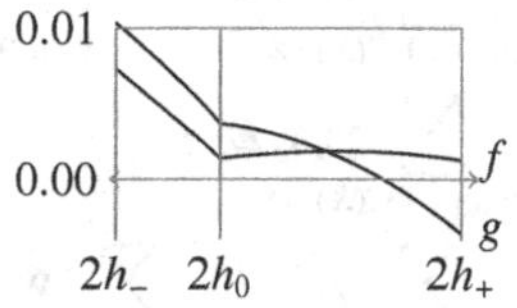

Figure 6.15 The functions g takes negative values, but the function f remains positive, as predicted by the cell cluster inequality. The nondifferentiability at $2h_0$ is inherited from the nondifferentiability of L.

The proof of the following lemma is deferred, because it relies on many computer calculations and is extremely long and complex. The non-computer parts of the proof take up most of the remainder of the book. In this chapter the lemma is treated as an unproved assertion.

Lemma* 6.95 *For any saturated packing* V *and any* $\mathbf{u}_0 \in V$,

$$\sum_{\mathbf{u}_1 \in V \,:\, h(\mathbf{u}_0, \mathbf{u}_1) \le h_0} L(h\{\mathbf{u}_0, \mathbf{u}_1\}) \le 12. \tag{6.96}$$

[6] [OXLZLEZ]

Lemma 6.97 *Inequality (6.96) implies that for every saturated packing V, there exists a negligible FCC-compatible function $G : V \to \mathbb{R}$.*

Remark 6.98 In light of Lemma 6.13, inequality 6.96 implies the Kepler conjecture.

Proof By Lemma 6.86, the proof reduces to showing that there exists a constant c_0 such that for all $r \geq 1$

$$\sum_{X \subset B(\mathbf{0},r)} \gamma(X, L) \geq c_0 r^2.$$

If a cell X does not belong to any cell cluster, then

$$\gamma(X, L) \geq 0$$

by Lemma 6.92. Note that the function $\beta(\varepsilon, X)$ averages to zero for any 4-cell X:

$$\sum_{\varepsilon \in \mathrm{EC}(X)} \beta(\varepsilon, X) = 0.$$

Hence, the terms involving β in sums may be disregarded in this proof. (These terms may be disregarded here, but they are needed in Lemma 6.93.)

Theorem 6.93 gives the required inequality for cell clusters. Again, using big O notation,

$$\begin{aligned}
\sum_{X \subset B(\mathbf{0},r)} \gamma(X, L) &= \sum_{X \subset B(\mathbf{0},r)\,:\,\mathrm{EC}(X) \neq \emptyset} \gamma(X, L) + \sum_{X \subset B(\mathbf{0},r)\,:\,\mathrm{EC}(X) = \emptyset} \gamma(X, L) \\
&\geq \sum_{X \subset B(\mathbf{0},r)\,:\,\mathrm{EC}(X) \neq \emptyset} \gamma(X, L) \\
&= \sum_{X \subset B(\mathbf{0},r)} \gamma(X, L) \sum_{\varepsilon \in \mathrm{EC}(X)} \mathrm{wt}(X) \\
&= \sum_{\varepsilon \subset B(\mathbf{0},r)} \sum_{X\,:\,\varepsilon \in \mathrm{EC}(X)} \gamma(X, L)\mathrm{wt}(X) \qquad + O(r^2) \\
&= \sum_{\varepsilon \subset B(\mathbf{0},r)} \Gamma(\varepsilon) \qquad + O(r^2) \\
&\geq O(r^2).
\end{aligned}$$

□

Definition 6.99 ($\mathcal{B}$) Let $\mathcal{B}$ be the *annulus* $\bar{B}(\mathbf{0}, 2h_0) \setminus B(\mathbf{0}, 2)$, where $\bar{B}(\mathbf{0}, r)$ is the closed ball of radius r.

Corollary 6.100 *If the Kepler conjecture is false, there exists a finite packing $V \subset \mathcal{B}$ with the following properties.*

$$\sum_{\mathbf{u} \in V} L(h\{\mathbf{0}, \mathbf{u}\}) > 12. \tag{6.101}$$

The proof of the Kepler conjecture proceeds by assuming that there is a counterexample to Inequality 6.96 and then deriving a contradiction. This corollary formulates the potential counterexample in slightly simpler terms.

Proof If the Kepler conjecture is false, Inequality 6.96 is violated for some packing V and some $\mathbf{u}_0 \in V$. After translating V to $V - \mathbf{u}_0$ and $\mathbf{u}_0$ to $\mathbf{0}$, it follows without loss of generality that $\mathbf{u}_0 = \mathbf{0} \in V$. After the replacement of V with the finite subset $V \cap \mathcal{B}$, it follows without loss of generality that the packing is a finite subset of $\mathcal{B}$. □

6.5 Counting Spheres

This section proves two estimates about a packing $V \subset \mathcal{B}$ that satisfies Inequality 6.101. The first estimate (Lemma 6.110) shows that the cardinality of V is thirteen, fourteen, or fifteen. The second estimate (See Lemma 6.112.) shows that no point $\mathbf{v} \in V$ can be strongly isolated from the other points of V. To prove these two estimates, we need a formula for the smallest possible area of a spherical polygon that contains a disk. This formula is developed in the first subsection.

6.5.1 solid angle

The following lemma is analogous to the Rogers decomposition of a polyhedron into simplices. The lemma constructs $2k$ points to be used to triangulate (a subset of) a polygon (Figure 6.16).

Lemma 6.102 *Let P be a two-dimensional bounded polyhedron in $\mathbb{R}^2$. Let k be the number of facets of P. Let $r > 0$. Suppose that*

$$\{\mathbf{v} \in \mathbb{R}^2 \ : \ \|v\| < r\} \subset P.$$

Then there exist nonzero points $\mathbf{w}_j \in P$ for $j = 0, \ldots, 2k-1$ such that

1. *The polar cycle on $\{\mathbf{w}_j \ : \ j\}$ is given by $\sigma(\mathbf{w}_j) = \mathbf{w}_{j+1}$, with indexing mod $2k$.*
2. *$\theta(\mathbf{w}_{2i}, \mathbf{w}_{2i+1}) = \theta(\mathbf{w}_{2i+1}, \mathbf{w}_{2i+2}) < \pi/2$, where θ denotes the relative polar coordinate of Lemma 2.89.*

3. $\|\mathbf{w}_{2i}\| = r$ *and* $\|\mathbf{w}_{2i+1}\| = r \sec \theta(\mathbf{w}_{2i}, \mathbf{w}_{2i+1})$, *for* $i = 0, \ldots, k-1$.
4. $\mathbf{w}_{2i} \cdot (\mathbf{w}_{2i\pm1} - \mathbf{w}_{2i}) = 0$.

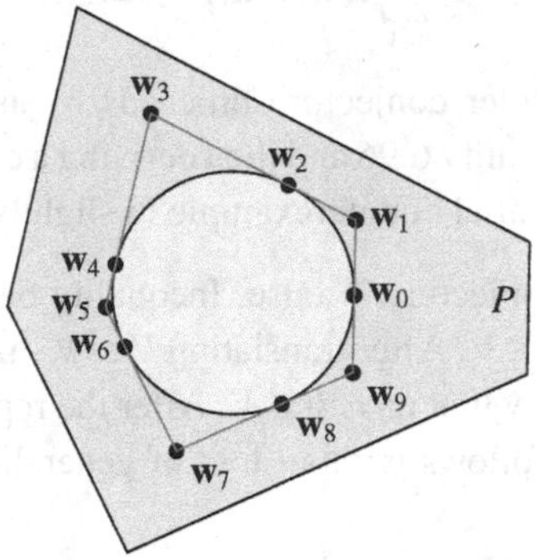

Figure 6.16 A set of reference points $\mathbf{w}_i$ can be constructed outside an open disk of radius r and inside a given polyhedron P in $\mathbb{R}^2$.

Proof Enumerate the distinct facets $F_1, \ldots, F_k$ of P, and for each one select a defining equation

$$F_i = P \cap \{\mathbf{p} \,:\, \mathbf{u}_i \cdot \mathbf{p} = b_i\}, \quad \text{where } \|\mathbf{u}_i\| = 1 \text{ and } b_i \geq 0.$$

We may assume that ordering of facets by increasing subscripts is the ordering by the polar cycle on $\mathbf{u}_i \in \mathbb{R}^2$. The assumption that P contains an open disk of radius r gives $r \leq b_i$.

We claim that $r\mathbf{u}_i \in P$ *and does not lie in any facet except possibly* F_i. Otherwise, for some j,

$$r \leq b_j \leq (r\mathbf{u}_i) \cdot \mathbf{u}_j \leq r\|\mathbf{u}_i\|\,\|\mathbf{u}_j\| = r.$$

This is the case of equality in Cauchy–Schwarz, which implies that $\mathbf{u}_i = \mathbf{u}_j$. The definition of face implies that $b_i = b_j$, and $i = j$. The claim ensues.

We claim that $0 < \theta(\mathbf{u}_i, \mathbf{u}_{i+1}) < \pi$. Indeed, from the previous claim it follows that $0 < \theta(\mathbf{u}_i, \mathbf{u}_{i+1})$. By the boundedness of P, for any nonzero $\mathbf{v}$ orthogonal to $\mathbf{u}_i$, there exists $s > 0$ such that $r\mathbf{u}_i + s\mathbf{v}$ lies in some facet $F_j \neq F_i$. This gives $r\mathbf{u}_j \cdot \mathbf{u}_i + s\mathbf{u}_j \cdot \mathbf{v} = b_j$. The condition $r\mathbf{u}_i \in P \setminus F_j$ gives $r\mathbf{u}_j \cdot \mathbf{u}_i < b_j$. Hence $\mathbf{u}_j \cdot \mathbf{v} > 0$. For an appropriate choice of sign of $\mathbf{v}$, this gives $\theta(\mathbf{u}_i, \mathbf{u}_{i+1}) \leq \theta(\mathbf{u}_i, \mathbf{u}_j) < \pi$.

Suppressing the subscript i, we write $\psi = \theta(\mathbf{u}_i, \mathbf{u}_{i+1})/2$. Let $\mathbf{u}_i'$ be the point in the plane given in polar coordinates by

$$\|\mathbf{u}_i'\| = r \sec \psi, \quad \theta(\mathbf{u}_i, \mathbf{u}_i') = \theta(\mathbf{u}_i', \mathbf{u}_{i+1}) = \psi.$$

We claim that $\mathbf{u}'_i \in P$. Indeed, for every j, we have

$$\mathbf{u}_j \cdot \mathbf{u}'_i = \|\mathbf{u}_j\| \, \|\mathbf{u}'_i\| \cos\varphi = r \sec\psi \cos\varphi,$$

where $\varphi = \theta(\mathbf{u}'_i, \mathbf{u}_j)$. From the polar order, and the construction of $\mathbf{u}'_i$ along the bisector of $\mathbf{u}_i, \mathbf{u}_{i+1}$, it follows that

$$\psi \le \varphi \le 2\pi - \psi.$$

Hence, $\cos\varphi \le \cos\psi$. This gives

$$\mathbf{u}_j \cdot \mathbf{u}'_i \le r \le b_j.$$

This shows that $\mathbf{u}'_i$ satisfies all the defining conditions of P.

Set $\mathbf{w}_{2i} = \mathbf{u}_i$ and $\mathbf{w}_{2i+1} = \mathbf{u}'_i$. It is clear from construction that the polar cycle on $\{\mathbf{w}_j \ : \ j\}$ is compatible with the indexing. The enumerated properties of the lemma now follow from Lemma 2.89. The lemma ensues. □

Lemma 6.103 *Let P be a bounded polyhedron in $\mathbb{R}^3$ that contains $\mathbf{0}$ as an interior point. Let F be a facet of P, given by an equation*

$$F = \{\mathbf{p} \ : \ \mathbf{p} \cdot \mathbf{v} = b_0\} \cap P.$$

Let W_F be the corresponding topological component of $Y(V_P, E_P)$. Assume that W_F contains the right-circular cone

$$\text{rcone}^0(\mathbf{0}, \mathbf{v}, t) \subset W_F \tag{6.104}$$

for some t such that $0 < t < 1$. *Then*

$$\text{sol}(W_F) \ge 2\pi - 2k \arcsin\left(t \sin(\pi/k) \right),$$

where k is the number of edges of F.

Proof Project the facet F to $\mathbb{R}^2$ by projecting onto the coordinates of $\mathbf{e}_2$ and $\mathbf{e}_3$ of an orthonormal frame $(\mathbf{e}_1, \mathbf{e}_2, \mathbf{e}_3)$ adapted to $(\mathbf{0}, \mathbf{v}, \ldots)$. By the Pythagorean theorem, the hypothesis (6.104) implies that a disk of radius

$$b_1\sqrt{1 - t^2}/t$$

is contained in the projected face, where $b_1 = b_0/\|\mathbf{v}\|$ is the distance from $\text{aff}(F)$ to $\mathbf{0}$. Apply Lemma 6.102 to the projected face, and pull the points $\mathbf{w}_j$ back to points on F with the same names.

By the additivity of measure over measurable sets that are disjoint up to a null set, we may partition into wedges:

$$\begin{aligned}\mathrm{sol}(W_F) &= \sum_j \mathrm{sol}(W_F \cap W(\mathbf{0}, \mathbf{v}, \mathbf{w}_j, \mathbf{w}_{j+1}))\\ &\geq \sum_j \mathrm{sol}(\mathrm{aff}_+^0(\mathbf{0}, \{\mathbf{v}, \mathbf{w}_j, \mathbf{w}_{j+1}\})).\end{aligned}$$

The solid triangles that appear in the last sum are primitive volumes, which are computed in terms of dihedral angles in Chapter 3. Set

$$\beta_{2j} = \beta_{2j+1} = \mathrm{dih}_V(\{\mathbf{0}, \mathbf{v}\}, \{\mathbf{w}_{2j}, \mathbf{w}_{2j\pm1}\}) \text{ and } a = \mathrm{arc}_V(\mathbf{0}, \{\mathbf{v}, \mathbf{w}_{2j}\}) = \arccos t.$$

The three vectors $\mathbf{v}$, $\mathbf{w}_{2j} - \mathbf{v}$, and $\mathbf{w}_{2j+1} - \mathbf{w}_{2j}$ are mutually orthogonal by the final claim of Lemma 6.102. Lemma 2.68 gives

$$\mathrm{dih}_V(\{\mathbf{0}, \mathbf{w}_{2j}\}, \{\mathbf{v}, \mathbf{w}_{2j+1}\}) = \pi/2,$$

because

$$\begin{aligned}(\mathbf{w}_{2j} \times \mathbf{v}) \cdot (\mathbf{w}_{2j} \times \mathbf{w}_{2j+1}) &= (\mathbf{w}_{2j} \times \mathbf{v}) \cdot (\mathbf{w}_{2j} \times (\mathbf{w}_{2j+1} - \mathbf{w}_{2j}))\\ &= (\mathbf{w}_{2j+1} - \mathbf{w}_{2j}) \cdot ((\mathbf{w}_{2j} \times \mathbf{v}) \times \mathbf{w}_{2j})\\ &= 0.\end{aligned}$$

Consider a spherical triangle with sides a, b, c and opposite angles α, β, γ. If $\gamma = \pi/2$, then by Girard's formula, the area of the triangle is

$$\alpha + \beta - \pi/2,$$

and by the spherical law of cosines (Lemma 2.71)

$$\cos\alpha = \cos a \sin\beta.$$

This determines the area $g(a, \beta)$ of the triangle as a function of a and β:

$$g(a, \beta) = \beta - \arcsin(\cos a \sin\beta).$$

The solid angle of W_F is at least sum of the areas of the triangles:

$$\sum_{j=0}^{2k-1} g(a, \beta_j),$$

with angle sum

$$\sum_{j=0}^{2k-1} \beta_j = 2\pi.$$

The second partial of g with respect to β is

$$\frac{\partial^2 g(a,\beta)}{\partial\beta^2} = \frac{\cos a \sin^2 a \sin\beta}{\sin^2 \alpha} \geq 0.$$

Thus, the function is convex in β. By convexity, the minimum area occurs when all angles are equal $\beta = \beta_j = \pi/k$.

The solid angle bound of the lemma is equal to

$$2kg(a, \pi/k)$$

where $\cos a = t$. □

6.5.2 a polyhedral bound

Definition 6.105 (weakly saturated) Let r and r' be real numbers such that $2 \leq r \leq r'$. Define a set $V \subset \mathbb{R}^3 \setminus B(\mathbf{0}, 2)$ to be *weakly saturated* with parameters (r, r') if for every $\mathbf{p} \in \mathbb{R}^3$

$$2 \leq \|\mathbf{p}\| \leq r' \quad \Longrightarrow \quad \exists \mathbf{u} \in V.\ \|\mathbf{u} - \mathbf{p}\| < r.$$

Lemma 6.106 *Fix r and r' such that $2 \leq r \leq r'$. Let V be a weakly saturated finite packing with parameters (r, r'). For any $g : V \to \mathbb{R}$, let $P(V, g)$ be the polyhedron given by the intersection of half-spaces*

$$\{\mathbf{p} : \mathbf{u} \cdot \mathbf{p} \leq g(\mathbf{u})\}, \quad \mathbf{u} \in V.$$

Then $P(V, g)$ is bounded.

Proof Assume for a contradiction that $P = P(V, g)$ is unbounded and there exists $\mathbf{p} \in P$ such that $\|\mathbf{p}\| > g(\mathbf{u})r'/2$ for all $\mathbf{u} \in V$. Let $\mathbf{v} = r'\mathbf{p}/\|\mathbf{p}\|$ so that $r' = \|\mathbf{v}\|$. By the weak saturation of V, there exists $\mathbf{u} \in V$ such that $\|\mathbf{v} - \mathbf{u}\| < r$. Then,

$$\begin{aligned}
\|\mathbf{p}\| &> g(\mathbf{u})r'/2 \geq \mathbf{u} \cdot (r'\mathbf{p})/2 = \|\mathbf{p}\|\mathbf{u} \cdot \mathbf{v}/2 \\
&= \|\mathbf{p}\|(\|\mathbf{u}\|^2 + \|\mathbf{v}\|^2 - \|\mathbf{u} - \mathbf{v}\|^2)/4 \\
&> \|\mathbf{p}\|(4 + r'^2 - r^2)/4 \\
&\geq \|\mathbf{p}\|.
\end{aligned}$$

This contradiction shows that P is bounded. □

Lemma 6.107 *Let*

$$g(h) = \arccos(h/2) - \pi/6.$$

Then

$$\mathrm{arc}(2h, 2h', 2) \geq g(h) + g(h'), \tag{6.108}$$

for all $h, h' \in [1, h_0]$.

Proof The function g can be rewritten as

$$g(h) = \mathrm{arc}(2h, 2, 2) - \mathrm{arc}(2, 2, 2)/2.$$

It is enough to prove a more general inequality in symmetrical form

$$f(a_2, b_2) - f(a_1, b_2) - f(a_2, b_1) + f(a_1, b_1) \geq 0, \tag{6.109}$$

when

$$2 \leq a_1 \leq a_2 \leq 2h_0, \text{ and } 2 \leq b_1 \leq b_2 \leq 2h_0,$$

where $f(a, b) = \mathrm{arc}(a, b, 2)$. A calculation gives

$$\frac{\partial^2 f(a, b)}{\partial a\, \partial b} = \frac{32ab}{\upsilon(a^2, b^2, 4)^{3/2}} > 0.$$

Thus, by holding a fixed, $\partial f/\partial a$ is increasing in b:

$$\frac{\partial f(a, b_2)}{\partial a} - \frac{\partial f(a, b_1)}{\partial a} \geq 0.$$

This shows that with b_1 and b_2 fixed, $f(a, b_2) - f(a, b_1)$ is increasing in a. Equation 6.109 ensues. □

Since $L(h) \leq 1$ when $h \geq 1$, it is clear that a finite packing V that satisfies Inequality 6.101 has cardinality greater than twelve. The following lemma also gives an upper bound on the cardinality of V.

Lemma 6.110 *If $V \subset \mathcal{B}$ is a packing that satisfies Inequality 6.101, then the cardinality of V is thirteen, fourteen, or fifteen.*

Proof (Following Marchal.) Consider a finite packing $V = \{\mathbf{u}_1, \ldots, \mathbf{u}_N\} \subset \mathcal{B}$ satisfying Inequality 6.101. The packing V contains more than twelve points because otherwise Inequality 6.101 cannot hold, as $L(h) \leq 1$.

By adding points as necessary, the packing becomes weakly saturated in the sense of Definition 6.105, with $r = 2$ and $r' = 2h_0$. It is enough to show that this enlarged set has cardinality less than sixteen. Let

$$g(h) = \arccos(h/2) - \pi/6,$$

and let $h_i = \|\mathbf{u}_i\|/2$. Then $h_i \leq h_0 = 1.26$. Consider the spherical disks D_i of radii $g(h_i)$, centered at $\mathbf{u}_i/\|\mathbf{u}_i\|$ on the unit sphere. These disks do not overlap by Lemma 6.107.

For each i, the plane through the circular boundary of D_i bounds a half-space containing the origin. The intersection of these half-spaces is a polyhedron P, which is bounded by Lemma 6.106. (See Figure 6.17.) Lemma 5.61 associates a fan (V_P, E_P) with P. (The set V_P is dual to V; the set V_P is in bijection with extreme points of P, whereas V is in bijection with the facets of P.) There are natural bijections between the following sets.

1. $V = \{\mathbf{u}_1, \ldots, \mathbf{u}_N\}$.
2. The facets of P.
3. The set of topological components of $Y(V_P, E_P)$.
4. The set of faces in the hypermap $\mathrm{hyp}(V_P, E_P)$.

The first conclusion of Lemma 5.54 gives the bijection of the first two sets. Lemmas 5.62 and 5.42 give the other bijections.

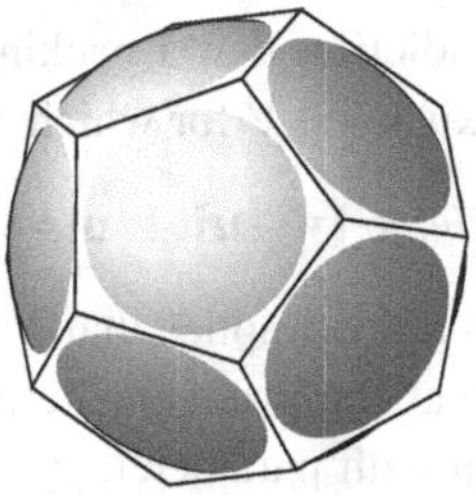

Figure 6.17 A polyhedron is constructed by extending planes through the circular boundaries of disks D_i on the unit sphere.

By Lemma 5.66, the number of edges of the facet i is k_i, the cardinality of the corresponding face in $\mathrm{hyp}(V_P, E_P)$. By Lemma 6.103, the solid angle of the topological component W_i of $Y(V_P, E_P)$ is at least $\mathrm{reg}(g(h_i), k_i)$, where

$$\mathrm{reg}(a, k) = 2\pi - 2k(\arcsin(\cos(a)\sin(\pi/k))).$$

By a *computer calculation*[7] [21]

$$\mathrm{reg}(g(h), k) \geq c_0 + c_1 k + c_2 L(h), \qquad \text{for all } k = 3, 4, \ldots, \qquad 1 \leq h \leq h_0, \quad (6.111)$$

where

$$c_0 = 0.591, \quad c_1 = -0.0331, \quad c_2 = 0.506.$$

The sum $\sum_i k_i$ is the number of darts in $\mathrm{hyp}(V_P, E_P)$ by Lemma 5.61. By

[7] [BIEFJHU] This is a linear lower bound on the area of a regular polygon.

Lemma 4.22, $\sum_i k_i \le (6N - 12)$. Summing over i, an estimate on N follows:

$$\begin{aligned} 4\pi &= \sum_i \mathrm{sol}(W_i) \\ &\ge \sum_i \mathrm{reg}(g(h_i), k_i) \\ &\ge c_0 N + c_1 \sum_i k_i + c_2 \sum L(h_i) \\ &\ge c_0 N + c_1(6N - 12) + c_2 12. \end{aligned}$$

This gives $16 > N$. □

Lemma 6.112 *Assume that $V \subset \mathcal{B}$ is a packing that satisfies Inequality 6.101. Then for every $\mathbf{v} \in V$ such that $\|\mathbf{v}\| = 2$, there exists $\mathbf{u} \in V$ such that $0 < \|\mathbf{v} - \mathbf{u}\| < 2h_0$.*

Proof Assume for a contradiction that a packing V exists that satisfies the inequality for which there exists $\mathbf{v} \in V$ for which

$$2h_0 \le \|\mathbf{v} - \mathbf{u}\|, \quad \mathbf{u} \ne \mathbf{v}. \tag{6.113}$$

The assumption that (6.101) holds implies that $N \ge 13$. Create one large disk D'_1 centered at $\mathbf{v}/2$ and repeat the proof of the previous lemma. Extend the packing to a weak saturation with parameters $r = r' = 2h_0$. This can be done in a way that maintains the assumptions on $\mathbf{v}$. By Lemma 6.106, the polyhedron is bounded. By a *computer calculation*[8] [21]

$$a' = 0.797 < \mathrm{arc}(2, 2h, 2h_0) - g(h) \quad \text{for } 1 \le h \le h_0.$$

By (6.113), we may take a' for the arcradius of the large disk D'_1. By a *computer calculation*[9] [21]

$$\mathrm{reg}(a', k) \ge c_0 + c_1 k + c_2 L(1) + c_3, \quad k = 3, 4, \ldots \tag{6.114}$$

8 [WAZLDCD]
9 [UKBRPFE]

where $c_3 = 1$. Then

$$\begin{aligned} 4\pi &= \sum_{i=1}^{N} \mathrm{sol}(W_i) \\ &\geq \mathrm{reg}(a', k_1) + \sum_{i=2}^{N} \mathrm{reg}(g(h_i), k_i) \\ &\geq c_0 N + c_1 \sum_{i=1}^{N} k + c_2 \sum_{i=1}^{N} L(h_i) + c_3 \\ &\geq c_0 N + c_1(6N - 12) + c_2 12 + c_3. \end{aligned}$$

This gives a contradiction $13 > N \geq 13$. □

7

Local Fan

Summary. *The difficult technical estimates that we need for the proof of the Kepler conjecture are found in this chapter. The standard form of the main estimate (Theorem 7.43) takes the form*

$$\tau(V, E, F) \geq d(k), \quad k = \mathrm{card}(V) \in \{3, 4, 5, 6\}. \tag{7.1}$$

Here (V, E) is a fan satisfying various technical conditions, and F is a face of the hypermap of (V, E). The function d is defined by a table of real numbers. Heuristically, the real-valued function τ measures the looseness of a packing. Large values of τ indicate that the points of V are loosely arranged around the face and small values of τ indicate a tight packing. The main estimate gives limits to the tightness of a packing, with the eventual aim of showing that no packing can have density greater than the FCC packing.

This chapter also proves the well-know result that the perimeter of a geodesically convex spherical polygon is never greater than 2π, the length of a great circle.

7.1 Localization

The *localization* of a fan along a face discards everything but the part of the fan near the face. The localization is used to focus attention on a single face in a fan. We also introduce a notion of convexity that is suitable for local fans.

7.1.1 basics

Roughly speaking, a local fan is to a fan what a polygon is to a biconnected plane graph.

Definition 7.2 (local fan) A triple (V, E, F) is a *local fan* if the following conditions hold.

1. (FAN) (V, E) is a fan.
2. (FACE) F is a face of $H = \text{hyp}(V, E)$.
3. (DIHEDRAL) H is isomorphic to Dih_{2k}, where $k = \text{card}(F)$.

A local fan (V, E, F) is said to be *nonreflexive*[1] if the following additional conditions hold.

4. (ANGLE) $\text{azim}(x) \le \pi$ for all darts $x \in F$.
5. (WEDGE) $V \subset W_{\text{dart}}(x)$ for all $x \in F$.

In the proof of the Kepler conjecture in this chapter and the next, all local fans are nonreflexive. Local fans (that are reflexive) appear in applications to other packing problems in Section 8.6.

Remark 7.3 (visualization) If (V, E, F) is a local fan, the intersection of $X(V, E)$ with the unit sphere is a spherical polygon, which gives a visual representation of the fan (Figure 7.1). In the first part of the chapter, the lengths $\|\mathbf{v}\|$, for $\mathbf{v} \in V$, have little importance, and the spherical polygon captures the relevant features of the local fan. The choice of F distinguishes the interior of the polygon from its exterior. If the local fan is nonreflexive, then the interior of the polygon is geodesically convex.

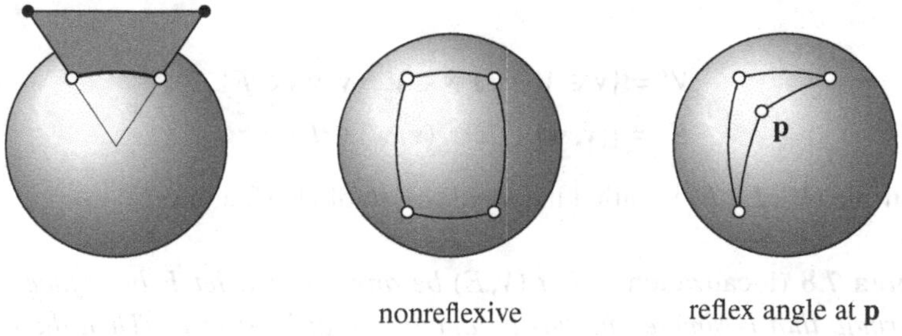

Figure 7.1 To represent a local fan, each blade is intersected with the unit sphere to give a spherical polygon. The spherical polygon of a nonreflexive local fan has no reflex angles.

[1] An angle greater than π is a *reflex angle*.

Lemma 7.4 *For any local fan (V, E, F), there is a bijection from F onto V given by*

$$(\mathbf{v}, \mathbf{w}) \mapsto \mathbf{v}.$$

Moreover, write $\mathbf{v} \mapsto (\mathbf{v}, \rho\mathbf{v})$ for the inverse map. Then $\rho : V \to V$ is a cyclic permutation. That is, the orbit of each $\mathbf{v}$ under ρ is V.

Proof The map from a face to the set of nodes is a bijection for the dihedral hypermap Dih_{2k}. It is a also bijection for a fan isomorphic to Dih_{2k}.

For all $(\mathbf{v}, \rho\mathbf{v}) \in F$,

$$f(\mathbf{v}, \rho\mathbf{v}) = (\rho\mathbf{v}, \rho^2\mathbf{v}),$$

so that the orbit of $\mathbf{v}$ under ρ in V corresponds under the bijection to the orbit of a dart under f on F, which F. Thus, ρ is a cyclic permutation of V of order $k = \mathrm{card}(V)$. □

Definition 7.5 (ρ, node) For any local fan (V, E, F), write $\rho = \rho_{V,E,F} : V \to V$ and node : $F \to V$ for the bijections of the preceding lemma.

Definition 7.6 (interior angle, $\angle$, W_{dart}) For any local fan (V, E, F), write

$$\angle(\mathbf{v}) = \mathrm{azim}((\mathbf{v}, \rho\mathbf{v})),$$

for all $\mathbf{v} \in V$. This is the *interior* angle of the local fan at $\mathbf{v}$. Also, write

$$W^0_{\mathrm{dart}}(F, \mathbf{v}) = W^0_{\mathrm{dart}}((\mathbf{v}, \rho\mathbf{v})), \quad W_{\mathrm{dart}}(F, \mathbf{v}) = W_{\mathrm{dart}}((\mathbf{v}, \rho\mathbf{v})).$$

Definition 7.7 (localization) Let (V, E) be a fan and let F be a face of $\mathrm{hyp}(V, E)$. Let

$$V' = \{\mathbf{v} \in V : \exists\, \mathbf{w} \in V.\ (\mathbf{v}, \mathbf{w}) \in F\}.$$
$$E' = \{\{\mathbf{v}, \mathbf{w}\} \in E : (\mathbf{v}, \mathbf{w}) \in F\}.$$

The triple (V', E', F) is called the *localization* of (V, E) along F.

Lemma 7.8 (localization) *Let (V, E) be any fan and let F be a face of its hypermap that is simple and has cardinality of at least three. Then the localization (V', E', F) is a local fan. Moreover, the angle* $\mathrm{azim}(x)$ *and the wedges $W_{dart}(x)$ and $W^0_{dart}(x)$ do not depend on whether they are computed relative to* $\mathrm{hyp}(V, E)$ *or to* $\mathrm{hyp}(V', E')$ *for all $x \in F$.*

Proof The proof that (V', E') is a fan consists of various simple verifications based on the techniques of Remark 5.6. The details are left to the reader.

The dart set D' of $\text{hyp}(V', E')$ is naturally identified with the disjoint union $F \coprod F'$, where $F = \{(\mathbf{v}, \rho\mathbf{v}) : \mathbf{v} \in V\}$ and $F' = \{(\mathbf{v}, \rho^{-1}\mathbf{v}) : \mathbf{v} \in V\}$. Under this identification, F is a face of $\text{hyp}(V', E')$. The face, node, and edge permutations have orders k, 2, and 2, respectively. By Lemma 4.54, this bijection extends to an isomorphism of hypermaps Dih_{2k} onto $\text{hyp}(V', E')$.

The proof that $\text{azim}(x)$ and $W^0_{\text{dart}}(x)$ do not depend on the choice of fan is a consequence of their definitions:

$$\text{azim}(x) = \text{azim}(\mathbf{0}, \mathbf{v}, \mathbf{w}, \sigma(\mathbf{v}, \mathbf{w})), \quad \text{and}$$
$$W^0_{\text{dart}}(x) = W^0_{\text{dart}}(\mathbf{0}, \mathbf{v}, \mathbf{w}, \sigma(\mathbf{v}, \mathbf{w})).$$

where $x = (\mathbf{v}, \mathbf{w})$. It is enough to check that $\sigma(\mathbf{v}, \mathbf{w}) \in E'(\mathbf{v})$. But $\{\sigma(\mathbf{v}, \mathbf{w}), \mathbf{v}\} \in F$, so this is indeed the case. □

Lemma 7.9 *Let (V, E) be any fan and let F be a face of its hypermap. Let (V', E') be the localization of (V, E) along F. Assume that (V, E) is fully surrounded. Then (V', E', F) is a nonreflexive local fan.*

Proof Lemma 7.8 gives all the properties of a nonreflexive local fan except for property (WEDGE): $V' \subset W_{\text{dart}}(x)$ for every dart $x \in F$. By Lemma 5.42, $U_F \subset W^0_{\text{dart}}(x)$. The wedge $W_{\text{dart}}(x)$ is closed and contains $W^0_{\text{dart}}(x)$. Hence, the closure $\bar{U}_F$ is contained in $W_{\text{dart}}(x)$. Let $\mathbf{v} \in V'$ and choose $\mathbf{w} \in V$ such that $y = (\mathbf{v}, \mathbf{w}) \in F$. Since the dart y leads into U_F, every neighborhood of $\mathbf{v}$ meets U_F. Thus, $\mathbf{v} \in \bar{U}_F \subset W_{\text{dart}}(x)$. This completes the proof. □

7.1.2 geometric type

Definition 7.10 (generic, lunar, circular) A local fan (V, E, F) is *generic* if it is nonreflexive and if for every $\{\mathbf{v}, \mathbf{w}\} \in E$ and every $\mathbf{u} \in V$,

$$C\{\mathbf{v}, \mathbf{w}\} \cap C^0_-\{\mathbf{u}\} = \emptyset.$$

A local fan is *circular* if it is nonreflexive and if there exists $\mathbf{u} \in V$ and $\{\mathbf{v}, \mathbf{w}\} \in E$ such that

$$C^0\{\mathbf{v}, \mathbf{w}\} \cap C^0_-\{\mathbf{u}\} \neq \emptyset.$$

A nonreflexive local fan is *lunar* with pole $\{\mathbf{v}, \mathbf{w}\} \subset V$ if it is nonreflexive, if it is not circular, if $\mathbf{v} \neq \mathbf{w}$, and if $\{\mathbf{v}, \mathbf{w}\}$ is a parallel set (Figure 7.2).

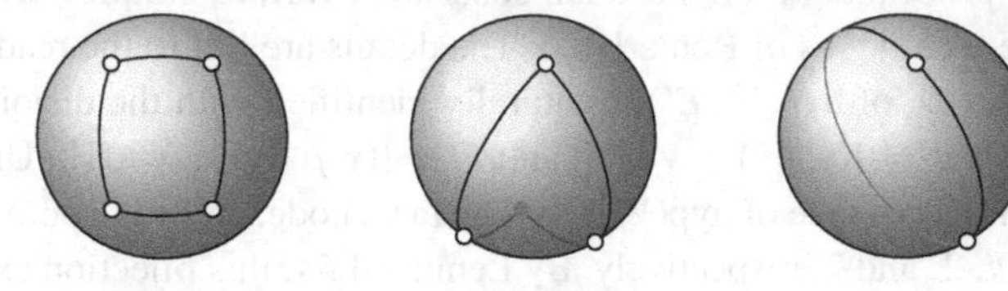

Figure 7.2 A nonreflexive local fan is generic, lunar, or circular. The figures follow the conventions of Remark 7.3.

Lemma 7.11 (trichotomy) *Every nonreflexive local fan is either generic, lunar, or circular. Moreover, these three properties are mutually exclusive.*

Proof If (V, E, F) is not generic, select some $\{\mathbf{v}, \mathbf{w}\} \in E$ and some $\mathbf{u} \in V$ such that

$$C\{\mathbf{v}, \mathbf{w}\} \cap C^0_-\{\mathbf{u}\} \neq \varnothing. \tag{7.12}$$

Now $C\{\mathbf{v}, \mathbf{w}\} = C^0\{\mathbf{v}, \mathbf{w}\} \cup C\{\mathbf{v}\} \cup C\{\mathbf{w}\}$. If, for some such triple $(\mathbf{u}, \mathbf{v}, \mathbf{w})$, the intersection (7.12) meets $C^0\{\mathbf{v}, \mathbf{w}\}$, then the nonreflexive local fan is circular. Otherwise, the nonreflexive local fan is lunar. □

Definition 7.13 (straight) Let (V, E, F) be a local fan. If $\angle(\mathbf{v}) = \pi$, then $\mathbf{v}$ is *straight.*

Lemma 7.14 *Assume that* $\{\mathbf{0}, \mathbf{u}, \mathbf{w}\}$ *and* $\{\mathbf{0}, \mathbf{u}, \mathbf{v}\}$ *are not collinear sets. Then* $\operatorname{azim}(\mathbf{0}, \mathbf{u}, \mathbf{v}, \mathbf{w}) = \pi$ *if and only if there exists a plane A such that* $\{\mathbf{0}, \mathbf{u}, \mathbf{v}, \mathbf{w}\} \subset A$ *and such that the line* $\operatorname{aff}\{\mathbf{0}, \mathbf{u}\}$ *separates* $\mathbf{v}$ *from* $\mathbf{w}$ *in A.*

Proof The given azimuth angle is π if and only if $\operatorname{dih}(\{\mathbf{0}, \mathbf{u}\}, \{\mathbf{v}, \mathbf{w}\}) = \pi$. This holds exactly when $\{\mathbf{0}, \mathbf{u}, \mathbf{v}, \mathbf{w}\}$ is coplanar, and the line $\operatorname{aff}\{\mathbf{0}, \mathbf{u}\}$ separates $\mathbf{v}$ from $\mathbf{w}$ in A. □

Lemma 7.15 *Let (V, E, F) be a nonreflexive local fan. Let $k = \operatorname{card}(F)$. Assume that for some $0 < r \leq k - 1$ and some $\mathbf{v} \in V$, the set $U = \{\mathbf{v}, \rho\mathbf{v}, \ldots, \rho^r\mathbf{v}\}$ is contained in a plane A passing through* $\mathbf{0}$. *Let* $\mathbf{e}$ *be the unit normal to A in the direction* $\mathbf{v} \times \rho\mathbf{v}$. *Then the set U is cyclic with respect to* $(\mathbf{0}, \mathbf{e})$, *and the azimuth cycle σ on U is*

$$\sigma\mathbf{u} = \begin{cases} \rho\mathbf{u}, & \mathbf{u} \neq \rho^r\mathbf{v}, \\ \mathbf{v}, & \mathbf{u} = \rho^r\mathbf{v}. \end{cases}$$

Furthermore, for all $0 \le i \le r-1$,

$$(\rho^i \mathbf{v} \times \rho^{i+1}\mathbf{v}) \cdot \mathbf{e} > 0.$$

Proof We write $\mathbf{v}_i = \rho^i \mathbf{v}$ for $i = 0, \ldots, r$.

We claim that $(\mathbf{v}_i \times \mathbf{v}_{i+1}) \cdot \mathbf{e} > 0$ *for all* $i \le r-1$. Indeed, the base case $(\mathbf{v}_0 \times \mathbf{v}_1) \cdot \mathbf{e} > 0$ of an induction argument holds by assumption. Assume for a contradiction that the inequality holds for i, but not for $i+1$. Then

$$\mathrm{aff}^0_+(\{\mathbf{0}, \mathbf{v}_{i+1}\}, \mathbf{v}_i) = \mathrm{aff}^0_+(\{\mathbf{0}, \mathbf{v}_{i+1}\}, \mathbf{v}_{i+2}).$$

This forces $C^0\{\mathbf{v}_i, \mathbf{v}_{i+1}\}$ to meet $C^0\{\mathbf{v}_{i+1}, \mathbf{v}_{i+2}\}$, which is contrary to the definition of a fan. Thus, the claim holds.

The fact that U is cyclic follows trivially from the fact that U is contained in a plane A through $\mathbf{0}$ and that $\mathbf{e}$ is orthogonal to A.

For all $0 \le i \le r-1$, $\sigma \mathbf{v}_i = \mathbf{v}_{i+1}$. Otherwise, there is some

$$\mathbf{u} \in (U \setminus \{\mathbf{v}_i, \mathbf{v}_{i+1}\}) \cap W^0(\mathbf{0}, \mathbf{e}, \mathbf{v}_i, \mathbf{v}_{i+1}) \cap A.$$

However, by the claim, this intersection is a subset of $C^0\{\mathbf{v}_i, \mathbf{v}_{i+1}\}$, and $\mathbf{u} \in C^0\{\mathbf{v}_i, \mathbf{v}_{i+1}\}$ is contrary to the property (INTERSECTION) of fans. The result ensues. □

Lemma 7.16 *Let* (V, E, F) *be a nonreflexive local fan and let* $\mathbf{u}, \mathbf{v}, \mathbf{w} \in V$ *satisfy the following conditions.*

1. $\{\mathbf{0}, \mathbf{u}, \mathbf{v}, \mathbf{w}\}$ *is contained in a plane* A.
2. $\mathbf{u}, \mathbf{w} \notin \mathrm{aff}\{\mathbf{0}, \mathbf{v}\}$.
3. $\mathrm{aff}^0_+(\{\mathbf{0}, \mathbf{v}\}, \mathbf{u}) \ne \mathrm{aff}^0_+(\{\mathbf{0}, \mathbf{v}\}, \mathbf{w})$.

Then $\mathbf{v}$ *is straight. Moreover,* $\rho\mathbf{v}, \rho^{-1}\mathbf{v} \in A$.

Proof Let $x = (\mathbf{v}, \rho\mathbf{v}) \in F$. Order $\mathbf{u}$ and $\mathbf{w}$ so that

$$\mathrm{azim}(\mathbf{0}, \mathbf{v}, \rho\mathbf{v}, \mathbf{u}) \le \mathrm{azim}(\mathbf{0}, \mathbf{v}, \rho\mathbf{v}, \mathbf{w}).$$

By the definition of nonreflexive local fan, by the conditions $\mathbf{u}, \mathbf{w} \in W_{\mathrm{dart}}(x)$, and by Lemma 7.14,

$$\begin{aligned}0 &\le \mathrm{azim}(\mathbf{0}, \mathbf{v}, \rho\mathbf{v}, \mathbf{u})\\ &= \mathrm{azim}(\mathbf{0}, \mathbf{v}, \rho\mathbf{v}, \mathbf{w}) - \mathrm{azim}(\mathbf{0}, \mathbf{v}, \mathbf{u}, \mathbf{w})\\ &= \mathrm{azim}(\mathbf{0}, \mathbf{v}, \rho\mathbf{v}, \mathbf{w}) - \pi\\ &\le \mathrm{azim}(\mathbf{0}, \mathbf{v}, \rho\mathbf{v}, \rho^{-1}\mathbf{v}) - \pi\\ &= \mathrm{azim}(x) - \pi\\ &= \angle(\mathbf{v}) - \pi\\ &\le 0.\end{aligned}$$

Hence, each inequality is equality. In particular, $\mathbf{v}$ is straight. In particular, $0 = \mathrm{azim}(\mathbf{0}, \mathbf{v}, \rho\mathbf{v}, \mathbf{u})$, so that

$$\rho\mathbf{v} \in \mathrm{aff}_+(\{\mathbf{0}, \mathbf{v}\}, \mathbf{u}) \subset A.$$

Similarly,

$$\rho^{-1}\mathbf{v} \in \mathrm{aff}_+(\{\mathbf{0}, \mathbf{v}\}, \mathbf{w}) \subset A.$$

□

If Lemma 7.16 can be applied once to a set of vectors, then it can often be applied repeatedly along a chain of vectors. For example, the conclusion of the lemma implies that $\rho^{-1}\mathbf{v} \in A$. In fact, by the definition of fan,

$$\rho^{-1}\mathbf{v} \in A \setminus \mathrm{aff}\{\mathbf{0}, \mathbf{v}\} = \mathrm{aff}_+^0(\{\mathbf{0}, \mathbf{v}\}, \mathbf{u}) \cup \mathrm{aff}_+^0(\{\mathbf{0}, \mathbf{v}\}, \mathbf{w}).$$

Suppose that $\rho^{-1}\mathbf{v}$ lies in the second term of the union. If $\mathbf{w} \ne \rho^{-1}\mathbf{v}$, then the assumptions of the lemma are met for $\{\mathbf{w}, \rho^{-1}\mathbf{v}, \mathbf{v}\}$, giving the conclusions that $\rho^{-1}\mathbf{v}$ is straight and that $\rho^{-2}\mathbf{v} \in A$. Repeating the argument on a new set of vectors, we obtain a chain

$$\pi = \angle(\mathbf{v}) = \angle(\rho^{-1}\mathbf{v}) = \cdots,$$

with $\mathbf{v}, \rho^{-1}\mathbf{v}, \ldots \in A$. Another chain $\mathbf{v}, \rho\mathbf{v}, \ldots$ of vectors can be constructed in the other direction. This process of chaining gives the following lemma.

Lemma 7.17 (circular geometry) *Let (V, E, F) be a circular fan. Then*

1. *$\mathbf{v}$ is straight for all $\mathbf{v} \in V$.*
2. *The set V lies in a plane A through $\mathbf{0}$.*
3. *For some choice of unit vector $\mathbf{e}$ orthogonal to A, the set V is cyclic with respect to $(\mathbf{0}, \mathbf{e})$, and the azimuth cycle on V coincides with $\rho : V \to V$.*
4. $\mathrm{azim}(\mathbf{0}, \mathbf{e}, \mathbf{v}, \rho\mathbf{v}) = \mathrm{dih}(\{\mathbf{0}, \mathbf{e}\}, \{\mathbf{v}, \rho\mathbf{v}\}) = \mathrm{arc}_V(\mathbf{0}, \{\mathbf{v}, \rho\mathbf{v}\}) < \pi$.

Proof Let $\mathbf{v}, \mathbf{u} \in V$ be such that $C^0\{\mathbf{u}, \rho\mathbf{u}\}$ meets $C^0_-\{\mathbf{v}\}$. Apply Lemma 7.16 to $\{\mathbf{u}, \mathbf{v}, \rho\mathbf{u}\}$ to conclude that $\mathbf{v}$ is straight, and that some plane A contains $\{\mathbf{0}, \mathbf{u}, \rho\mathbf{u}, \mathbf{v}, \rho\mathbf{v}, \rho^{-1}\mathbf{v}\}$. If $\mathbf{w} \in V \cap A$, then there exists $\mathbf{w}_1, \mathbf{w}_2 \in (V \cap A) \setminus \{\mathbf{w}\}$ for which the assumptions of Lemma 7.16 hold for $\mathbf{w}_1, \mathbf{w}, \mathbf{w}_2$. Then $\mathbf{w}$ is straight, and $\rho\mathbf{w} \in V \cap A$. The set $V \cap A$ is therefore preserved by ρ. By observing that V is the only nonempty subset of V that is preserved by ρ, it follows that $V \subset A$ and that $\mathbf{w}$ is straight for all $\mathbf{w} \in V$.

By Lemma 7.15, V is cyclic with respect to a unit vector $\mathbf{e}$ orthogonal to A. The azimuth cycle on V is $\mathbf{v} \mapsto \rho\mathbf{v}$.

We turn to the final conclusion. By the final conclusion of Lemma 7.15 and Lemma 2.81, the azimuth angle is less than π. Under this constraint, the azimuth angle equals the dihedral angle by Lemma 2.80. By definition, the dihedral angle is the angle $\operatorname{arc}_V(\mathbf{0}, *)$ of an orthogonal projection of $\{\mathbf{v}, \rho\mathbf{v}\}$ to a plane with normal $\mathbf{e}$. But $\{\mathbf{v}, \rho\mathbf{v}\}$ is already a subset of the plane A, so that the projection is the identity map, and the dihedral angle is $\operatorname{arc}_V(\mathbf{0}, \{\mathbf{v}, \rho\mathbf{v}\})$. □

Lemma 7.18 (lunar geometry) *Let (V, E, F) be a lunar fan with pole $\{\mathbf{v}, \mathbf{w}\} \subset V$. Then*

1. *$\mathbf{u}$ is straight for all $\mathbf{u} \in V \setminus \{\mathbf{v}, \mathbf{w}\}$.*
2. $0 < \angle(\mathbf{v}) = \angle(\mathbf{w}) \leq \pi$.
3. $V \cap \operatorname{aff}_+(\{\mathbf{0}, \mathbf{v}\}, \rho\mathbf{v}) = \{\mathbf{v}, \rho\mathbf{v}, \ldots, \mathbf{w}\}$.
4. $V \cap \operatorname{aff}_+(\{\mathbf{0}, \mathbf{v}\}, \rho^{-1}\mathbf{v}) = \{\mathbf{w}, \rho\mathbf{w}, \ldots \mathbf{v}\}$.

Proof Set $V_1 = \{\mathbf{v}, \rho\mathbf{v}, \ldots, \mathbf{w}\}$ and $V_2 = \{\mathbf{w}, \rho\mathbf{w}, \ldots, \mathbf{v}\}$. Let $\mathbf{u} \in V \setminus \{\mathbf{v}, \mathbf{w}\}$ be arbitrary. Apply Lemma 7.16 to the set $\{\mathbf{v}, \mathbf{u}, \mathbf{w}\}$ to find that $\mathbf{u}$ is straight and that $\{\mathbf{0}, \mathbf{u}, \rho\mathbf{u}, \rho^{-1}\mathbf{u}\}$ belongs to a plane $A(\mathbf{u})$. Now $A(\mathbf{u})$ and $A(\rho\mathbf{u})$ are both the unique plane containing $\{\mathbf{0}, \mathbf{u}, \rho\mathbf{u}\}$; hence, $A(\mathbf{u}) = A(\rho\mathbf{u})$ when $\rho\mathbf{u} \notin \{\mathbf{v}, \mathbf{w}\}$. By induction, there are planes A_1, A_2 such that $V_i \subset A_i$. There is an azimuth cycle σ_i on V_i such that $\sigma_i\mathbf{u} = \rho\mathbf{u}$, when $\mathbf{u} \in A_i \setminus \{\mathbf{v}, \mathbf{w}\}$.

The angles $\angle(\mathbf{v})$ and $\angle(\mathbf{w})$ are both equal to the dihedral angle between the half-planes $\operatorname{aff}_+(\{\mathbf{0}, \mathbf{v}\}, \rho^{\pm}\mathbf{v})$. In particular, $0 < \angle(\mathbf{v}) = \angle(\mathbf{w}) \leq \pi$. □

Lemma 7.19 (monotonicity) *Let (V, E, F) be a nonreflexive local fan and let k be the cardinality of F. Fix $\mathbf{v}_0 \in V$. Assume that $\{\mathbf{0}, \mathbf{v}_0, \mathbf{u}\}$ is not collinear for any $\mathbf{u} \in V \setminus \{\mathbf{v}_0\}$. For all i, set $\mathbf{v}_i = \rho^i\mathbf{v}_0$ and $\beta(i) = \operatorname{azim}(\mathbf{0}, \mathbf{v}_0, \mathbf{v}_1, \mathbf{v}_i)$. Then*

$$0 = \beta(1) \leq \beta(2) \leq \cdots \leq \beta(k-1) \leq \pi.$$

(See Figure 7.3.) Moreover, if $\beta(i) = 0$ for some $1 < i \leq k - 1$, then

$$\angle(\mathbf{v}_1) = \cdots = \angle(\mathbf{v}_{i-1}) = \pi,$$

and $\{\mathbf{v}_1, \ldots, \mathbf{v}_i\} \subset \text{aff}^0_+(\{\mathbf{0}, \mathbf{v}_0\}, \mathbf{v}_1)$. *Finally, if* $\beta(i) = \beta(k-1)$ *for some* $1 \le i < k-1$, *then*

$$\angle(\mathbf{v}_{i+1}) = \cdots = \angle(\mathbf{v}_{k-1}) = \pi, \quad \text{and } \{\mathbf{v}_i, \ldots, \mathbf{v}_{k-1}\} \subset \text{aff}^0_+(\{\mathbf{0}, \mathbf{v}_0\}, \mathbf{v}_{k-1}).$$

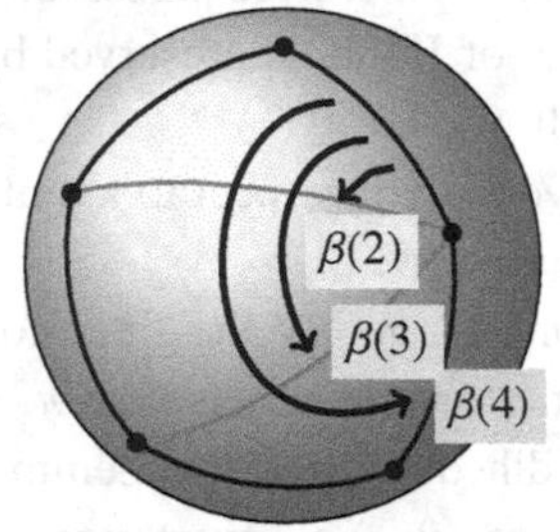

Figure 7.3 The angles $\beta(k)$ are increasing in k on a nonreflexive local fan. The three arrows mark the angles $\beta(2)$, $\beta(3)$, and $\beta(4)$.

Proof With respect to a frame, the points $\mathbf{v}_j$ can be represented in spherical coordinates (r_j, θ_j, ϕ_j). In an appropriate frame, $\phi_0 = 0$ and $\theta_j = \beta(j)$ for all j. From $\mathbf{v}_j \in W_{\text{dart}}(F, \mathbf{v}_0)$ and $\angle(\mathbf{v}_0) \le \pi$, it follows that $0 \le \theta_j \le \theta_{k-1} \le \pi$ when $0 \le j \le k-1$.

One may assume the induction hypothesis that $0 \le \beta(1) \le \cdots \le \beta(i)$. The condition

$$\mathbf{v}_0 \in W_{\text{dart}}(F, \mathbf{v}_i)$$

implies that

$$0 \le \text{azim}(\mathbf{0}, \mathbf{v}_i, \mathbf{v}_{i+1}, \mathbf{v}_0) \le \text{azim}(\mathbf{0}, \mathbf{v}_i, \mathbf{v}_{i+1}, \mathbf{v}_{i-1}) \le \pi.$$

By Lemma 2.81, the resulting inequality

$$\sin(\text{azim}(\mathbf{0}, \mathbf{v}_i, \mathbf{v}_{i+1}, \mathbf{v}_0)) \ge 0$$

reduces to a triple-product:

$$(\mathbf{v}_0 \times \mathbf{v}_i) \cdot \mathbf{v}_{i+1} \ge 0.$$

In spherical coordinates, this inequality becomes

$$r_0 r_i r_{i+1} \sin\phi_i \sin\phi_{i+1} \sin(\theta_{i+1} - \theta_i) \ge 0. \tag{7.20}$$

Under the noncollinearity assumption, $\sin\phi_i \sin\phi_{i+1} \ne 0$ (when $0 < i < k-1$). Once we deal with the degenerate case $\theta_{i+1} = 0, \theta_i = \pi$, these inequalities give $\theta_i \le \theta_{i+1}$, and the result follows by induction.

Turn to the degenerate case $\theta_{i+1} = 0, \theta_i = \pi$. In this case, the set $\{\mathbf{0}, \mathbf{v}_0, \mathbf{v}_i, \mathbf{v}_{i+1}\}$

is coplanar. Let $C_+^0 = C_+^0\{\mathbf{v}_i, \mathbf{v}_{i+1}\}$. The values of the angles θ_i and θ_{i+1} imply that C_+^0 meets the line $\mathrm{aff}\{\mathbf{0}, \mathbf{v}_0\}$. In particular, $\epsilon\mathbf{v}_0 \in C_+^0$ for some choice of sign $\epsilon \in \{\pm 1\}$. The definition of a fan implies $\mathbf{v}_0 \notin C_+^0$. Hence $-\mathbf{v}_0 \in C_+^0$. By the definition of a circular fan, (V, E, F) is circular. The lemma follows in this case from the explicit description of circular fans in Lemma 7.17. This completes the proof of the first statement of the lemma.

Assume that $\beta(i) = \theta_i = 0$ for some $1 < i \le k-1$. Then by the first conclusion, $\beta(j) = 0$ for $0 \le j \le i$. That is, $\mathbf{v}_1, \ldots, \mathbf{v}_i$ all lie in the half-plane $\mathrm{aff}_+^0(\{\mathbf{0}, \mathbf{v}_0\}, \mathbf{v}_1)$. In particular, they are coplanar. A chaining argument based on Lemma 7.16 gives the result.

The final conclusion follows by a similar chaining argument. □

7.2 Modification

7.2.1 deformation

This subsection develop a theory of deformations of a nonreflexive local fan (V, E, F), including sufficient conditions for the deformation of a nonreflexive local fan to remain a nonreflexive local fan.

Definition 7.21 (deformation) A *deformation* of a nonreflexive local fan (V, E, F) over an interval $I \subset \mathbb{R}$ is a function $\varphi : V \times I \to \mathbb{R}^3$ with the following properties.

1. $\varphi(v, *) : I \to \mathbb{R}^3$ is continuous for each $v \in V$,
2. $0 \in I$, and
3. $\varphi(\mathbf{v}, 0) = \mathbf{v}$ for all $\mathbf{v} \in V$.

Notation 7.22 Beware of the notational distinction between the zenith angle ϕ and the deformation φ. When a deformation φ is given, write $\mathbf{v}(t)$ as an abbreviation of $\varphi(\mathbf{v}, t)$ for $t \in I$. Also, set

$$V(t) = \{\mathbf{v}(t) \;:\; \mathbf{v} \in V\},$$
$$E(t) = \{\{\mathbf{v}(t), \mathbf{w}(t)\} \;:\; \{\mathbf{v}, \mathbf{w}\} \in E\},$$
$$F(t) = \{(\mathbf{v}(t), \mathbf{w}(t)) \;:\; (\mathbf{v}, \mathbf{w}) \in F\}.$$

A deformation does not require $(V(t), E(t), F(t))$ to be a nonreflexive local fan for all $t \in I$, although this is often the case. The permutation $\rho : V \to V$ gives $\varphi(\rho\mathbf{v}, t) \in V(t)$ for every $\mathbf{v} \in V$.

The following three lemmas give conditions ensuring that properties of fans are preserved under deformation.

Lemma 7.23 (local deformation) *Let (φ, V, I) be a deformation of a local fan (V, E, F) over an interval I. Then $(V(t), E(t), F(t))$ is a local fan for all sufficiently small $t \in I$.*

Lemma 7.24 (lunar deformation) *Let (φ, V, I) be a deformation of a lunar fan (V, E, F) with pole $\{\mathbf{v}, \mathbf{w}\}$ over an interval I. Assume that $\angle(\mathbf{w}) < \pi$. Suppose that there is an index i such that $\mathbf{v}_i$ is straight and such that $\mathbf{v}_j$ remains fixed for $j \neq i$. Suppose that $\mathbf{v}_i(t)$ remains in the plane through $\{\mathbf{0}, \mathbf{v}, \mathbf{w}, \mathbf{v}_i\}$. Then $(V(t), E(t), F(t))$ is a lunar fan for all sufficiently small $t \in I$.*

Lemma 7.25 (generic deformation) *Let (φ, V, I) be a deformation of a generic local fan (V, E, F) over an interval I. Assume that the azimuth angle of $\mathbf{v}_i(t)$ is at most π for all sufficiently small $t \in I$, whenever $\mathbf{v}_i$ is straight. Then $(V(t), E(t), F(t))$ is a generic local fan for all sufficiently small $t \in I$.*

Proof (7.23) We examine in turn each of the defining properties of a local fan.

(CARDINALITY) The set $V(t)$ is the image of V and is therefore finite and nonempty.

(ORIGIN) Since φ is continuous and $\mathbf{0} \notin V$, it follows that $\mathbf{0} \notin V(t)$ for sufficiently small t.

(NONPARALLEL) If $\mathbf{v}, \mathbf{w}$ are nonparallel, then $\mathbf{v}(t)$ and $\mathbf{w}(t)$ are nonparallel for sufficiently small t.

(INTERSECTION) If $\varepsilon \cap \varepsilon' = \emptyset$, then $C(\varepsilon) \cap S^2$ has a positive distance from $C(\varepsilon') \cap S^2$, where S^2 is a unit sphere. Hence, for sufficiently small times, the deformation of these sets remain disjoint. If $\varepsilon = \{\mathbf{u}, \mathbf{v}\}$ and $\varepsilon' = \{\mathbf{v}, \mathbf{w}\}$, where $\mathbf{u} \neq \mathbf{w}$, then again the deformation of $C(\varepsilon) \cap C(\varepsilon')$ is $C(\{\mathbf{v}(t)\})$ for sufficiently small t. The other cases are similar.

(FACE), (DIHEDRAL) The azimuth cycle on $E(\mathbf{v}(t))$ is preserved; hence, the combinatorial properties of the hypermap do not change when t is sufficiently small. □

proof (7.24) (7.25) The proofs of these two lemmas may be combined. The main part of the proof shows that the local fans obtained by deformation are nonreflexive. From there, the proof is completed in two cases as follows.

(LUNAR) By assumption, the poles $\{\mathbf{v}, \mathbf{w}\}$ of the lunar fan are not straight. The poles remain fixed poles under the deformation and remain unstraightened. This shows that the deformation of a lunar fan is lunar.

(GENERIC) Genericity is stated as a finite collection of open conditions $\mathbf{v} \notin C\{\mathbf{u}, \mathbf{w}\}$. These conditions continue to hold for sufficiently small t.

Now we return to the proof of nonreflexivity (page 195).

(ANGLE) If $\text{azim}(x) < \pi$, then the inequality remains strict for sufficiently small t. If $\text{azim}(x) = \pi$, then the straightness assumptions of the lemmas give the inequality $\text{azim}(x) \le \pi$ for all sufficiently small t.

(WEDGE) We consider two cases, depending on the type of the local fan.

If the fan (V, E, F) is lunar with pole $\{\mathbf{v}, \mathbf{w}\}$, the assumptions of the lemma make $W_{\text{dart}}(x(t))$ independent of t for every dart x of F. Then $\mathbf{v}_i(t)$ (for the special index i that appears in the statement of the lemma) remains in the half-plane $\text{aff}^0_+(\{\mathbf{v}, \mathbf{w}\}, \mathbf{v}_i)$, which is a subset of $W_{\text{dart}}(x)$, for every $x \in F$. The result ensues.

Finally consider a generic fan (V, E, F). For simplicity, write $W_{\text{dart}}(\mathbf{u})$ for $W_{\text{dart}}(x)$, where $x = (\mathbf{u}, \rho\mathbf{u}) \in F$. The property $\mathbf{v} \in W^0_{\text{dart}}(x)$ is an open condition. It holds for sufficiently small t. The proof then reduces to the case $\mathbf{v} \in W_{\text{dart}}(\mathbf{u}) \setminus W^0_{\text{dart}}(\mathbf{u})$, for some $\mathbf{u}, \mathbf{v} \in V$. By Lemma 7.19, $\mathbf{u}$ and $\mathbf{v}$ can be connected by a sequence

$$\rho\mathbf{u}, \rho^2\mathbf{u}, \ldots \rho^r\mathbf{u} \quad \text{or} \quad \rho\mathbf{v}, \rho^2\mathbf{v}, \ldots, \rho^r\mathbf{v},$$

where all intermediate terms are straight.

Thus, it suffices to prove the following statement by an induction on r: for all $\mathbf{v}, \mathbf{w} \in V$, we have $\mathbf{v}(t) \in W_{\text{dart}}(\mathbf{w}(t))$ and $\mathbf{w}(t) \in W_{\text{dart}}(\mathbf{v}(t))$, where $\mathbf{w} = \rho^r\mathbf{v}$, provided $\rho^i\mathbf{v}$ is straight for all $0 < i < r$. If $r = 1$, the statement is a triviality because the azimuth angle spanned by $W_{\text{dart}}(\mathbf{v}(t))$ is defined by the azimuth angle of $(\mathbf{0}, \mathbf{v}(t), \rho\mathbf{v}(t), \rho^{-1}\mathbf{v}(t))$.

Now assume the induction hypothesis holds for all numbers less than r. Fix $\mathbf{v}$ and write $\mathbf{v}_i = \rho^i\mathbf{v}$. We show that $\mathbf{v}_r \in W_{\text{dart}}(\mathbf{v}_0)$, leaving the symmetrical claim $\mathbf{v}_0 \in W_{\text{dart}}(\mathbf{v}_r)$ to the reader. We return to the coordinate representation that was used in the proof of Lemma 7.19, which applies to generic fans. With respect to a frame, the points $\mathbf{v}_j$ can be represented in spherical coordinates (r_j, θ_j, ϕ_j). In an appropriate frame, $\phi_0 = 0$ and $\theta_j = \beta(j)$ for all j.

From the induction hypothesis $\mathbf{v}_0(t) \in W_{\text{dart}}(\mathbf{v}_{r-1}(t))$ and the assumption $\text{azim}(x(t)) \le \pi$ of the lemma, we find

$$0 \le \text{azim}(\mathbf{0}, \mathbf{v}_{r-1}(t), \mathbf{v}_r(t), \mathbf{v}_0(t)) \le \text{azim}(\mathbf{0}, \mathbf{v}_{r-1}(t), \mathbf{v}_r(t), \mathbf{v}_{r-2}(t)) \le \pi.$$

By relation (7.20), this gives

$$\sin(\theta_r(t) - \theta_{r-1}(t)) \ge 0. \tag{7.26}$$

When $t = 0$, we have $\theta_j(0) = \beta(j) = 0$, for $j = 1, \ldots, r$. By continuity, when t is sufficiently small, $\theta_r(t) - \theta_{r-1}(t)$ is near zero, so that (7.26) gives

$\theta_r(t) \geq \theta_{r-1}(t)$. The induction hypothesis $\mathbf{v}_{r-1}(t) \in W_{\text{dart}}(\mathbf{v}_0(t))$ gives $\theta_{r-1}(t) \geq 0$. Hence $\theta_r(t) \geq 0$.

The inequality $\theta_r(t) < \theta_{k-1}(t)$ holds at $t = 0$ and by continuity for small t. We obtain

$$0 \leq \theta_r(t) \leq \theta_{k-1}(t).$$

This is precisely the desired relation $\mathbf{v}_r(t) \in W_{\text{dart}}(\mathbf{v}_0(t))$, expressed in spherical coordinates. □

7.2.2 slicing

This subsection shows that a nonreflexive local fan can be sliced along a internal blade to divide it into two nonreflexive local fans.

Lemma 7.27 *Let (V, E, F) be a nonreflexive local fan. If $\mathbf{v}, \mathbf{w} \in V$ are nonparallel, then $C\{\mathbf{v}, \mathbf{w}\} \subset W_{dart}(x)$ for any dart $x \in F$.*

Proof This is an elementary consequence of the cone shape of $W_{\text{dart}}(x)$, the condition that $V \subset W_{\text{dart}}(x)$, and definitions. □

Definition 7.28 (slice) Let (V, E, F) be a nonreflexive local fan. Assume that $\mathbf{v}, \mathbf{w} \in V$ are nonparallel and that $(V, E') = (V, E \cup \{\{\mathbf{v}, \mathbf{w}\}\})$ is a fan. Let F' be the face of $\text{hyp}(V, E')$ containing the dart $(\mathbf{w}, \mathbf{v})$. Write

$$(V[\mathbf{v}, \mathbf{w}], E[\mathbf{v}, \mathbf{w}], F[\mathbf{v}, \mathbf{w}])$$

for the localization of (V, E') along F', where

$$\begin{aligned}
V[\mathbf{v}, \mathbf{w}] &= \{\mathbf{v}, \rho\mathbf{v}, \rho^2\mathbf{v}, \ldots, \mathbf{w}\},\\
E[\mathbf{v}, \mathbf{w}] &= \{\{\mathbf{v}, \rho\mathbf{v}\}, \ldots, \{\rho^{-1}\mathbf{w}, \mathbf{w}\}, \{\mathbf{w}, \mathbf{v}\}\},\\
F[\mathbf{v}, \mathbf{w}] &= \{(\mathbf{v}, \rho\mathbf{v}), (\rho\mathbf{v}, \rho^2\mathbf{v}), \ldots, (\rho^{-1}\mathbf{w}, \mathbf{w}), (\mathbf{w}, \mathbf{v})\}.
\end{aligned}$$

The triple $(V[\mathbf{v}, \mathbf{w}], E[\mathbf{v}, \mathbf{w}], F[\mathbf{v}, \mathbf{w}])$ is called the *slice* of (V, E, F) along $(\mathbf{v}, \mathbf{w})$. See Figure 7.4.

To allow contexts with more than one nonreflexive local fan (V, E, F), we extend the notation, writing $\angle(H, \mathbf{v})$ for $\angle(\mathbf{v})$ in the hypermap H. Similarly, we write $W^0_{\text{dart}}(H, \mathbf{v})$ for $W^0_{\text{dart}}(x)$ and so forth.

Lemma 7.29 (slicing) *Let (V, E, F) be a nonreflexive local fan with hypermap H. Select $\mathbf{v}, \mathbf{w} \in V$ such that $\mathbf{v}$ is not parallel with $\mathbf{w}$. Assume that $C^0\{\mathbf{v}, \mathbf{w}\} \subset W^0_{dart}(x)$ for all darts $x \in F$. Then*

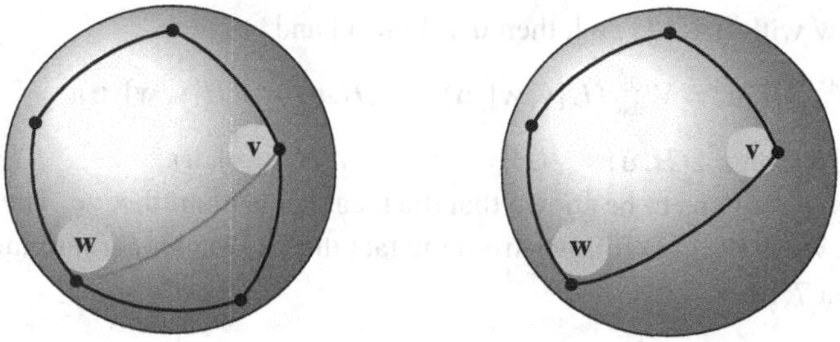

Figure 7.4 A slice of a fan along $(\mathbf{v}, \mathbf{w})$.

1. *$(V[\mathbf{v},\mathbf{w}], E[\mathbf{v},\mathbf{w}], F[\mathbf{v},\mathbf{w}])$ and $(V[\mathbf{w},\mathbf{v}], E[\mathbf{w},\mathbf{v}], F[\mathbf{w},\mathbf{v}])$ are nonreflexive local fans.*
2. *Let $H[\mathbf{v},\mathbf{w}]$ and $H[\mathbf{w},\mathbf{v}]$ be the hypermaps of these two nonreflexive local fans, respectively. Let $g : V \to \mathbb{R}$ be any function. Then*

$$\sum_{\mathbf{v}\in V} g(\mathbf{v})\angle(H,\mathbf{v}) = \sum_{\mathbf{v}\in V[\mathbf{v},\mathbf{w}]} g(\mathbf{v})\angle(H[\mathbf{v},\mathbf{w}],\mathbf{v}) + \sum_{\mathbf{v}\in V[\mathbf{w},\mathbf{v}]} g(\mathbf{v})\angle(H[\mathbf{w},\mathbf{v}],\mathbf{v}).$$

Proof *For each $\mathbf{u} \in \{\mathbf{v},\mathbf{w}\}$, the node $\mathbf{u}$ is not parallel with any element of* $V \setminus \{\mathbf{u}\}$. Otherwise, say that $\mathbf{u}$ and $\mathbf{v}$ are parallel. Then the fan is circular or lunar with pole $\{\mathbf{u},\mathbf{v}\}$. By Lemmas 7.17 and 7.18,

$$\mathbf{w} \in \text{aff}_+(\{\mathbf{0},\mathbf{v}\},\rho\mathbf{v}) \cup \text{aff}_+(\{\mathbf{0},\mathbf{v}\},\rho^{-1}\mathbf{v}).$$

This gives $C^0\{\mathbf{v},\mathbf{w}\} \not\subset W^0_{\text{dart}}(x)$, where x is the dart of F at $\mathbf{v}$. This contradicts an assumption.

(V, E') is a fan, where $E' = E \cup \{\{\mathbf{v},\mathbf{w}\}\}$. Indeed, except for the intersection property, all of the properties of a fan follow trivially from the fact that (V, E) is a fan and that $\mathbf{v}$ and $\mathbf{w}$ are nonparallel. (Note the similarity with Lemma 5.22.) The intersection property also is trivial except in the case $\varepsilon = \{\mathbf{v},\mathbf{w}\}$ and $\varepsilon' \setminus \varepsilon \neq \emptyset$. Select $\mathbf{u} \in \varepsilon' \setminus \varepsilon$. It follows from the node partition of Lemma 5.15 that

$$\begin{aligned} C(\varepsilon) \cap C(\varepsilon') &= (C(\mathbf{v}) \cap C(\varepsilon')) \cup (C(\mathbf{w}) \cap C(\varepsilon')) \\ &= C(\{\mathbf{v}\} \cap \varepsilon') \cup C(\{\mathbf{w}\} \cap \varepsilon') \\ &= C(\{\mathbf{v},\mathbf{w}\} \cap \varepsilon'). \end{aligned}$$

The intersection property thus holds and (V, E') is a fan.

It follows by Lemma 7.8 that $(V[\mathbf{v},\mathbf{w}], E[\mathbf{v},\mathbf{w}], F[\mathbf{v},\mathbf{w}])$ is a local fan.

The second conclusion of the lemma follows from the following identities.

If $\mathbf{u} \neq \mathbf{v}, \mathbf{w}$ with $\mathbf{u} \in V[\mathbf{v}, \mathbf{w}]$, then $\mathbf{u} \notin V[\mathbf{w}, \mathbf{v}]$ and

$$W^0_{\text{dart}}(H, \mathbf{u}) = W^0_{\text{dart}}(H[\mathbf{v}, \mathbf{w}], \mathbf{u}), \quad \angle(H, \mathbf{u}) = \angle(H[\mathbf{v}, \mathbf{w}], \mathbf{u}). \tag{7.30}$$

If $\mathbf{u} \in \{\mathbf{v}, \mathbf{w}\}$, then $\angle(H, \mathbf{u}) = \angle(H[\mathbf{v}, \mathbf{w}], \mathbf{u}) + \angle(H[\mathbf{w}, \mathbf{v}], \mathbf{u})$.

Finally, it remains to be shown that the local fan is nonreflexive. The conclusion $V[\mathbf{v}, \mathbf{w}] \subset W_{\text{dart}}(x)$ follows from the fact that the angles $\beta(i)$ are increasing in Lemma 7.19. □

Definition 7.31 When (V, E, F) is a local fan, set

$$\text{sol}(V, E, F) = 2\pi + \sum_{x \in F} (\text{azim}(x) - \pi) = 2\pi + \sum_{\mathbf{v} \in V} (\angle(\mathbf{v}) - \pi).$$

When $\text{card}(V) = 3$, this definition reduces to Girard's formula for the solid angle of a triangle. For conforming fans, the definition reduces to the (SOLID ANGLE) formula of U_F for conforming fans (page 126). However, we cannot apply the results from Chapter 5 directly, because a local fan is not conforming: (V, E) is not fully surrounded. Instead, we rely on the following lemma.

Lemma 7.32 *Let (V, E, F) be a generic nonreflexive local fan. Then*

$$\text{sol}(V, E, F) \geq 0.$$

Proof We argue by complete induction on $k = \text{card}(V)$. The base case of the induction is $k = 3$. In this case, the formula reduces to Girard's formula (Lemma 3.23) for the solid angle of the triangle $\text{aff}^0_+(\mathbf{0}, V)$, which is certainly nonnegative.

When $k \geq 4$, if we can find distinct $\mathbf{v}, \mathbf{w} \in V$ such that $C^0\{\mathbf{v}, \mathbf{w}\} \subset W^0_{\text{dart}}(x)$, then we may apply Lemma 7.29 and use the induction hypothesis to write $\text{sol}(V, E, F)$ as the sum of two nonnegative terms, to obtain the lemma. We may therefore assume that for all $\mathbf{v}, \mathbf{w} \in V$, $C^0\{\mathbf{v}, \mathbf{w}\} \not\subset W^0_{\text{dart}}(x)$.

We may assume that some $\mathbf{v} \in V$ is not straight. Otherwise, it follows trivially from definitions that $\text{sol}(V, E, F) = 2\pi > 0$.

Select any $\mathbf{v} \in V$ that is not straight, and write $\mathbf{v}_i = \rho^i \mathbf{v}$. By Lemma 7.19 and the assumptions, it follows that there is some $i : 1 \leq i \leq k - 2$ such that $\mathbf{v}_j$ is straight, for all $j \neq 0, i, i + 1$. In particular, there are at most three unstraightened elements of V. By Girard's formula, the value of $\text{sol}(V, E, F)$ is equal to the solid angle of

$$\text{aff}^0_+(\mathbf{0}, \{\mathbf{v}_0, \mathbf{v}_i, \mathbf{v}_{i+1}\}),$$

which is nonnegative. □

7.3 Polarity

7.3.1 construction

This section constructs a polar fan (V', E', F') from a local fan (V, E, F).

Definition 7.33 (polar) Let (V, E, F) be a local fan, with permutation $\rho = \rho_{V,E,F} : V \to V$. Define a map $(') : V \to \mathbb{R}^3$ by $\mathbf{v} \mapsto \mathbf{v}' = \mathbf{v} \times \rho\mathbf{v}$. Set

$$V' = \{\mathbf{v}' \;:\; \mathbf{v} \in V\}$$
$$E' = \{\{\mathbf{v}', (\rho\mathbf{v})'\} \;:\; \mathbf{v} \in V\}$$
$$F' = \{(\mathbf{v}', (\rho\mathbf{v})') \;:\; \mathbf{v} \in V\}.$$

The triple (V', E', F') is called the *polar* of (V, E, F). See Figure 7.5.

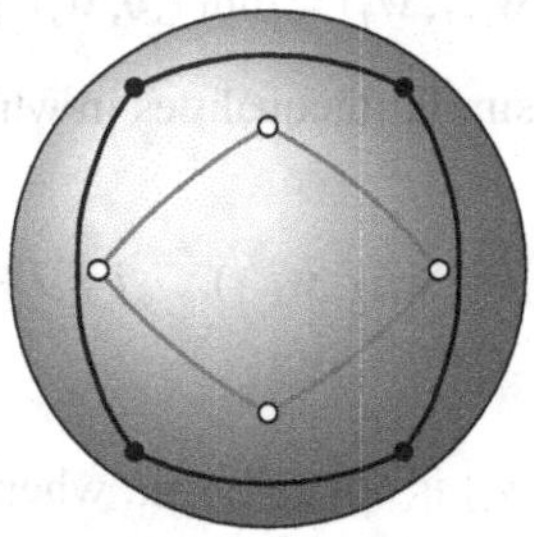

Figure 7.5 A quadrilateral and its polar.

The properties of the polar are established in the following lemma.

Lemma 7.34 *Let (V, E, F) be a generic nonreflexive local fan. Assume that $\angle(\mathbf{v}) < \pi$ for every $\mathbf{v} \in V$. Then (V', E', F') is a generic nonreflexive local fan satisfying* $\text{card}(V') = \text{card}(V)$. *Moreover, for every $\mathbf{v} \in V$,*

$$\text{arc}_V(\mathbf{0}, \{\mathbf{v}', (\rho\mathbf{v})'\}) = \pi - \angle(\rho\mathbf{v}) \in (0, \pi),$$
$$\text{arc}_V(\mathbf{0}, \{\mathbf{v}, \rho\mathbf{v}\}) = \pi - \angle'(\mathbf{v}') \in (0, \pi),$$

where $\angle'$ is the azimuth angle function on the nodes of the polar fan (V', E', F').

Proof Fix $\mathbf{v} \in V$ and write $\mathbf{v}_i = \rho^i\mathbf{v}$ and $\mathbf{w}_i = \mathbf{v}_i \times \mathbf{v}_{i+1}$. In a generic fan, $\{\mathbf{0}, \mathbf{v}_i, \rho\mathbf{v}_i\}$ is not collinear. By Lemma 2.64, $\mathbf{w}_i \neq \mathbf{0}$.

The proof makes repeated use of the cross product identities of Lemmas 2.65 and 2.81, without further mention. Recall that $(\sim)$ is the equivalence relation on $\mathbb{R}$ given by $x \sim y$ when there exists $t > 0$ such that $x = ty$. Abbreviate $(\mathbf{u} \times \mathbf{v}) \cdot \mathbf{w}$ to $[\mathbf{u}, \mathbf{v}, \mathbf{w}]$.

The sign of $\sin(\mathrm{azim}(\mathbf{0}, \mathbf{w}_i, \mathbf{w}_{i+1}, \mathbf{w}_j))$, when $j \neq i, i+1$, is determined by a calculation:

$$\begin{aligned}\sin(\mathrm{azim}(\mathbf{0}, \mathbf{w}_i, \mathbf{w}_{i+1}, \mathbf{w}_j)) &\sim (\mathbf{w}_i \times \mathbf{w}_{i+1}) \cdot \mathbf{w}_j \\ &= ((\mathbf{v}_i \times \mathbf{v}_{i+1}) \times (\mathbf{v}_{i+1} \times \mathbf{v}_{i+2})) \cdot \mathbf{w}_j \\ &= (\mathbf{v}_{i+1}[\mathbf{v}_i, \mathbf{v}_{i+1}, \mathbf{v}_{i+2}]) \cdot (\mathbf{v}_j \times \mathbf{v}_{j+1}) \\ &= [\mathbf{v}_i, \mathbf{v}_{i+1}, \mathbf{v}_{i+2}][\mathbf{v}_j, \mathbf{v}_{j+1}, \mathbf{v}_{i+1}] \\ &\sim \sin(\mathrm{azim}(\mathbf{0}, \mathbf{v}_{i+1}, \mathbf{v}_{i+2}, \mathbf{v}_i)) \sin(\mathrm{azim}(\mathbf{0}, \mathbf{v}_j, \mathbf{v}_{j+1}, \mathbf{v}_{i+1})) \\ &\sim \sin \angle(\mathbf{v}_{i+1}) \sin(\mathrm{azim}(\mathbf{0}, \mathbf{v}_j, \mathbf{v}_{j+1}, \mathbf{v}_{i+1})) \\ &\sim \sin(\mathrm{azim}(\mathbf{0}, \mathbf{v}_j, \mathbf{v}_{j+1}, \mathbf{v}_{i+1})).\end{aligned}$$

By Lemma 7.19, this final term is positive. We conclude by Lemma 2.80 that

$$0 < \mathrm{azim}(\mathbf{0}, \mathbf{w}_i, \mathbf{w}_{i+1}, \mathbf{w}_j) = \mathrm{dih}_V(\{\mathbf{0}, \mathbf{w}_i\}, \{\mathbf{w}_{i+1}, \mathbf{w}_j\}) < \pi.$$

When $j \neq i-1, i$, we have similar inequalities in which i is replaced with $i-1$. This implies that

$$\mathbf{w}_j \in W^0(\mathbf{0}, \mathbf{w}_i, \mathbf{w}_{i+1}, \mathbf{w}_{i-1}), \quad j \neq i-1, i, i+1. \tag{7.35}$$

We claim that

$$\{\mathbf{0}, \mathbf{w}_i, \mathbf{w}_j\} \text{ is not collinear, when } i \neq j. \tag{7.36}$$

Indeed, by Lemma 2.64, it is enough to show that $\mathbf{w}_i \times \mathbf{w}_j \neq \mathbf{0}$. We compute

$$\mathbf{w}_i \times \mathbf{w}_j = (\mathbf{v}_i \times \mathbf{v}_{i+1}) \times (\mathbf{v}_j \times \mathbf{v}_{j+1}) = [\mathbf{v}_i, \mathbf{v}_j, \mathbf{v}_{j+1}]\mathbf{v}_{i+1} - [\mathbf{v}_{i+1}, \mathbf{v}_j, \mathbf{v}_{j+1}]\mathbf{v}_i.$$

The coefficients $[\mathbf{v}_i, \mathbf{v}_j, \mathbf{v}_{j+1}]$ and $[\mathbf{v}_{i+1}, \mathbf{v}_j, \mathbf{v}_{j+1}]$ are nonzero by the preceding calculations. If $\mathbf{w}_i \times \mathbf{w}_j = \mathbf{0}$, then $\{\mathbf{0}, \mathbf{v}_i, \mathbf{v}_{i+1}\}$ is collinear. This contradicts the defining properties of the fan (V, E, F). This establishes the claim.

The calculation of $\mathrm{arc}_V(\mathbf{0}, \{\mathbf{w}_i, \mathbf{w}_{i+1}\})$ relies on Lemma 2.68:

$$\begin{aligned}\mathrm{arc}_V(\mathbf{0}, \{\mathbf{w}_i, \mathbf{w}_{i+1}\}) &= \mathrm{arc}_V(\mathbf{0}, \{\mathbf{v}_i \times \mathbf{v}_{i+1}, \mathbf{v}_{i+1} \times \mathbf{v}_{i+2}\}) \\ &= \pi - \mathrm{arc}_v(\mathbf{0}, \{\mathbf{v}_{i+1} \times \mathbf{v}_i, \mathbf{v}_{i+1} \times \mathbf{v}_{i+2}\}) \\ &= \pi - \mathrm{dih}_V(\{\mathbf{0}, \mathbf{v}_{i+1}\}, \{\mathbf{v}_i, \mathbf{v}_{i+1}\}) \\ &= \pi - \angle(\mathbf{v}_{i+1}).\end{aligned}$$

The calculation of $\angle'(\mathbf{w}_{i+1})$ also relies on Lemma 2.68:

$$\begin{aligned}\angle'(\mathbf{w}_{i+1}) &= \mathrm{dih}_V(\{\mathbf{0}, \mathbf{w}_{i+1}\}, \{\mathbf{w}_i, \mathbf{w}_{i+2}\})\\ &= \mathrm{arc}_V(\mathbf{0}, \{\mathbf{w}_{i+1} \times \mathbf{w}_i, \mathbf{w}_{i+1} \times \mathbf{w}_{i+2}\})\\ &= \mathrm{arc}_V(\mathbf{0}, \{(\mathbf{v}_{i+1} \times \mathbf{v}_{i+2}) \times (\mathbf{v}_i \times \mathbf{v}_{i+1}), (\mathbf{v}_{i+1} \times \mathbf{v}_{i+2}) \times (\mathbf{v}_{i+2} \times \mathbf{v}_{i+3})\})\\ &= \mathrm{arc}_V(\mathbf{0}, \{-\mathbf{v}_{i+1}[\mathbf{v}_i, \mathbf{v}_{i+1}, \mathbf{v}_{i+2}], \mathbf{v}_{i+2}[\mathbf{v}_{i+1}, \mathbf{v}_{i+2}, \mathbf{v}_{i+3}]\})\\ &= \mathrm{arc}_V(\mathbf{0}, \{-\mathbf{v}_{i+1}, \mathbf{v}_{i+2}\})\\ &= \pi - \mathrm{arc}_V(\mathbf{0}, \{\mathbf{v}_{i+1}, \mathbf{v}_{i+2}\}).\end{aligned}$$

It is a routine verification to check that (V', E', F') is a local fan. The verification uses the calculations (7.35) and (7.36) to give the separation properties (ORIGIN), (NONPARALLEL), and (INTERSECTION) in the definition of fan. This fan cannot be lunar, because (7.35) precludes poles. It cannot be circular because the azimuth angles $\angle'(\mathbf{v}')$ are less than π. It must be generic. We leave the remaining verifications to the reader. □

7.3.2 perimeter

The perimeter bound of 2π for convex spherical polygons is classical [37, p. 100]. This section proves the bound 2π on the perimeter of a nonreflexive local fan (Lemma 7.38). A great circle has perimeter 2π. The proof shows that the perimeter of the fan is related to the solid angle of the polar fan by duality. The upper bound on the perimeter is equivalent to the nonnegativity of the solid angle of the polar.

Definition 7.37 (perimeter) Let (V, E, F) be a nonreflexive local fan. Set

$$\mathrm{per}(V, E, F) = \sum_{i=0}^{k-1} \mathrm{arc}_V(\mathbf{0}, \{\rho^i\mathbf{v}, \rho^{i+1}\mathbf{v}\}),$$

where $k = \mathrm{card}(F)$. The right-hand side of this formula is easily seen to be independent of the choice of $\mathbf{v} \in V$. Call per the *perimeter* of the nonreflexive local fan. If $\mathbf{v}, \mathbf{w} \in V$ are distinct nodes, define the *partial perimeter*

$$\mathrm{per}(V, E, F, \mathbf{v}, \mathbf{w}) = \sum_{i=0}^{r-1} \mathrm{arc}_V(\mathbf{0}, \{\rho^i\mathbf{v}, \rho^{i+1}\mathbf{v}\}),$$

where r is chosen so that $\mathbf{w} = \rho^r\mathbf{v}$ and $0 < r \le k - 1$.

Lemma 7.38 (perimeter majorization) *The perimeter of every nonreflexive local fan is at most 2π.*

Proof *If the nonreflexive local fan (V, E, F) is circular, then its perimeter is* $\mathrm{per}(V, E, F) = 2\pi$. Indeed, by Lemma 7.17, the arcs making up the perimeter all lie in a common plane. The azimuth cycle on V coincides with $\rho : V \to V$. The sum of the terms in the formula defining the perimeter is the sum of the azimuth angles in the azimuth cycle. The sum is 2π by Lemma 2.94.

If the nonreflexive local fan is lunar, then its perimeter is $\mathrm{per}(V, E, F) = 2\pi$. Indeed, by Lemma 7.18, the set V is contained in the union of two half-planes. The perimeter is the sum of arcs in a half-circle in the first half-plane plus the sum of arcs in a half-circle in the second half-plane. This sum is 2π.

Finally, assume that the nonreflexive local fan is generic. Suppose for a contradiction that the lemma is false. Consider all counterexamples that minimize the cardinality of V.

A nonreflexive local fan (V, E, F) is determined by V and the cyclic permutation $\rho : V \to V$: $E = \{\{\mathbf{v}, \rho\mathbf{v}\} : \mathbf{v} \in V\}$ and $F = \{(\mathbf{v}, \rho\mathbf{v}) : \mathbf{v} \in V\}$.

In such a counterexample, if there is any straight dart $x = (\mathbf{v}, \mathbf{w}) \in F$, then there is a new nonreflexive local fan (V', E', F') with $V' = V \setminus \{\mathbf{v}\}$ and $\rho' : V' \to V'$ given by

$$\rho'(\mathbf{u}) = \begin{cases} \rho(\mathbf{u}), & \text{if } \rho(\mathbf{u}) \neq \mathbf{v}, \\ \rho(\mathbf{v}), & \text{if } \rho(\mathbf{u}) = \mathbf{v}. \end{cases}$$

This is a nonreflexive local fan with the same perimeter, contrary to the presumed minimality of the counterexample. Thus, in the minimal counterexample $\mathrm{azim}(x) < \pi$ for all $x \in F$.

Let (V', E', F') be the polar fan. Its solid angle, which is nonnegative, is given by Lemma 7.32. We find by Lemma 7.34 and the definition of perimeter that

$$\begin{aligned} 0 &\leq \mathrm{sol}(V', E', F') \\ &= 2\pi + \sum_{\mathbf{v}' \in V'} (\angle'(\mathbf{v}') - \pi) \\ &= 2\pi - \sum_{\mathbf{v} \in V} \mathrm{arc}_V(\mathbf{0}, \{\mathbf{v}, \rho\mathbf{v}\}) \\ &= 2\pi - \mathrm{per}(V, E, F). \end{aligned}$$

The lemma ensues. □

7.4 Main Estimate

Our aim becomes single-minded throughout the rest of the chapter; we wish to give a proof of the main estimate (Theorem 7.43). This, the most intricate proof in the book, requires substantial preparation. Assuming the existence of some counterexample to the main estimate, a compactness argument gives the existence of a minimal counterexample. The properties of minimal counterexamples are developed in a long sequence of lemmas. Eventually, enough properties of a minimal counterexample are established to conclude that it cannot exist.

7.4.1 statement of results

This subsection states the main results of the chapter.

Definition 7.39 (h_0, τ, dih_i) Let (V, E, F) be a nonreflexive local fan. Recall that $h_0 = 1.26$ and $L(h) = (h_0 - h)/(h_0 - 1)$, when $h \le h_0$. Set

$$\rho_0(y) = 1 + \frac{\mathrm{sol}_0}{\pi} \cdot \frac{y-2}{2h_0 - 2} = 1 + \frac{\mathrm{sol}_0}{\pi}(1 - L(y/2)),$$

$$\tau(V, E, F) = \sum_{x \in F} \rho_0(\|\mathrm{node}(x)\|)\mathrm{azim}(x) + (\pi + \mathrm{sol}_0)(2 - k(F)),$$

where $\mathrm{sol}_0 = 3\arccos(1/3) - \pi \approx 0.551$ is the solid angle of a spherical equilateral triangle of side $\pi/3$, and $k(F)$ is the cardinality of F. Let

$$\tau_{tri}(y_1, y_2, y_3, y_4, y_5, y_6) = \sum_{i=1}^{3} \rho_0(y_i)\, \mathrm{dih}_i(y_1, \ldots, y_6) - (\pi + \mathrm{sol}_0), \qquad (7.40)$$

where

$$\begin{aligned}
\mathrm{dih}_1(y_1, y_2, y_3, y_4, y_5, y_6) &= \mathrm{dih}(y_1, y_2, y_3, y_4, y_5, y_6),\\
\mathrm{dih}_2(y_1, y_2, y_3, y_4, y_5, y_6) &= \mathrm{dih}(y_2, y_3, y_1, y_5, y_6, y_4), \text{ and}\\
\mathrm{dih}_3(y_1, y_2, y_3, y_4, y_5, y_6) &= \mathrm{dih}(y_3, y_1, y_2, y_6, y_4, y_5). \qquad (7.41)
\end{aligned}$$

Definition 7.42 (standard, protracted, diagonal) Let (V, E) be a fan. We write $\|\varepsilon\|$ for $\|\mathbf{v} - \mathbf{w}\|$, when $\varepsilon = \{\mathbf{v}, \mathbf{w}\} \subset V$. We say that ε is *standard* if

$$2 \le \|\varepsilon\| \le 2h_0.$$

We say that ε is *protracted* if

$$2h_0 \le \|\varepsilon\| \le \sqrt{8}.$$

If $\mathbf{v}, \mathbf{w} \in V$ are distinct, and $\varepsilon = \{\mathbf{v}, \mathbf{w}\}$ is not an edge in E, then we call ε a *diagonal* of the fan.

Theorem 7.43 (main estimate) *Let (V, E, F) be a nonreflexive local fan (Definition 7.2). We make the following additional assumptions on (V, E, F).*

1. (PACKING) *V is a packing. That is, for every $\mathbf{v}, \mathbf{w} \in V$, if $\|\mathbf{v} - \mathbf{w}\| < 2$, then $\mathbf{v} = \mathbf{w}$.*
2. (ANNULUS) *$V \subset \mathcal{B}$.*
3. (DIAGONAL) *For all distinct elements $\mathbf{v}, \mathbf{w} \in V$, if $\{\mathbf{v}, \mathbf{w}\} \notin E$, then*
$$\|\mathbf{v} - \mathbf{w}\| \geq 2h_0.$$
4. (CARD) *Let $k = \text{card}(E) = \text{card}(F)$. Then $3 \leq k \leq 6$.*

In this context, we have the following conclusions.

1. *Assume $k \geq 4$. If every edge of E is standard, then*
$$\tau(V, E, F) \geq d(k), \text{ where } d(k) = \begin{cases} 0.206, & \text{if } k = 4, \\ 0.4819, & \text{if } k = 5, \\ 0.712, & \text{if } k = 6. \end{cases}$$
2. *Assume $k = 5$. Assume that every edge of E is standard. Assume that every diagonal ε of the fan satisfies $\|\varepsilon\| \geq \sqrt{8}$. Then*
$$\tau(V, E, F) \geq 0.616.$$
3. *Assume $k = 5$. Assume there exists some protracted edge in E and that the other four are standard. Then*
$$\tau(V, E, F) \geq 0.616.$$
4. *Finally, assume that $k = 4$. Assume that there exists some protracted edge in E and that the other three are standard. Assume that both diagonals ε of the fan satisfy $\|\varepsilon\| \geq \sqrt{8}$. Then*
$$\tau(V, E, F) \geq 0.477.$$

There are two related inequalities that we will prove separately. For that reason, we state them as a separate lemma.

Lemma 7.44 *Let (V, E, F) be a nonreflexive local fan. Under the same hypotheses on (V, E, F) as in Theorem 7.43,*

1. *Assume $k = 3$. Then*
$$\tau(V, E, F) \geq 0.$$

2. *Assume $k = 4$. Assume that every edge of E is standard. Assume that both diagonals ε of the fan satisfy $\|\varepsilon\| \geq 3$. Then*

$$\tau(V, E, F) \geq 0.467.$$

The proof of the main estimate occupies the rest of the chapter. We refer to the first conclusion of the theorem as the *standard main estimate*.

The main estimate and Lemma 7.44 are obtained by computer calculation, proving nonlinear inequalities by interval arithmetic. Two difficulties arise in the proof of the main estimate. First, nonlinear optimization is in general NP hard; and our calculations in particular rapidly become more difficult to carry out as the dimension increases. When $k = 3$, the set $V = \{\mathbf{v}_1, \mathbf{v}_2, \mathbf{v}_3\}$ is six-dimensional (nine spacial coordinates minus a three-dimensional group of rotational symmetries). These calculations in six dimensions are relatively simple. However, by the time $k = 6$, the dimension of V has reached fifteen, which is beyond our computational capacity. We are forced to prove a sequence of lemmas, showing that any configuration (V, E, F) that minimizes τ lies in an explicit low-dimensional subset of this set of local nonreflexive fans, where low-dimensional means anything small enough to be treated directly by a computer calculation.

The second source of difficulty comes from numerical instabilities. For numerical stability, we insist on using analytic functions on compact domains. One of our favorite strategies is to slice along internal blades to cut local fans into smaller fans, and inductively build up the desired estimates from the smaller fans. However, when we slice along an internal blade, it is very difficult to avoid computations on simplices that flatten into simplices of zero volume. The functions defining τ are not analytic at flat simplices. They behave as $\sqrt{\Delta}$, with Δ tending to 0 from above. Concerns such as these force us to use relatively short diagonals when we slice. The general heuristic we use is that degeneracies are avoided when $\|\varepsilon\| < 3.106\ldots$, and calculations become stable when $\|\varepsilon\| < 3.01$ (see the proof of Lemma 7.52).

We do not present a complete proof of the main estimate in the text, because much of it is done by computer. In the rest of this chapter, we describe how the local fans of large dimension (especially, the case $k = 6$) can be reduced to much lower dimension. From there, the reader must trust that the small calculations have been executed, or turn directly to the computer implementation for details.

7.4.2 constraints

Let (V, E, F) be a nonreflexive local fan that satisfies all the assumptions of the main estimate. The main estimate takes the form of a collection of bounds

$$\tau(V, E, F) > d, \tag{7.45}$$

assuming various length constraints on the edges and diagonals of the fan. In building up these estimates inductively (by slicing into smaller fans), we will need to consider further estimates of the same general form (7.45), under many different length constraints on edges and diagonals. With that in mind, we introduce a *constraint system*.

Definition 7.46 (torsor, adjacent) Let $k > 1$ be an integer. A *torsor* is a set I with a given simply transitive action of $\mathbb{Z}/k\mathbb{Z}$ on I. We write the application of $j \in \mathbb{Z}/k\mathbb{Z}$ to $i \in I$ as $j + i$ or $i + j$. We also write $j + i$ for the application of the image of $j \in \mathbb{Z}$ in $\mathbb{Z}/k\mathbb{Z}$ to $i \in I$. Note that each choice of base point $i_0 \in I$ gives a bijection $i \mapsto i + i_0$ between $\mathbb{Z}/k\mathbb{Z}$ and I. We say that i and j are not *adjacent* if $i \neq j \pm 1$. An *isomorphism of torsors* is a bijection that respects the action.

We use the constant $c_{\text{stab}} = 3.01$ to make the constraint systems numerically stable. Its use will become apparent in Lemma 7.52.

Definition 7.47 (constraint system, stable) A *constraint system* s consists of the following data:

1. a natural number $k \in \{3, 4, 5, 6\}$,
2. a $\mathbb{Z}/k\mathbb{Z}$-torsor I,
3. a real number d,
4. real constants a_{ij}, b_{ij}, satisfying $a_{ij} = a_{ji}$, $b_{ij} = b_{ji}$, $a_{ij} \le b_{ij}$, for $i, j \in I$, and
5. a subset $J \subset \{\{i, 1 + i\} \ : \ i \in I\}$, such that $\text{card}(J) + k \le 6$.

We say that a constraint system s is *stable* if the following additional properties hold.

1. $$0 = a_{ii} \text{ and } 2 \le a_{ij} \text{ for all } i, j \in I \text{ such that } i \neq j.$$
2. Also, $$b_{i,i+1} \le c_{\text{stab}}.$$
3. If $\{i, j\} \in J$, then $[a_{ij}, b_{ij}] = [\sqrt{8}, c_{\text{stab}}]$.

Remark 7.48 The number k represents the number of edges in a given local fan. In practice, the set I is an indexing set for the set of nodes $V = \{\mathbf{v}_i : i \in I\}$ of a local fan (V, E, F), indexed such that $\rho_{V,E,F}\mathbf{v}_i = \mathbf{v}_{i+1}$. The constants a_{ij} and b_{ij} prescribe the lower and upper bounds on the edges and diagonals of a fan (V, E, F) with V:

$$a_{ij} \leq \|\mathbf{v}_i - \mathbf{v}_j\| \leq b_{ij}.$$

A constant d appears in (7.45). The set J is used to make minor adjustments to the estimates, and will be explained later. In most cases, we can take $J = \varnothing$.

For each constraint system s, we write $k(s)$, $d(s)$, $I(s)$, $a_{ij}(s)$, and so forth for the associated parameters.

Example 7.49 *The constants in the conclusions of the main estimate (Theorem 7.43) can be packaged into stable constraint systems. For example, the standard main estimate for $k = 6$ gives the constraint system $d = 0.712$, $J = \varnothing$, I an indexing set of cardinality six, and*

$$a_{ij} = \begin{cases} 0, & i = j, \\ 2, & j = i \pm 1, \\ 2h_0, & \text{otherwise}, \end{cases} \qquad b_{ij} = \begin{cases} 0, & i = j, \\ 2h_0, & j = i \pm 1, \\ 4h_0^+, & \text{otherwise}, \end{cases}$$

where h_0^+ is any constant greater than h_0. The upper bound $4h_0$ on any diagonal comes from the triangle inequality: $\|\mathbf{v}_i - \mathbf{v}_j\| \leq \|\mathbf{v}_i\| + \|\mathbf{v}_j\| \leq 4h_0$.

We write S_{main} for the set of stable constraint systems s, with a fixed choice of torsor for each k, for all cases of the main estimate.

Example 7.50 (ear) *We have a stable constraint system s given by $k = \mathrm{card}(I) = 3$, $d = 0.11$, J a singleton, and*

$$[a_{ij}, b_{ij}] = \begin{cases} [0, 0], & \text{if } i = j, \\ [\sqrt{8}, c_{stab}], & \text{if } \{i, j\} \in J, \\ [2, 2h_0], & \text{otherwise}. \end{cases}$$

We call s an ear (by analogy with an ear in a triangulation of a polygon, which is a triangle that has two of its edges in common with the polygon).

Next we associate a set $\mathcal{B}_s$ with each constraint system s.

Definition 7.51 ($\mathcal{B}_s$) For every constraint system s, and every function $\mathbf{v} : I(s) \to \mathcal{B}$, let $V_\mathbf{v} \subset \mathcal{B}$ be the image of $\mathbf{v}$. Let $E_\mathbf{v}$ be the image of $i \mapsto \{\mathbf{v}_i, \mathbf{v}_{i+1}\}$. Let $F_\mathbf{v}$ be the image of $i \mapsto (\mathbf{v}_i, \mathbf{v}_{i+1})$. Let $\mathcal{B}_s$ be the set of all functions $\mathbf{v}$ that have the following properties.

1. $a_{ij}(s) \le \|\mathbf{v}_i - \mathbf{v}_j\| \le b_{ij}(s)$, for all $i, j \in I(s)$.
2. $(V_{\mathbf{v}}, E_{\mathbf{v}}, F_{\mathbf{v}})$ is a nonreflexive local fan.

Lemma 7.52 *Let s be a stable constraint system. Then $\mathcal{B}_s$ is compact (as a subset of $\mathcal{B}^k \subset \mathbb{R}^{3k}$).*

Proof The set $\mathcal{B}$ is defined as a closed subset of a closed ball in $\mathbb{R}^3$. It is compact. By taking products of a compact set, $\mathcal{B}^k$ is compact. The set $\mathcal{B}_s$ is defined by two conditions. The first enumerated condition in the definition of $\mathcal{B}_s$ is a closed constraint. It is enough to check that the condition the second condition is also a closed constraint. That is, it is enough to show that the set of functions $\mathbf{v}$ such that $(V_{\mathbf{v}}, E_{\mathbf{v}}, F_{\mathbf{v}})$ is a nonreflexive local fan is closed in $\mathcal{B}^k$.

For this, we run through each defining property of fan, local, and nonreflexive in turn, and check that they are all closed conditions. For that purpose, consider a function

$$\mathbf{v} : \{1, \ldots, k\} \to \mathcal{B}.$$

that lies in the closure of functions in $\mathcal{B}_s$. We show that the limit $(V_{\mathbf{v}}, E_{\mathbf{v}}, F_{\mathbf{v}})$ is also a nonreflexive local fan. By the stability condition $2 \le \|\mathbf{v}_i - \mathbf{v}_j\|$, when $i \ne j$, we see that $\mathbf{v}$ is an injective function on the domain $\{1, \ldots, k\}$. We find that $V_{\mathbf{v}}$ is a subset of $\mathcal{B}$ of cardinality k. In particular, it is a nonempty finite set such that $\mathbf{0} \notin V_{\mathbf{v}}$. This verifies the first two defining properties of a fan.

The condition (NONPARALLEL) of a fan follows from the estimates based on stability.

$$2 \le \|\mathbf{v}_i - \mathbf{v}_{i+1}\| \le c_{\text{stab}},$$

If $\mathbf{v}_i$ and $\mathbf{v}_{i+1}$ are parallel, we get a contradiction:

$$\|\mathbf{v}_i - \mathbf{v}_{i+1}\| = |\,\|\mathbf{v}_i\| \pm \|\mathbf{v}_{i+1}\|\,|$$

which is at least $4 > c_{\text{stab}}$ or no greater than $2h_0 - 2 < 2$.

We turn to the condition (INTERSECTION). This is the most tedious part of the proof, because there are several cases involved in showing that for all $\varepsilon, \varepsilon' \in E \cup \{\{\mathbf{w}\} : \mathbf{w} \in V\}$,

$$C(\varepsilon) \cap C(\varepsilon') = C(\varepsilon \cap \varepsilon').$$

We leave most of these routine verifications to the reader. Two cases are noteworthy. (1) Suppose that ε and ε' are disjoint sets of cardinality two, such that the data for $\mathbf{v}$ gives a nonempty intersection $C^0(\varepsilon) \cap C^0(\varepsilon') \ne \varnothing$. The intersection of these two blades is an open condition, so that this failure to satisfy the fan constraint is open, and satisfaction of the constraint is therefore closed. (2) Suppose that $\varepsilon = \{\mathbf{u}\}$ and $\varepsilon' \in E_{\mathbf{v}}$ is disjoint from ε. Suppose for a contradiction that $C^0(\varepsilon')$ meets $C(\varepsilon)$. We obtain a planar quadrilateral with diagonals ε' and $\{\mathbf{0}, \mathbf{u}\}$. By contracting the diagonal ε', we obtain a rhombus of side 2. By vector geometry, the two diagonals d_1 and d_2 of the rhombus satisfy

$$d_1^2 = 16 - d_2^2. \tag{7.53}$$

We have $d_2 \le 2h_0$ because $\mathbf{u}$ is an element of the annulus $\mathcal{B}$, and ε' satisfies an upper bound coming from the stability conditions:

$$d_1^2 = \|\varepsilon'\|^2 \le c_{\text{stab}}^2 < 3.106^2 < 16 - (2h_0)^2 \le 16 - d_2^2 = d_1^2. \tag{7.54}$$

This is a contradiction.

The defining properties of local fan are combinatorial, and depend only on s. In the definition of nonreflexive, the condition (ANGLE) is given as a closed condition on the azimuth angle. The condition (WEDGE) is also given as a closed condition. This completes the proof. □

The following lemma is based on the same methods as the previous lemma. It tells us that sufficiently short blades are necessarily internal.

Lemma 7.55 *Let s be a stable constraint system, and let $\mathbf{v} \in \mathcal{B}_s$. Let $\mathbf{u}, \mathbf{w} \in V_{\mathbf{v}}$ satisfy $2 \le \|\mathbf{u} - \mathbf{w}\| \le c_{stab}$ where $\{\mathbf{u}, \mathbf{w}\} \notin E_{\mathbf{v}}$. Then $\mathbf{u}$ and $\mathbf{w}$ are nonparallel. Moreover, $C^0\{\mathbf{u}, \mathbf{w}\} \subset W^0_{dart}(x)$ for all $x \in F$.*

Proof The proof that $\mathbf{u}$ and $\mathbf{w}$ are nonparallel is identical to the proof in the previous lemma that showed $\mathbf{v}_i$ and $\mathbf{v}_{i+1}$ are not parallel.

We turn to the second conclusion of the lemma. Assume for a contradiction that the second conclusion of the lemma is false. We can find an edge $\{\mathbf{u}_1, \mathbf{u}_2\} \in E_{\mathbf{v}}$ such that

$$C^0\{\mathbf{u}, \mathbf{w}\} \cap C\{\mathbf{u}_1, \mathbf{u}_2\} \ne \emptyset.$$

Moreover, $C^0\{\mathbf{u}, \mathbf{w}\}$ lies in a half-space with boundary $\operatorname{aff}\{\mathbf{0}, \mathbf{u}_1, \mathbf{u}_2\}$. This forces

$$C^0\{\mathbf{u}, \mathbf{w}\} \cap \{\mathbf{u}_1, \mathbf{u}_2\} \ne \emptyset.$$

To be definite, assume that $\mathbf{u}_1 \in C^0\{\mathbf{u}, \mathbf{w}\}$. We obtain a planar quadrilateral with diagonals $\{\mathbf{0}, \mathbf{u}_1\}$ and $\{\mathbf{u}, \mathbf{w}\}$. We obtain the same contradiction as in the proof of case (2) of (INTERSECTION) in the previous lemma, by deforming the quadrilateral to a rhombus. □

The set J is used to make a small correction $d(s, \mathbf{v})$ to the constant $d(s)$.

Definition 7.56 ($d(s, \mathbf{v})$) Set $\sigma(s) = 1$ when s is an ear; $\sigma = -1$, otherwise. Let $V = \{\mathbf{v}_i \,:\, i \in I(s)\}$ be a set of points in $\mathbb{R}^3$. Write

$$d(s, \mathbf{v}) = d(s) + 0.1\,\sigma(s) \sum_{\{i,j\} \in J(s)} (c_{\text{stab}} - \|\mathbf{v}_i - \mathbf{v}_j\|). \tag{7.57}$$

This correction to $d(s)$ makes it a bit easier to prove inequalities when $\sigma(s) = -1$, at the cost of slightly more difficult inequalities for ears. The set $J(s)$ is empty for $s \in S_{\text{main}}$, so this correction does not directly affect the main estimates:

$$d(s, \mathbf{v}) = d(s), \text{ for all } s \in S_{\text{main}}.$$

Definition 7.58 (τ^*) Let s be a stable constraint system. Define

$$\tau^* : \{(s, \mathbf{v}) \ :\ \mathbf{v} \in \mathcal{B}_s\} \to \mathbb{R}$$

by

$$(s, \mathbf{v}) \mapsto \tau(V_{\mathbf{v}}, E_{\mathbf{v}}, F_{\mathbf{v}}) - d(s, \mathbf{v}).$$

Lemma 7.59 (continuity) *Let s be a stable constraint system. Then the function*

$$\mathbf{v} \mapsto \tau^*(s, \mathbf{v})$$

is a continuous function on $\mathcal{B}_s$. Moreover, if $\mathcal{B}_s$ is nonempty, then the function attains a minimum.

Proof The function τ^* is a polynomial in $\|\mathbf{v}_i\|$ and $\text{azim}(\mathbf{0}, \mathbf{v}_i, \mathbf{v}_{i+1}, \mathbf{v}_{i-1})$. The norm and azimuth angle are both continuous functions of $\mathbf{v}$. Moreover, a continuous function on a compact space attains its minimum. □

The largest constant $d(s)$ that arises in our calculations will be $d(s) = 0.712$ (in the standard main estimate for $k = 6$). In particular, the following lemma allows us to assume in all that follows that the fan of a counterexample is not circular.

Lemma 7.60 *Let s be a stable constraint system. Let $\mathbf{v} \in \mathcal{B}_s$. Suppose that*

$$d(s) \le 0.9 \text{ and } \text{sol}(V_{\mathbf{v}}, E_{\mathbf{v}}, F_{\mathbf{v}}) \ge \pi.$$

Then $\tau^*(s, \mathbf{v}) > 0$. In particular, if $d(s) \le 0.9$ and $\tau^*(s, \mathbf{v}) \le 0$, then $(V_{\mathbf{v}}, E_{\mathbf{v}}, F_{\mathbf{v}})$ is not a circular local fan.

Proof We have

$$\begin{aligned}
\tau(V,E,F) &= (\pi + \mathrm{sol}_0)(2-k) + \sum_{x\in F} \mathrm{azim}(x)\left(1 + \frac{\mathrm{sol}_0}{\pi}\frac{\|v\|-2}{2h_0-2}\right)\\
&\geq (\pi + \mathrm{sol}_0)(2-k) + \sum_{x\in F} \mathrm{azim}(x)\\
&= \mathrm{sol}(V_{\mathbf{v}}, E_{\mathbf{v}}, F_{\mathbf{v}}) + (2-k)\,\mathrm{sol}_0\\
&\geq \pi + (2-6)\,\mathrm{sol}_0\\
&> 0.92.
\end{aligned}$$

$$\begin{aligned}
d(s,\mathbf{v}) &= d(s) + 0.1\,\sigma \sum_{\{i,j\}\in J(s)} (c_{\mathrm{stab}} - \|\mathbf{v}_i - \mathbf{v}_j\|)\\
&\leq d(s) + 0.1(c_{\mathrm{stab}} - \sqrt{8})\\
&\leq d(s) + 0.02\\
&\leq 0.92.
\end{aligned}$$

Hence $\tau^*(s,\mathbf{v}) = \tau(V,E,F) - d(s,\mathbf{v}) > 0$. When the fan is circular,

$$\mathrm{sol}(V_{\mathbf{v}}, E_{\mathbf{v}}, F_{\mathbf{v}}) = 2\pi.$$

The result ensues. □

7.4.3 minimality

We slice torsors as we did earlier with local fans (Figure 7.6).

Definition 7.61 (slice) Let I be a $\mathbb{Z}/k\mathbb{Z}$-torsor, with action given by $(j,i) \mapsto j+i$, for $i \in I$. Let $p, q \in I$ that are not adjacent. Set

$$I[p,q] = \{p, 1+p, 2+p, \ldots, q\} \subset I.$$

Note that the cardinality of $I[p,q]$ is

$$m = 1 + \min\{m \in \mathbb{N} \;:\; m + p = q\}.$$

We make $I[p,q]$ into a $\mathbb{Z}/m\mathbb{Z}$-torsor with action $(j,i) \mapsto j +' i$, given by the iterates of

$$1 +' i = \begin{cases} 1+i & \text{if } i \neq q,\\ p & \text{if } i = q.\end{cases}$$

The $\mathbb{Z}/m\mathbb{Z}$-torsor $I[p,q]$ is called the *slice* of I along (p,q).

To prove the main estimate, we use a finite set S of constraint systems that includes S_{main}. To obtain the main estimate by induction by slicing fans into

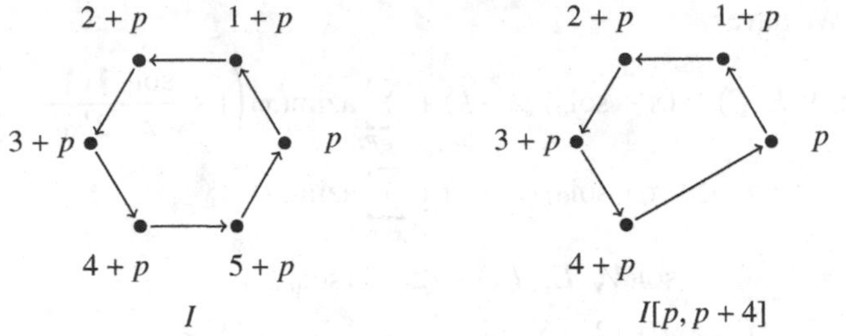

Figure 7.6 Given $p, q \in I$, the slice $I[p, q]$ follows the cyclic order through I from p to q, then returns directly from q to p.

pieces, the constraint systems must be compatible. We use the following coherence conditions.

Definition 7.62 (diagonal cover) Let s be a constraint system and let $p, q \in I(s)$ where p and q are not adjacent. (In particular, $k(s) = \text{card}(I(s)) > 3$.) We say that a pair $\{s', s''\}$ of constraint systems *covers* the diagonal $\{p, q\}$ of s, if the following conditions hold.

1. $I(s') = I[p, q]$ and $I(s'') = I[q, p]$, up to isomorphisms of torsors.
2. $d(s) \le d(s') + d(s'')$.
3. $J(t) \subset J(s) \cup \{\{p, q\}\}$, for $t = s', s''$. Also, $\{p, q\} \in J(s')$ if and only if $\{p, q\} \in J(s'')$, if and only if s' or s'' is an ear.
4. For $t = s', s''$,
$$a_{ij}(t) = a_{ij}(s) \text{ and } b_{ij}(t) = b_{ij}(s),$$
when $i, j \in I(t)$, provided $\{i, j\} \neq \{p, q\}$.
5. For $t = s', s''$,
$$a_{pq}(s), b_{pq}(s) \in [a_{pq}(t), b_{pq}(t)].$$

A cover of a diagonal $\{p, q\}$ is used when we slice a fan of cardinality k into two smaller fans with cardinalities $k(s')$ and $k(s'')$. All of the edge length constraints are to be preserved under slicing, with a mild compatibility condition on the new edge created by the slice.

If $\{s', s''\}$ covers a diagonal $\{p, q\}$, then we can use the inclusions $I(s') \subset I(s)$ and $I(s'') \subset I(s)$ to restrict an element $\mathbf{v} : I(s) \to \mathcal{B}$ to $\mathbf{v}' : I(s') \to \mathcal{B}$ and $\mathbf{v}'' : I(s'') \to \mathcal{B}$. The inequality $d(s) \le d(s') + d(s'')$ of a covered diagonal implies a related inequality.

Lemma 7.63 *Let s be a stable constraint system with diagonal $\{p, q\}$ that is covered by a pair of stable constraint systems $\{s', s''\}$. Let $\mathbf{v} \in \mathcal{B}_s$ and let $\mathbf{v}'$ and*

$\mathbf{v}''$ be constructed from $\mathbf{v}$ as above. Assume that if $\{i,j\} \in J(s) \cup J(s') \cup J(s'')$, then $\|\mathbf{v}_i - \mathbf{v}_j\| \le c_{stab}$. Then

$$d(s, \mathbf{v}) \le d(s', \mathbf{v}') + d(s'', \mathbf{v}'') \tag{7.64}$$

and

$$\tau^*(s, \mathbf{v}) \ge \tau^*(s, \mathbf{v}') + \tau^*(s, \mathbf{v}''). \tag{7.65}$$

Moreover, $\mathbf{v}' \in \mathcal{B}_{s'}$ and $\mathbf{v}'' \in \mathcal{B}_{s''}$.

Proof The existence of a diagonal forces $k(s) > 3$. In particular, s is not an ear, so that $\sigma(s) = -1$. Every element of $J(s)$, by the definition of a constraint system has the form $\{i, i+1\}$. In particular, $\{p, q\} \notin J(s)$. Let $\delta = 1$ if $\{p, q\} \in J(s')$ and $\delta = 0$, otherwise. By covering properties, $\{p, q\} \in J(s')$, if and only if $\{p, q\} \in J(s'')$. If $\delta = 1$, then at least one of $\sigma(s'), \sigma(s'')$ is 1. Abbreviate $c_{ij}(\mathbf{v}) = c_{\text{stab}} - \|\mathbf{v}_i - \mathbf{v}_j\|$, for $\{i, j\} \in J(s) \cup \{\{p, q\}\}$. We have $c_{ij}(\mathbf{v}) \ge 0$, for all $\{i, j\} \in J(s) \cup J(s') \cup J(s'')$, by assumption. The sets $J(s')$ $J(s'')$ are disjoint if $\delta = 0$ and meet in the singleton set $\{\{p, q\}\}$ if $\delta = 1$. These observations give

$$\begin{aligned}
&d(s', \mathbf{v}') + d(s'', \mathbf{v}'') - d(s, \mathbf{v}) \\
&\quad \ge (d(s') + d(s'') - d(s)) \quad + 0.1(\sigma(s') + \sigma(s''))\,\delta\, c_{pq}(\mathbf{v}) \quad + 0.1 \sum_{\{i,j\} \in J(s) \setminus (J(s') \cup J(s''))} c_{ij}(\mathbf{v}) \\
&\quad \ge 0 + 0 + 0.
\end{aligned}$$

The statement about τ^* follows from Lemma 7.29. □

The proof of the main estimate has the following structure. In an iterative process, we construct an explicit finite set S of constraint systems that includes the set S_{main} of constraint systems appearing in the main estimate. We give a proof that for every $s \in S$ and every $\mathbf{v} \in \mathcal{B}_s$

$$\tau^*(s, \mathbf{v}) > 0. \tag{7.66}$$

Definition 7.67 (level function, minimal counterexample) Let S be a set of constraint systems. We say that $\ell : S \to \mathbb{N}$ is a *level* function on S if for every $s, s' \in S$, $k(s') < k(s)$ implies $\ell(s') < \ell(s)$. We say that $(s, \mathbf{v}) \in S \times \mathcal{B}_s$ is a *minimal counterexample* relative to S and ℓ if the following conditions hold.

1. $\mathbf{v}$ minimizes the function $\mathbf{v} \mapsto \tau^*(s, \mathbf{v})$ over $\mathcal{B}_s$.
2. $\tau^*(s, \mathbf{v}) \le 0$.
3. If $s' \in S$ is any constraint system such that $\ell(s') < \ell(s)$, then for all $\mathbf{v}' \in \mathcal{B}_{s'}$, we have $\tau^*(s', \mathbf{v}') > 0$.

Lemma 7.68 *Let S be a set of constraint systems with level function ℓ. Assume that τ^* attains its minimum on each nonempty $\mathcal{B}_s$, $s \in S$. If (7.66) fails to hold for some $s \in S$, then there exists a minimal counterexample relative to S and ℓ.*

Proof Select some s that minimizes $\ell(s)$, from the set $S_1 \subset S$ of constraint systems s, such that $\mathcal{B}_s$ contains points violating (7.66). Let $\mathbf{v}$ minimize τ^* on $\mathcal{B}_s$. By construction, it satisfies the defining properties of a minimal counterexample. □

Lemma 7.69 (minimality criteria) *Let S be a set of constraint systems. Let $s \in S$. Suppose that for every $\mathbf{w} \in \mathcal{B}_s$, one of the following criteria holds.*

1. (NUMERICAL POSITIVITY) *A calculation*[2] *shows $\tau^*(s, \mathbf{w}) > 0$.*
2. (DEFORMATION) *For some $\epsilon > 0$, there exists a continuous function $\mathbf{v}$: $[0, \epsilon) \to \mathcal{B}_s$ such that $\mathbf{v}(0) = \mathbf{w}$ and*
$$\tau(s, \mathbf{v}(t)) < \tau(s, \mathbf{w}),$$
for all sufficiently small positive numbers t.
3. (DIAGONAL) *The system s is stable. Also, there exist $p, q \in I(s)$ and stable $s', s'' \in S$ such that $\{s', s''\}$ covers the diagonal $\{p, q\}$ and such that*
$$\|\mathbf{v}_p - \mathbf{v}_q\| \le \min(b_{pq}(s'), b_{pq}(s'')),$$
4. (TRANSFER) *The constraint system s is not an ear, and there exists $s' \in S$ such that the following conditions hold.*
   - a. $\mathbf{w} \in \mathcal{B}_{s'}$;
   - b. $\ell(s') < \ell(s)$;
   - c. $I(s') = I(s)$ *(up to isomorphism);*
   - d. $d(s') \ge d(s)$; *and*
   - e. $J(s') \subset J(s)$.

Then there does not exist an element $\mathbf{w} \in \mathcal{B}_s$ such that $(s, \mathbf{w})$ is a minimal counterexample relative to S and ℓ.

Proof The first two criteria are clearly incompatible with minimality. The third is incompatible with minimality by Lemma 7.63. If the final criterion holds at $\mathbf{w}$, then $k(s) = k(s')$, and $\tau(V_{\mathbf{w}}, E_{\mathbf{w}}, F_{\mathbf{w}})$ is the same for both s and s'. Also, $d(s', \mathbf{w}) \ge d(s, \mathbf{w})$ and $\tau^*(s', \mathbf{w}) \le \tau^*(s, \mathbf{w})$. Then $(s', \mathbf{w})$ certifies that $(s, \mathbf{w})$ is not minimal. □

[2] In practice, these are computer calculations.

7.4.4 reducing dimension

As we pointed out at the beginning of this section, the dimension of $\mathcal{B}_s$ is so large that it gives computational difficulties. The dimension of $\mathcal{B}_s$ is $3k - 3 \leq 15$, and to obtain reasonable performance, we prefer to restrict our computer calculations to at most six or seven dimensions. This subsection gives a collection of lemmas that show that some minimal counterexample (for suitable S) must lie in a subset of $\mathcal{B}_s$ of small dimension. This will allow us to use computers to complete the verifications of the main estimate.

Throughout this subsection we examine a constraint system s. With the (DEFORMATION) criterion of Lemma 7.69 in mind, we specifically avoid minimal counterexamples that will be treated later with the (DIAGONAL) criterion of the same lemma. We say that $(s, \mathbf{v})$ is *free*, if $a_{ij}(s) < \|\mathbf{v}_i - \mathbf{v}_j\| < b_{ij}(s)$ for all diagonals $\{i, j\} \notin E_{\mathbf{v}}$. We limit our discussion in this section to free pairs $(s, \mathbf{v})$.

We consider differentiable curves

$$\mathbf{v} : (-\epsilon, \epsilon) \to \mathcal{B}^{k(s)}.$$

If we show that $\tau^*(s, \mathbf{v}(t)) < \tau^*(s, \mathbf{v}(0))$ and $\mathbf{v}(t) \in \mathcal{B}_s$, whenever t is positive and sufficiently small, then $(s, \mathbf{v}(0))$ is not a minimal counterexample. For simplicity, we will start our study with curves that move a single point:

$$\mathbf{v}_j(t) = \mathbf{w}_j \text{ if } j \neq i, \quad \mathbf{v}_i(0) = \mathbf{w}_i, \tag{7.70}$$

for some index $i \in I(s)$ and some fixed $\mathbf{w} \in \mathcal{B}_s$.

In the next three lemmas, we make the implicit assumption that the local fan $(V_{\mathbf{w}}, E_{\mathbf{w}}, F_{\mathbf{w}})$ is generic, or that it is a lunar fan satisfying the assumptions of Lemma 7.24.

Lemma 7.71 *If $(s, \mathbf{w})$ is a free minimal counterexample relative to S and ℓ, then for all i, one of the following constraints hold.*

1. *$\|\mathbf{w}_i - \mathbf{w}_{i+1}\|$ attains its lower bound $a_{i,i+1}(s)$.*
2. *$\|\mathbf{w}_i - \mathbf{w}_{i-1}\|$ attains its lower bound $a_{i,i-1}(s)$.*
3. *$\|\mathbf{w}_i\|$ attains its lower bound 2.*
4. *There exists j adjacent to i such that $\{i, j\} \in J(s)$.*

Proof Fix i. Assume for a contradiction that none of the constraints hold. The function τ^* is decreasing along the curve of the form (7.70) such that $\mathbf{v}_i(t) = (1 - t)\mathbf{w}_i$. That is, we push the point $\mathbf{w}_i$ radially towards the origin. Explicitly, along this deformation, up to a constant, τ^* is equal to a positive constant times $\|\mathbf{v}_i(t)\|$. If none of the constraints of the lemma are satisfied, then $\mathbf{v}(t) \in \mathcal{B}_s$ for all t positive and sufficiently small. □

Recall that we call $\mathbf{w}_i$ *straight* if $\angle(\mathbf{w}_i) = \pi$ in the local fan $(V_\mathbf{w}, E_\mathbf{w}, F_\mathbf{w})$.

Lemma 7.72 *Let $(s, \mathbf{w})$ be a free minimal counterexample relative to S and ℓ, and let $i \in I(s)$. If $\mathbf{w}_i$ is straight, then one of the following constraints hold.*

1. *$\|\mathbf{w}_i - \mathbf{w}_{i+1}\|$ attains its lower bound $a_{i,i+1}(s)$, and $\|\mathbf{w}_i - \mathbf{w}_{i-1}\|$ attains its lower bound $a_{i,i-1}(s)$.*
2. *$\|\mathbf{w}_i\|$ attains its lower bound 2.*
3. *There exists j adjacent to i such that $\{i, j\} \in J(s)$.*

Proof The set $\{\mathbf{0}, \mathbf{w}_{i-1}, \mathbf{w}_i, \mathbf{w}_{i+1}\}$ lies in a plane A. Assume for a contradiction that none of the constraints holds. By the previous lemma one of the norm constraints is satisfied, say

$$\|\mathbf{w}_i - \mathbf{w}_{i+1}\| = a_{i,i+1}(s).$$

We consider a curve $\mathbf{v}$ of the form (7.70). We let the curve $\mathbf{v}_i$ describes a circle through $\mathbf{w}_i$ with center $\mathbf{w}_{i+1}$ in the plane A. Parameterize the curve so that as t increases, the norm $\|\mathbf{v}_i(t)\|$ decreases. The function $\tau^*(s, \mathbf{v})$ is decreasing in t. Explicitly, the function again depends linearly on $\|\mathbf{v}_i(t)\|$, because the azimuth angles remain fixed. The result ensues. □

The following lemma allows us to propagate a lower bound constraint from one edge to an adjacent one.

Lemma 7.73 *Let S be a set of constraint systems and ℓ a level function. Let $s \in S$, and let $i \in I(s)$. Assume that $(s, \mathbf{w})$ is a free minimal counterexample relative to S and ℓ such that $\mathbf{w}_i$ is straight and $\|\mathbf{w}_i - \mathbf{w}_{i+1}\| = a_{i,i+1}(s)$. Assume $a_{i,i+1}(s) < b_{i,i+1}(s)$. Then one of the following conclusions hold.*

1. *There exists j adjacent to i such that $\{i, j\} \in J(s)$.*
2. *$\|\mathbf{w}_i - \mathbf{w}_{i-1}\| = a_{i,i-1}(s)$.*
3. *There exists a free minimal counterexample $(s, \mathbf{u})$ relative to S and ℓ that differs from $\mathbf{w}$ only at index i, such that $\mathbf{u}_i$ is straight, and such that*

$$\|\mathbf{u}_i - \mathbf{u}_{i-1}\| > a_{i,i-1}(s) \text{ and } \|\mathbf{u}_i - \mathbf{u}_{i+1}\| > a_{i,i+1}(s).$$

Moreover, the corresponding result holds with $i + 1$ and $i - 1$ interchanged.

Proof Let $(s, \mathbf{w})$ be a free minimal counterexample as described. Assume that neither of the first two conclusions hold. In particular, $\|\mathbf{w}_i - \mathbf{w}_{i-1}\| > a_{i,i-1}(s)$. We consider a curve $\mathbf{v}$ of the form (7.70) that moves $\mathbf{v}_i$ in a circular arc with center $\mathbf{0}$ through the point $\mathbf{w}_i$ and in the fixed plane determined by $\{\mathbf{0}, \mathbf{w}_i, \mathbf{w}_{i+1}\}$. The function τ^* is constant along this curve. We orient the curve to increase $\|\mathbf{w}_i - \mathbf{w}_{i+1}\|$. For sufficiently, small t, we find that $\mathbf{v}(t) \in \mathcal{B}_s$ satisfies the third conclusion. □

Remark 7.74 (lateral motion) We continue to study curves $\mathbf{v}$ of the form (7.70), with motion confined to a single index $i \in I$. Let $j, k \in I$ be the two indices adjacent to i, so that $\{j, k\} = \{i-1, i+1\}$. We consider a curve $\mathbf{v}_i$ in $\mathbb{R}^3$ with parameter t that describes the circle through $\mathbf{w}_i$ at fixed distance from $\mathbf{0}$ and $\mathbf{w}_j$ (Figure 7.7). By Lemma 7.29 applied to τ^*, up to a term that is constant along the curve, the function $\tau^*(s, \mathbf{v})$ depends on $\mathbf{v}$ only through the three points $\mathbf{w}_j$, $\mathbf{v}_i$, and $\mathbf{w}_k$. The function τ^* is invariant under orthogonal transformations. The dependence on $\mathbf{v}$ can be expressed through the function τ_{tri} in Definition 7.39, evaluated at the six edge lengths of the simplex $\{\mathbf{0}, \mathbf{w}_j, \mathbf{v}_i, \mathbf{w}_k\}$. The derivative of τ^* along the curve is given by the partial derivative of τ_{tri} with respect to a single variable, say $y_4 = \|\mathbf{v}_i(t) - \mathbf{w}_k\|$:

$$\frac{\partial \tau^*(s, \mathbf{v}(t))}{\partial t} = \frac{\partial \tau_{tri}}{\partial y_4} \frac{\partial y_4}{\partial t}. \tag{7.75}$$

Even when the dimension of $\mathcal{B}_s$ is large, the right-hand side of this equation is a function of just six variables, and can be estimated by computer. By a *computer calculation*[3] [21] we can show that under rather general conditions on s, the function τ_{tri} is increasing in y_4. More generally, when the partial derivative of τ_{tri} of y_4 vanishes, computer calculations of the second derivative show that the τ_{tri} has a local maximum (again under mild restrictions on the domain), so that there are no interior point local minima as a function of y_4. By applying lateral motions at one index after another, the dimension of the configuration space can be significantly reduced.

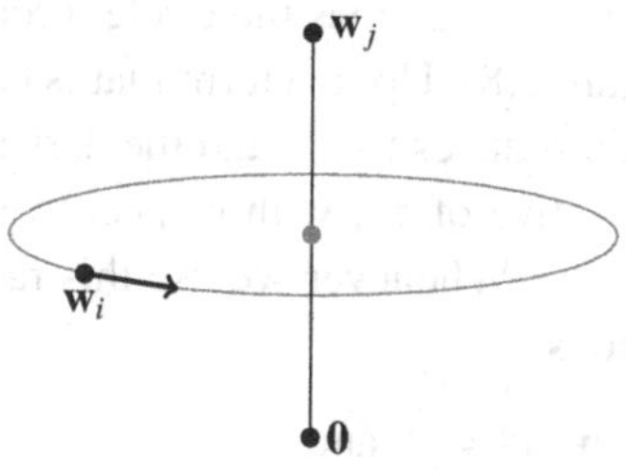

Figure 7.7 A lateral motion of $\mathbf{w}_i$ follows a circular path at fixed distance from $\mathbf{0}$ and $\mathbf{w}_j$, for some pair of adjacent indices i and j.

Remark 7.76 (straight node strategies) A complication occurs in lateral motions. Consider the lateral motion of $\mathbf{w}_i$ at fixed distance from $\mathbf{w}_j$ and $\mathbf{0}$, where j and k are the indices adjacent to i. The curve $\mathbf{v}$ is required to remain inside $\mathcal{B}_s$ for t sufficiently small and positive. But when $\mathbf{w}_i$, $\mathbf{w}_j$, or $\mathbf{w}_k$ is straight, the curve

[3] [UPONLFY]

does not generally remain inside $\mathcal{B}_s$: nonreflexivity can fail. We use three different strategies to get around this complication. In practice, these three strategies are sufficient to handle all the situations that arise.

First, we can try a lateral motion for a different ordered triple of pairs (i, j, k), such as a different index i altogether, or the triple (i, k, j), which interchanges the roles of j and k.

Second, we can restrict the direction of the lateral motion to decrease the angle at the straight node(s). Sometimes a computer calculation shows that for a given s, the azimuth angle of $\mathbf{w}_i$ is obtuse. When the azimuth angle at $\mathbf{w}_i$ is obtuse, basic trigonometry shows that if the lateral motion is directed to decrease the angle at $\mathbf{w}_j$, then it also decreases the angle at $\mathbf{w}_k$, and the curve remains initially inside $\mathcal{B}_s$.

Finally, to be concrete assume that $(j, k) = (i - 1, i + 1)$. Select ℓ so that $\mathbf{w}_{i-1}, \mathbf{w}_{i-2}, \ldots, \mathbf{w}_{\ell+1}$ are straight, but $\mathbf{w}_\ell$ is not. The point $\mathbf{0}$ and collection $W = \{\mathbf{w}_i, \ldots, \mathbf{w}_\ell\}$ lie in a common plane A. We consider a curve $\mathbf{v}$ that fixes all coordinates, except those in W, and isometrically moves the set W by a rotation of the plane A about the line through $\{\mathbf{0}, \mathbf{w}_\ell\}$. (This is the one place where we consider a curve $\mathbf{v}$ that moves more that one component $\mathbf{v}_i$ at once.) In this case, the dependence of τ^* on the curve factors through the six edges of the simplex $\{\mathbf{0}, \mathbf{v}_i, \mathbf{w}_{i+1}, \mathbf{w}_\ell\}$. The local minima of this function can be studied by computer by using the right-hand side of (7.75).

Remark 7.77 (radial motion) We continue to study curves $\mathbf{v}$ of the form (7.70), with motion confined to a single index i. We consider a curve $\mathbf{v}_i$ in $\mathcal{B} \subset \mathbb{R}^3$ with parameter t that describes the circle through $\mathbf{w}_i$ at fixed distance from $\mathbf{w}_{i-1}$ and $\mathbf{w}_{i+1}$ (Figure 7.8). Up to a term that is constant along the curve, the function $\tau^*(s, \mathbf{v})$ again reduces to τ_{tri} and the derivative with respect to t is given by the partial derivative of τ_{tri} with respect to $y_1 = \|\mathbf{v}_i\|$, provided we use parameterization $t = y_1$. Whenever we use this radial motion, we impose the following preconditions.

1. $\|\mathbf{w}_{i-1} - \mathbf{w}_i\| = \|\mathbf{w}_i - \mathbf{w}_{i+1}\| = 2$, and
2. $\|\mathbf{w}_{i-1} - \mathbf{w}_{i+1}\| \geq c_{\text{stab}}$.

Under these conditions a computer calculation of the first and second derivatives of τ_{tri} shows that it has no local minimum, provided $\mathbf{w}_i$ is not straight. To maintain nonreflexivity along the curve, we must also assume that neither of $\mathbf{w}_{i-1}, \mathbf{w}_{i+1}$ is straight, As a consequence of radial motion, any free minimal counterexample $(s, \mathbf{w})$ that satisfies this and the preconditions must have an extremal norm:

$$\|\mathbf{w}_i\| = 2, \text{ or } \|\mathbf{w}_i\| = 2h_0, \text{ or } \mathbf{w}_i \text{ is straight.} \tag{7.78}$$

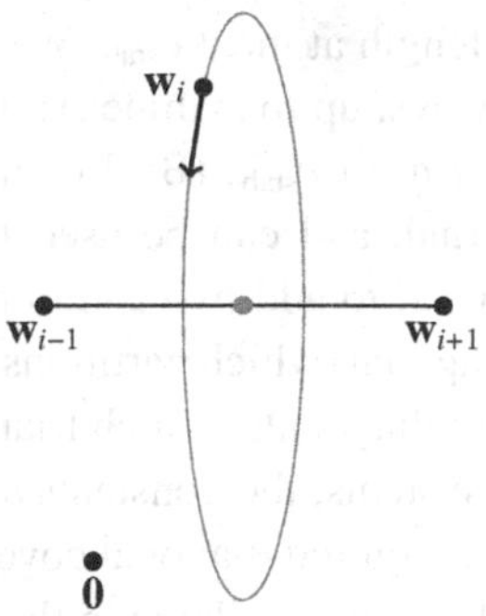

Figure 7.8 A radial motion of $\mathbf{w}_i$ towards or away from $\mathbf{0}$ follows a circular path at fixed distance from $\mathbf{w}_{i-1}$ and $\mathbf{w}_{i+1}$.

7.4.5 computer proof of main estimate

In this subsection, we sketch the computer proof of the main estimate. Before turning to the general case, we illustrate the methods, by proving Lemma 7.44.

computer proof (7.44) When $k = 3$, the space of configurations has six dimensions. The inequality $\tau(V, E, F) \geq 0$ is a simple direct computer calculation. When $k = 4$, the assumptions of the lemma give that both diagonals have length at least 3. By a rigorous computer estimate of dihedral angles, all nodes of a quadrilateral with diagonals at least 3 must have azimuth angle less than π. We laterally contract edges (by Remark 7.74) until both diagonals are precisely 3 or all four edges reach the lower bound 2. However, the rhombus diagonal inequality (7.53) shows that both diagonals become 3 before the four edges reach the lower bound 2. Thus, it is enough to consider the case when both diagonals are 3. By adding these two constraints, we have reduced the dimension from nine to seven. This seven-dimensional inequality is within reach of direct computer calculation. □

construction of S

We now sketch the proof of the main estimate, and describe the construction of set S of constraint systems and level set[4] ℓ, starting with S_{main}. We always wish

[4] We may replace the range of ℓ with any well-ordered set that has the same ordinal as $\mathbb{N}$. In particular, it is more convenient to use a lexicographic order on $\mathbb{N} \times Z$, for some fixed finite ordered set Z, with the property that the first coordinate of the pair is $k(s)$ and the second coordinate orders constraint systems with given k however we please in the construction that follows.

to slice along diagonals of length at most c_{stab}. We saturate S by adding diagonal covers in all possible ways, up to symmetry, of slicing systems $s \in S_{\text{main}}$ along diagonals of length at most c_{stab}. (By Lemma 7.55, the corresponding blades are necessarily internal, and can be used to slice the local fan.) For example, when $k = 5$, we need to add two additional constraint systems, one corresponding to a single diagonal (which partitions the pentagon into a quadrilateral and triangle), and two diagonals (which triangulates the pentagon). For these additional constraint systems, the constants d have been determined experimentally, and satisfy the required diagonal covering conditions. We do not list all the constants here. They are available in the computer code. We define $J(s)$ to be as large as possible, subject to the diagonal covering conditions.

For every s in this set S_{main}, we add a constraint system s' to S, where s' has the same parameters as s, except that for all non-adjacent indices i and j, we set

$$a_{ij}(s') = \max(a_{ij}(s), c_{\text{stab}}), \quad b_{ij}(s') = \max(b_{ij}(s), c_{\text{stab}}). \tag{7.79}$$

so that

$$c_{\text{stab}} \le a_{ij}(s') \le b_{ij}(s'). \tag{7.80}$$

At this point, a minimal counterexample necessarily has diagonals greater than c_{stab}. By a transfer to new constraint system s' with these parameters, and choosing $\ell(s') < \ell(s)$, we may assume that if $(s', \mathbf{v})$ is a minimal counterexample, then the system s' satisfies (7.80).

In general, whenever we add any constraint system to S, we recursively add further constraint systems, corresponding to slicing along diagonals, to maintain the diagonal covering properties. We assume this is done in what follows, without further mention.

When $s \in S$ satisfies $k(s) > 3$, and $J(s) \neq \emptyset$, we transfer to an additional $s' \in S$ that is identical to s, except that $J(s') = \emptyset$ and $\ell(s') < \ell(s)$.

triangles and quadrilaterals

The computer proof that there are no minimal counterexamples $(s, \mathbf{v})$ with $k(s) = 3$ goes as follows. Up to rotational invariance, the function τ^* can be expressed in terms of the function τ_{tri} of six variables. The rigorous nonlinear minimization of τ^* is easily done by computer, and we find that for each $s \in S$, and every $\mathbf{v} \in \mathcal{B}_s$, we have $\tau^*(s, \mathbf{v}) > 0$.

The computer proof that there are no minimal counterexamples $(s, \mathbf{v})$ with $k(s) = 4$ is not much more difficult.

pentagons

We consider a transfer that combines all the cases with $k = 5$ (with diagonals greater than c_{stab}) into a single additional constraint system s of level lower than any other pentagon. For this we put the standard constraint on all edges but one

$$(a_{i,i+1}(s), b_{i,i+1}(s)) = \begin{cases} (2, 2h_0), & \text{if } i \neq 0 \\ (2, c_{\text{stab}}), & \text{if } i = 0, \end{cases}$$

where $I(s) = \mathbb{Z}/5\mathbb{Z}$. We use the usual modified constraint on all diagonals:

$$(a_{ij}(s), b_{ij}(s)) = (c_{\text{stab}}, 4h_0), \quad \text{if } i \text{ and } j \text{ are not adjacent.}$$

For the constant $d(s)$ we take the maximum of the constant $d(s')$ as s' runs over cases with $k(s') = 5$.

By combining all pentagonal cases into one, we may give a uniform proof. By lateral motions (Remark 7.74), a computer assisted argument shows that every minimal counterexample $(s, \mathbf{w})$ satisfies

$$\| \mathbf{w}_i - \mathbf{w}_{i+1} \| = 2,$$

for all i. We use the following lemma.

Lemma 7.81 *Consider any skew pentagon in $\mathbb{R}^3$ whose five edges equal 2. Then there are two diagonals of the pentagon with a common endpoint whose lengths are at most*

$$1 + \sqrt{5} \approx 3.23607.$$

Proof Cut the pentagon into a triangle and skew quadrilateral along the shortest diagonal, of length t. By the triangle inequality $t \leq 4$. The shortest diagonal of the skew quadrilateral is maximized, when the quadrilateral is planar, with equal diagonals. The length of this diagonal is given by the largest root u of

$$\Delta(u^2, 4, 4, u^2, 4, t^2) = 0.$$

Solve for u in terms of t to obtain

$$u = \sqrt{4 + 2t}.$$

If $t > 1 + \sqrt{5}$, this gives a contradiction $u < t$. So $t \leq 1 + \sqrt{5}$. Then also

$$u = \sqrt{4 + 2t} \leq 1 + \sqrt{5}.$$

□

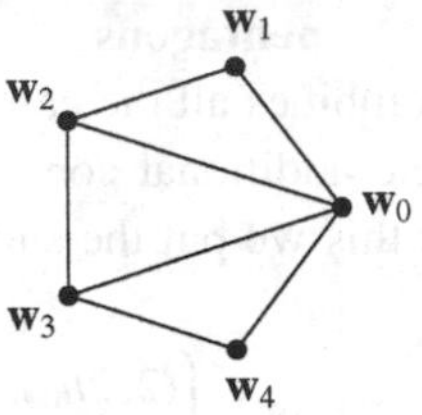

Figure 7.9 A pentagon is triangulated by two diagonals of length at most $1 + \sqrt{5}$ as shown. The peripheral nodes $\mathbf{w}_1$ and $\mathbf{w}_4$ are rigidly determined up to finite ambiguity by the simplex with extreme points $\mathbf{0}$, $\mathbf{w}_0$, $\mathbf{w}_2$, and $\mathbf{w}_3$.

We assume that indexing is chosen so that $I = \mathbb{Z}/5\mathbb{Z}$ and $(s, \mathbf{w})$ is a minimal counterexample with two chosen diagonals (Figure 7.9).

$$c_{\text{stab}} < \|\mathbf{w}_0 - \mathbf{w}_i\| \leq 1 + \sqrt{5}, \quad i = 2, 3.$$

Dihedral angle computer calculations show that under these constraints, none of $\{\mathbf{w}_0, \mathbf{w}_2, \mathbf{w}_3\}$ is straight. This allows us to apply radial motion (Remark 7.77), to show that $\mathbf{w}_1$ and $\mathbf{w}_4$ are extremal, in the sense of (7.78). This determines $\mathbf{w}_1$ and $\mathbf{w}_4$ (up to three cases each) as a function of $\mathbf{w}_0$, $\mathbf{w}_2$, $\mathbf{w}_3$. The calculations reduce in this way to a single simplex $\{\mathbf{0}, \mathbf{w}_0, \mathbf{w}_2, \mathbf{w}_3\}$, which have been carried out by computer.

hexagons

Only one of the inequalities in the main estimate has $k = 6$. It asserts that $\tau(V, E, F) > 0.712$. By adding an additional hexagon of lower level, we may transfer to a constraint system whose diagonals satisfy

$$a_{ij}(s) = c_{\text{stab}}. \tag{7.82}$$

We assume that our constraint system has this property. By lateral motions, we reduce to the case

$$\|\mathbf{w}_i - \mathbf{w}_{i+1}\| = 2,$$

for all i. We may assume that the indexing set I is $\mathbb{Z}/6\mathbb{Z}$. We triangulate with three blades $\{\mathbf{w}_{2i}, \mathbf{w}_{2i+2}\}$, for $i = 0, 2, 4$. See Figure 7.10.

Lemma 7.83 *The norms $\|\mathbf{w}_{2i} - \mathbf{w}_{2i+2}\|$, for $\mathbf{w} \in \mathcal{B}_s$, are at least c_{stab} and at most* 3.915.

Proof The lower bound comes from (7.82). By Lemma 2.48, the squares of the edges x_{ij} of the simplex $\{\mathbf{0}, \mathbf{w}_{2i}, \mathbf{w}_{2i+1}, \mathbf{w}_{2i+2}\}$ gives a nonnegative value

$$\Delta(x_{ij}) \geq 0.$$

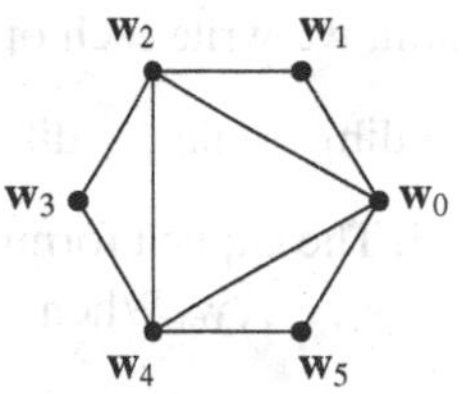

Figure 7.10 A hexagon is triangulated as shown. The peripheral nodes $\mathbf{w}_1$, $\mathbf{w}_3$, and $\mathbf{w}_5$ are rigidly determined up to finite ambiguity by the simplex with extreme points $\mathbf{0}, \mathbf{w}_0, \mathbf{w}_2$, and $\mathbf{w}_4$.

However, this polynomial is negative when $\|\mathbf{w}_{2i} - \mathbf{w}_{2i+2}\| > 3.915$. □

By radial motion, we may assume that (7.78) holds at each odd index $\mathbf{w}_{2i+1}$. We warn that these contraction arguments may produce nonreflexivity in the local fan at some of the even indices $\mathbf{v}_{2i}$. At this final stage, we abandon the nonreflexive condition. In fact, at this stage, we may abandon the geometry altogether, and view τ^* analytically as a sum of four terms τ_{tri}, indexed by the four triangles in the triangulation of the hexagon. After radial motion, the points $\mathbf{v}_{2i+1}$ are rigidly determined, up to three cases (7.78), in terms of the simplex $\{\mathbf{0}, \mathbf{v}_2, \mathbf{v}_4, \mathbf{v}_6\}$. We have reduced the calculations for $k = 6$ to a single simplex, which have been carried out by computer.

instabilities

We add a final lemma that we used to deal with the issue of numerical instability in the calculations when one of the simplices $\{\mathbf{0}, \mathbf{w}_{2i}, \mathbf{w}_{2i+1}, \mathbf{w}_{2i+2}\}$ is close to being planar.

Lemma 7.84 *Consider the function τ_{tri} on the domain*

$$y_1, y_2, y_3 \in [2, 2h_0], \quad y_4 \in [c_{stab}, 3.915], \quad y_5 = y_6 = 2, \quad \Delta(y_1^2, \ldots, y_6^2) \geq 0.$$

Then τ_{tri} has the following properties.

1. *If $y_1 = 2$, then $\tau_{tri}(y_1, y_2, y_3, y_4, y_5, y_6) \geq -\mathrm{sol}_0$.*
2. *If $y_1 = 2h_0$, then $\tau_{tri}(y_1, y_2, y_3, y_4, y_5, y_6) \geq 0$.*
3. *If $\mathrm{dih}_1(y_1, y_2, y_3, y_4, y_5, y_6) = \pi$, then*

$$\tau_{tri}(y_1, y_2, y_3, y_4, y_5, y_6) = \mathrm{sol}_0 \frac{y_1 - 2h_0}{2h_0 - 2}.$$

Proof The first claim is the trivial lower bound that we obtain by replacing $\rho_0(y)$ with 1, and the solid angle $-\pi + \sum_{i=1}^3 \mathrm{dih}_i$ with zero in the Definition 7.39 of τ_{tri}. The does not use the assumption that $y_1 = 2$.

To establish the second claim, we write each of the three terms

$$\pi - \text{dih}_1, \quad \text{dih}_2, \quad \text{dih}_3 .$$

in the form $f_i\sqrt{\Delta}$, for $i = 1, 2, 3$. The explicit formulas for dihedral angles show that f_i is an analytic function of $y_1, \ldots, y_6$. When $y_1 = 2h_0$, we obtain a formula for τ_{tri} of the general form

$$\tau_{tri}(y_1, \ldots, y_6) = f(y_1, \ldots, y_6)\sqrt{\Delta(y_1^2, \ldots, y_6^2)}$$

for some analytic function f. A *computer calculation*[5] [21] shows that $f \geq 0$ on the domain given in the lemma. Hence τ_{tri} is also nonnegative.

We turn to the third claim. If $\text{dih}_1 = \pi$, then $\text{dih}_2 = \text{dih}_3 = 0$. If we make these substitutions into the formula for τ_{tri}, the claim follows immediately. □

The lemma is used to avoid numerical instabilities as follows. The three statements of the lemma correspond to the three cases given by (7.78). We may assume that the simplex $\{\mathbf{0}, \mathbf{w}_{2i}, \mathbf{w}_{2i+1}, \mathbf{w}_{2i+2}\}$ falls into one these cases. When the simplex approaches a planar configuration, that is as Δ approaches zero, we replace the term τ_{tri} with the lower bound given by the lemma, to avoid computing a nonanalytic term directly. By doing this, all of the computer calculations go through without trouble.

[5] `[5202826650 a]`

8

Tame Hypermap

Summary. *This chapter is the last of the three core chapters on the proof of the Kepler conjecture. If V is a finite set of vectors in* $\mathbb{R}^3$*, let*

$$\mathcal{L}(V) = \sum_{\mathbf{v}\in V} L(\|\mathbf{v}\|/2).$$

Let $\mathcal{B}$ *be the annulus* $\bar{B}(\mathbf{0}, 2h_0) \setminus B(\mathbf{0}, 2)$*, where* $\bar{B}(\mathbf{0}, r)$ *is the closed ball of radius r. By Corollary 6.100, the Kepler conjecture holds if every packing V contained in* $\mathcal{B}$ *satisfies*

$$\mathcal{L}(V) \le 12. \tag{8.1}$$

In this chapter, we assume that there exists a counterexample V to this inequality and reach a contradiction. A subset of extremal counterexamples is selected that is particularly well-suited for further analysis. Every extremal counterexample gives rise to a fan and a corresponding hypermap. A detailed study of these hypermaps leads to a long list of properties that all such hypermaps must possess. A tame hypermap is defined by these properties. Tameness is thus an umbrella term that covers a long list of loosely related properties.

An earlier chapter on hypermaps gives an algorithm that generates all restricted hypermaps with a given bound on the number of nodes. Every tame hypermap is restricted and has at most fifteen nodes. Hence, a list of all tame hypermaps can be obtained by generating all restricted hypermaps and filtering out those that are not tame. This algorithm has been implemented and executed in computer code. The result is an explicit list that classifies tame hypermaps up to isomorphism. This classification solves a major step of the packing problem.

Each tame hypermap H gives rise to a nonlinear optimization problem to maximize $\mathcal{L}(V)$ subject to the constraint that the hypermap associated with V is isomorphic to H. This nonlinear optimization problem has a relaxation in the form of a linear program with a maximum that is at least as large as the maximum of the nonlinear program. Each linear program has been solved by computer. In every case, after branching into subcases, the maximum is less than 12. *Hence, Inequality 8.1 always holds, so that the Kepler conjecture is confirmed.*

8.1 Definition

This section gives the definition of a tame hypermap, which is object of study in this chapter. The definition depends on a large set of parameters, which have been determined by computer experimentation. On the one hand, the idea is to define a class of hypermaps that is finite and small enough to classify without much trouble. On the other hand, we have to prove that every counterexample V of (8.1) has a hypermap that is tame. The smaller the class of hypermaps, the more difficult it becomes to relate them to the counterexample V. In the end, we seek a balance between these contrary demands.

Definition 8.2 (triangle, quadrilateral) Faces of cardinality three in a hypermap are called *triangles* and those of cardinality four are called *quadrilaterals*.

Definition 8.3 (type, (p,q,r)) The *type* of a node is defined to be a triple of nonnegative integers (p,q,r), where p is the number of triangles meeting the node, q is the number of quadrilaterals meeting it, and r is the number of other faces meeting it, so that $p+q+r$ is the total number of faces meeting the node.

8.1.1 weight assignment

We call the constant $\text{tgt} = 1.541$, which arises repeatedly in this chapter, the *target*. The constant's name comes from its function as a measure of optimality. Below, in the definition of a tame hypermap (Definition 8.7), any hypermap that overshoots the target is not tame. We define two tables of constants: b and d. (There is no a or c.)

Definition 8.4 (b) Define $b : \mathbb{N}^2 \to \mathbb{R}$ by $b(p, q) = \text{tgt}$, except for the values in the following table.

	$q = 0$	1	2	3	4
$p = 0$	tgt	tgt	tgt	0.618	0.97
1	tgt	tgt	0.656	0.618	tgt
2	tgt	0.797	0.412	1.2851	tgt
3	tgt	0.311	0.817	tgt	tgt
4	0.347	0.366	tgt	tgt	tgt
5	0.04	1.136	tgt	tgt	tgt
6	0.686	tgt	tgt	tgt	tgt
7	1.450	tgt	tgt	tgt	tgt

Definition 8.5 (d) Define $d : \mathbb{N} \to \mathbb{R}$ by

$$d(k) = \begin{cases} 0 & k \leq 3, \\ 0.206 & k = 4, \\ 0.4819 & k = 5, \\ 0.712 & k = 6, \\ \text{tgt} = 1.541 & \text{otherwise.} \end{cases}$$

Definition 8.6 (weight assignment) A *weight assignment* of a hypermap H is a real-valued function τ on the set of faces of H. A weight assignment τ is *admissible* if the following properties hold.

1. (BOUND A) Let v be any node of type $(5, 0, 1)$ and let A be the set of triangles meeting that node. Then

$$\sum_{F \in A} \tau(F) \geq 0.63.$$

2. (BOUND B) If a node v has type $(p, q, 0)$, then

$$\sum_{F : v \cap F \neq \emptyset} \tau(F) \geq b(p, q).$$

3. (BOUND D) If the face F has cardinality k, then $\tau(F) \geq d(k)$.

The sum $\sum_F \tau(F)$ (over all faces) is called the *total weight*.

8.1.2 hypermap property

Definition 8.7 (tame) A hypermap is *tame* (Figure 8.1) if it satisfies the following conditions.

1. (PLANAR) The hypermap is plain and planar.
2. (SIMPLE) The hypermap is connected and simple. In particular, each intersection of a face with a node contains at most one dart.
3. (NONDEGENERATE) The edge map e has no fixed points.
4. (NO LOOPS) The two darts of each edge lie in different nodes.
5. (NO DOUBLE JOINS) At most one edge meets any two (not necessarily distinct) nodes.
6. (FACE COUNT) The hypermap has at least three faces.
7. (FACE SIZE) The cardinality of each face is at least three and at most six.
8. (NODE COUNT) There are thirteen, fourteen, or fifteen nodes.
9. (NODE SIZE) The cardinality of every node is at least three and at most seven.
10. (NODE TYPES) If a node has type (p, q, r) with $p + q + r \geq 6$ and $r \geq 1$, then $(p, q, r) = (5, 0, 1)$.
11. (WEIGHTS) There exists an admissible weight assignment of total weight less than tgt $= 1.541$.

8.2 Contravening Hypermap

8.2.1 standard fan

Let $V \subset \mathcal{B}$ be a packing. Define $E_{std} = E_{std}(V)$ and $E_{ctc} = E_{ctc}(V)$ by

$$E_{std} = \{\{\mathbf{v}, \mathbf{w}\} \subset V \ : \ 0 < \|\mathbf{v} - \mathbf{w}\| \leq 2h_0\}, \qquad (8.8)$$

$$E_{ctc} = \{\{\mathbf{v}, \mathbf{w}\} \subset V \ : \ \|\mathbf{v} - \mathbf{w}\| = 2\} \subset E_{std}. \qquad (8.9)$$

Lemma 8.10 *Let $V \subset \mathcal{B}$ be a packing. If $E = E_{std}$ or $E = E_{ctc}$, then (V, E) is a fan.*

Definition 8.11 The fans (V, E_{std}) and (V, E_{ctc}) are called the *standard fan* and the *contact fan*, respectively.

Proof (V, E_{std}) is a fan by Lemma 8.12. Since $E_{ctc} \subset E_{std}$, it follows from Lemma 5.3 that (V, E_{ctc}) is a fan. □

Lemma 8.12 (standard fan) *Let $V \subset \mathcal{B}$ be a packing. Then (V, E_{std}) is a fan.*

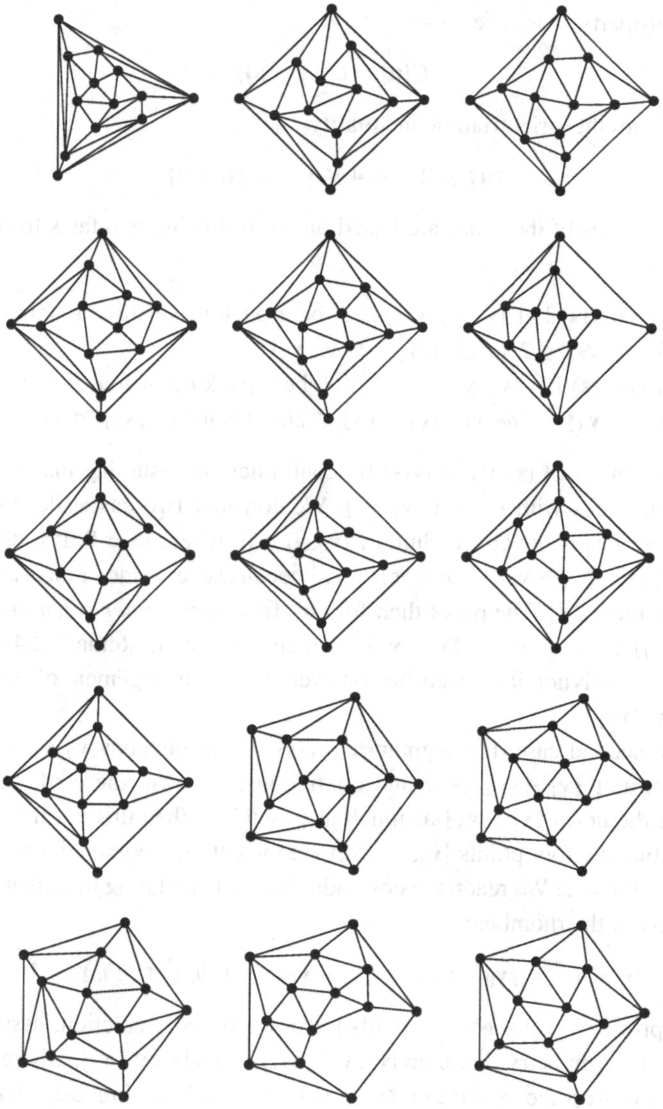

Figure 8.1 Here are some of the tens of thousands of planar graphs whose hypermaps are tame. The ones depicted here are the ones that are the most difficult to eliminate through linear programming.

Proof The properties (CARDINALITY), (ORIGIN), and (NONPARALLEL) (on page 113) follow by the methods of Remark 5.6.

(INTERSECTION) Some geometrical reasoning is required to establish the intersection property. The case

$$C\{\mathbf{u}\} \cap C\{\mathbf{v}\} = \{\mathbf{0}\}$$

follows from the strict triangle inequality

$$\|\mathbf{u}\| \le 2h_0 < 4 \le \|\mathbf{v}\| + \|\mathbf{u} - \mathbf{v}\|.$$

The other cases of the proof are based on the following two facts from Tarski arithmetic.

1. (THREE POINTS) Let $\{\mathbf{v}_1, \mathbf{v}_2, \mathbf{v}_3\} \subset \mathcal{B}$ be a packing of three points. Assume that $\|\mathbf{v}_1 - \mathbf{v}_3\| \le 2h_0$. Then $\mathbf{v}_2 \notin C\{\mathbf{v}_1, \mathbf{v}_3\}$.
2. (FOUR POINTS) Let $\{\mathbf{v}_1, \mathbf{v}_2, \mathbf{v}_3, \mathbf{v}_4\} \subset \mathcal{B}$ be a packing of four points. Assume that $\|\mathbf{v}_1 - \mathbf{v}_3\| \le 2h_0$ and $\|\mathbf{v}_4 - \mathbf{v}_2\| \le 2h_0$. Then $C\{\mathbf{v}_1, \mathbf{v}_3\} \cap C\{\mathbf{v}_4, \mathbf{v}_2\} = \{\mathbf{0}\}$.

We give a proof of (THREE POINTS) by contradiction, assuming that some configuration exists with $\mathbf{v}_2 \in C\{\mathbf{v}_1, \mathbf{v}_3\}$. We consider two cases, depending on whether $\mathbf{v}_2$ lies in the convex hull of $\{\mathbf{v}_0, \mathbf{v}_1, \mathbf{v}_3\}$, where $\mathbf{v}_0 = \mathbf{0}$. If it does lie in this hull, then $\|\mathbf{v}_2 - \mathbf{v}_i\| \le 2h_0$ for $i = 1, 3$ because every side of the triangle has length at most $2h_0$. The proof then follows from a *computer calculation*[1] [21] that $\Delta(x_{ij}) > 0$ for $x_{ij} = \|\mathbf{v}_i - \mathbf{v}_j\|^2$, which contradicts Remark 2.44, asserting that the polynomial Δ vanishes on every planar arrangement of four points $\{\mathbf{v}_0, \mathbf{v}_1, \mathbf{v}_3, \mathbf{v}_2\}$.

In the second case, the segment $\mathrm{conv}\{\mathbf{v}_1, \mathbf{v}_3\}$ meets $\mathrm{conv}\{\mathbf{v}_0, \mathbf{v}_2\}$. We drop the constraint $\|\mathbf{v}_1 - \mathbf{v}_3\| \ge 2$ and continuously deform the configuration to contract the norm $\|\mathbf{v}_1 - \mathbf{v}_3\|$ as much as possible. When this norm is as small as possible, the four points $\{\mathbf{v}_0, \ldots, \mathbf{v}_3\}$ are the extreme points of a rhombus of side length $r = 2$. We reach the contradiction by calculating the squares of the diagonals of the rhombus:

$$16 = 4r^2 = \|\mathbf{v}_0 - \mathbf{v}_2\|^2 + \|\mathbf{v}_1 - \mathbf{v}_3\|^2 \le (2h_0)^2 + (2h_0)^2 < 16.$$

The proof of (FOUR POINTS) is also a proof by contradiction, assuming as we may by symmetry that $\mathrm{conv}\{\mathbf{v}_1, \mathbf{v}_3\}$ meets $\mathrm{conv}\{\mathbf{v}_0, \mathbf{v}_4, \mathbf{v}_2\}$, where $\mathbf{v}_0 = \mathbf{0}$. Again, we drop the constraint $\|\mathbf{v}_1 - \mathbf{v}_3\| \ge 2$ and continuously deform the configuration to contract the norm $\|\mathbf{v}_1 - \mathbf{v}_3\|$ as much as possible. By (THREE POINTS), the continuous contraction does not pass through any configurations in which four of the points are coplanar. When the norm is as small as possible the lengths satisfy $\|\mathbf{v}_i - \mathbf{v}_j\| = 2$ for $i = 0, 2, 4$ and $j = 1, 3$. Let $\mathbf{v}_0^*$ be the reflection of $\mathbf{v}_0$ through the circumcenter of $\{\mathbf{v}_0, \mathbf{v}_2, \mathbf{v}_4\}$. The four points $\{\mathbf{v}_1, \mathbf{v}_3, \mathbf{v}_0, \mathbf{v}_0^*\}$ are

[1] [TVAWGDR] A certain configuration of four points cannot be coplanar.

the extreme points of a rhombus of side length $r = 2$. We reach the following contradiction by calculating the squares of the diagonals of the rhombus:

$$16 = 4r^2 = \|\mathbf{v}_0 - \mathbf{v}_0^*\|^2 + \|\mathbf{v}_1 - \mathbf{v}_3\|^2 \le (2\eta(2h_0, 2h_0, 2h_0))^2 + (2h_0)^2 < 16,$$

where $\eta(a, b, c)$ is the circumradius of a triangle with sides a, b, c. □

8.2.2 surrounded and isolated nodes

The purpose of this section is to select a special class of counterexamples to Inequality 8.1.

Definition 8.13 (isolated, surrounded) Let (V, E) be a fan. Say that $\mathbf{v} \in V$ is *isolated* in the fan if $E(\mathbf{v})$ is empty. Say that $\mathbf{v} \in V$ is *surrounded* in the fan if the azimuth angles of all darts at the node $\mathbf{v}$ are less than π. (In particular, the cardinality of $E(\mathbf{v})$ is at least three.)

The following lemma appears in Schütte and van der Waerden [37].

Lemma* 8.14 *Given any packing $V \subset \mathcal{B}$, there exists a bijection $\phi : V \to V'$ with a packing V' such that $\|\mathbf{v}\| = \|\phi(\mathbf{v})\|$ and such that every node $\mathbf{v}'$ in the contact fan of V' is either isolated or surrounded.*

Proof Consider all finite packings in bijective correspondence with V such that the bijection preserves distances to the origin. Among these packings, select one, V', with the largest number of isolated points in the contact graph.

We assume for a contradiction that there is a point $\mathbf{v} \in V'$ that is not isolated and not surrounded. It has a dart in the contact fan with angle at least π. Perturb $\mathbf{v}$ away from the contacts, making it isolated, while preserving its distance from the origin. The perturbed packing V'' has at least one more isolated point than V', in violation of the supposed maximality of V'. Hence, the conclusion of the lemma holds for V'. □

The following lemma shows that the corresponding result holds for the standard fan. Beware! The set of isolated and surrounded nodes depends on a choice of fan. The next proof makes frequent use of two different fans (V, E_{std}) and (V, E_{ctc}), which have different sets of isolated and surrounded nodes, even though the set V is the same in both cases.

Lemma 8.15 *Assume that there exists a counterexample to Inequality 8.1. Then there also exists a counterexample V to the inequality with the following properties.*

1. $V \subset \mathcal{B}$ is a packing.
2. $\mathcal{L}(V) > 12$, and no finite packing in $\mathcal{B}$ attains a value larger than $\mathcal{L}(V)$.
3. The cardinality of V is thirteen, fourteen, or fifteen.
4. For every $\mathbf{v} \in V$, the node $\mathbf{v}$ is surrounded or isolated in the standard fan (V, E_{std}).

Proof The cardinality of a counterexample is between thirteen and fifteen, by Lemma 6.110. The space of all packings $V \subset \mathcal{B}$ of cardinality thirteen, fourteen, or fifteen is a compact topological space. The function $V \mapsto \mathcal{L}(V)$ is continuous. Hence by real analysis, the maximum is achieved by some $W \subset \mathcal{B}$. The packing W has the first three properties of the lemma.

Let $\mathcal{V}$ be given by

$$\begin{aligned}\mathcal{V} = \{(V, \phi) \ :& \\ & V \subset \mathcal{B} \text{ is a packing,} \\ & \phi : W \to V \text{ is a bijection, and} \\ & \forall \mathbf{v} \in W.\ \|\phi(\mathbf{v})\| = \|\mathbf{v}\|. \\ \}. &\end{aligned}$$

The norm condition gives $\mathcal{L}(W) = \mathcal{L}(V)$ for all $(V, \phi) \in \mathcal{V}$. By fixing an enumeration $W = \{\mathbf{v}_0, \ldots, \mathbf{v}_{k-1}\}$, we identify $\mathcal{V}$ with a compact subspace of $\mathcal{B}^k \subset \mathbb{R}^{3k}$. Also, $\mathcal{V}$ is nonempty, because it contains (W, I_W). If $(V, \phi) \in \mathcal{V}$, then V has the first three properties of the lemma.

For any pair (V, ϕ) in $\mathcal{V}$ and $\mathbf{v} \in V$, let $c(V, \mathbf{v})$ be the minimum distance from $\mathbf{v}$ to a point in $V \setminus \{\mathbf{v}\}$. For $i = 0, \ldots, k-1$ define $c_i = c_i(V)$, by ordering the real numbers $c(V, \mathbf{v})$, $\mathbf{v} \in V$, in increasing order:

$$c_0 \le c_1 \le c_2 \le \cdots \le c_{k-1}.$$

The functions $c_i : \mathcal{V} \to \mathbb{R}$ are continuous on the compact space $\mathcal{V}$. There is a nonempty compact subset $\mathcal{V}_0$ of $\mathcal{V}$ on which c_0 attains its maximum value. Continuing recursively, there is a nonempty compact subset $\mathcal{V}_{i+1}$ of $\mathcal{V}_i$ on which c_{i+1} attains its maximum value.

Any $(V, \phi) \in \mathcal{V}_n$ has the fourth enumerated property of the lemma. Otherwise, there exists some node $\mathbf{v} \in V$ that is neither isolated nor surrounded in the standard fan. There exist i and j such that

$$\{k \ : \ c_k(V) = c(V, \mathbf{v})\} = \{k \ : \ i \le k \le j\}.$$

As $\mathbf{v}$ is not isolated in the standard fan, it follows that $c(V, \mathbf{v}) \le 2h_0$. As $\mathbf{v}$ is not surrounded in the standard fan in the cyclic order on

$$\{\mathbf{w} \in V \ : \ \|\mathbf{w} - \mathbf{v}\| = c(V, \mathbf{v})\},$$

some azimuth angle is at least π. Thus, there is a direction in which $\mathbf{v}$ can be perturbed that fixes $c_0, \ldots, c_{i-1}$, does not decrease $c_i, \ldots, c_{j-1}$, and increases c_j. This is contrary to the defining property of $(V, \phi) \in \mathcal{V}_n \subset \mathcal{V}_j \subset \mathcal{V}_i$. This establishes the claim. □

Lemma 8.16 *Assume that there exists a counterexample to Inequality 8.1. Then there also exists a counterexample V to the inequality with the following properties.*

1. *$V \subset \mathcal{B}$ is a packing.*
2. *$\mathcal{L}(V) > 12$, and no finite packing in $\mathcal{B}$ attains a value larger than $\mathcal{L}(V)$.*
3. *The cardinality of V is thirteen, fourteen, or fifteen.*
4. *Every node $\mathbf{v}$ is surrounded in the standard fan (V, E_{std}).*
5. *Every node $\mathbf{v}$ that is not surrounded in the contact fan (V, E_{ctc}) satisfies $\|\mathbf{v}\| = 2$.*

Proof By Lemma 8.15, some counterexample V satisfies the first three properties and in which every node is surrounded or isolated in the standard fan. By Lemma 6.112, if there are any isolated nodes in the standard fan, then it is not a counterexample. Hence, every node is surrounded in the standard fan.

A node $\mathbf{v}$ that is not surrounded in the contact fan satisfies $\|\mathbf{v}\| = 2$. Otherwise, the counterexample does not maximize $\mathcal{L}$. In fact, the packing that replaces $\mathbf{v}$ with $(1 - \epsilon)\mathbf{v}$ for sufficiently small $\epsilon > 0$ does better. □

Definition 8.17 (contravening) A finite packing V is a *contravening* packing if it satisfies the properties of Lemma 8.16. The hypermap $\text{hyp}(V, E_{std})$ is also said to be *contravening* when V is contravening.

8.3 Contravention is Tame

This section and the next one prove that every contravening hypermap is tame.

Let V be a contravening packing with standard fan $(V, E) = (V, E_{std})$ and let $H = \text{hyp}(V, E) = (D, e, n, f)$ be the hypermap attached to (V, E). The fan (V, E) is fully surrounded and a conforming fan by Lemma 5.42. We recall some of the properties of conforming fans from Section 5.3.2. The hypermap H is plain, planar, connected, and simple. The set of topological components of $Y(V, E)$ is in bijection with the set of faces of H. For each face of H, the corresponding component U_F is eventually radial with solid angle

$$\text{sol}(U_F) = 2\pi + \sum_{x \in F} (\text{azim}(x) - \pi). \tag{8.18}$$

Recall that

$$\sum_F \mathrm{sol}(U_F) = 4\pi. \tag{8.19}$$

Set

$$h(x) = \|\mathrm{node}(x)\|/2,$$

where as usual, node : $F \to V$ is given by

$$x \mapsto \mathrm{node}(x); \quad x = (\mathrm{node}(x), \ldots).$$

Define the weight function

$$\begin{aligned}
\tau(V, E, F) &= \sum_{x \in F} \mathrm{azim}(x)\left(1 + \frac{\mathrm{sol}_0}{\pi}(1 - L(h(x)))\right) + (\pi + \mathrm{sol}_0)(2 - k(F)) \\
&= \mathrm{sol}(U_F) + (2 - k(F))\,\mathrm{sol}_0 - \frac{\mathrm{sol}_0}{\pi}\sum_{x \in F} \mathrm{azim}(x)(L(h(x)) - 1) \\
&= \mathrm{sol}(U_F)\left(1 + \frac{\mathrm{sol}_0}{\pi}\right) - \frac{\mathrm{sol}_0}{\pi}\sum_{x \in F} \mathrm{azim}(x)(L(h(x))),
\end{aligned} \tag{8.20}$$

where sol_0 is the solid angle of a spherical equilateral triangle with a side of arclength $\pi/3$, and $k(F)$ is the cardinality of F. These formulas are equivalent by Equation 8.18. The first expression for $\tau(V, E, F)$ is particularly convenient because it expresses τ as a sum of local contributions from each dart. The main conjecture may be expressed in the following alternative form.

Lemma 8.21 (target) *Let V be a contravening packing. Then*

$$\sum_F \tau(V, E_{std}, F) < 4\pi - 20\,\mathrm{sol}_0 .$$

The sum runs over the faces F of $\mathrm{hyp}(V, E_{std})$.

Proof Use the formula (8.20). The sum of the solid angles U_F is 4π, and the sum of the azimuth angles at each node is 2π. Thus,

$$\sum_F \tau(V, E_{std}, F) = 4\pi\left(1 + \frac{\mathrm{sol}_0}{\pi}\right) - \left(\frac{\mathrm{sol}_0}{\pi}\right) 2\pi \sum_V L(\|\mathbf{v}\|/2) \tag{8.22}$$

$$= (4\pi - 20\,\mathrm{sol}_0) + 2\,\mathrm{sol}_0(12 - \mathcal{L}(V)). \tag{8.23}$$

In a contravening packing, $2\,\mathrm{sol}_0(12 - \mathcal{L}(V)) < 0$. The result ensues. □

Remark 8.24 The significance of the constant tgt = 1.541 is that it is a convenient rational approximation to the constant $4\pi - 20\,\mathrm{sol}_0 \approx 1.54065$, which appears in Lemma 8.21.

The theorem that follows is one of the main results of this chapter. The subsequent sections present the proof in a sequence of steps.

Theorem 8.25 *Let V be a contravening packing. Then the weight assignment $F \mapsto \tau(V, E_{std}, F)$ on $H = \text{hyp}(V, E_{std})$ is admissible. Moreover, the hypermap H is tame with weight assignment $\tau(V, E_{std}, *)$.*

8.3.1 general properties

Many of the properties of tameness are trivial or have been established in earlier sections. The following lemma quickly disposes of many of the properties of tameness.

Lemma 8.26 *A contravening hypermap H satisfies properties* (PLANAR), (SIMPLE), (NONDEGENERATE), (NO LOOPS), (NO DOUBLE JOINS), (FACE COUNT), (NODE COUNT), *and the first part of* (NODE SIZE) *of tameness on page 238.*

Proof The hypermap is plain, planar, connected, and simple by the general results established in the chapter on fan. That chapter also shows that the hypermap attached to a fan satisfies properties (NONDEGENERATE), (NO LOOPS), and (NO DOUBLE JOINS).

Properties (FACE COUNT) *and the first half of property* (NODE SIZE) *hold.* Indeed, every node is surrounded, meaning that the azimuth angles of the darts at the node are less than π. As the angles around the node sum to 2π, there are at least three darts in the node. Each of the darts in the node leads into a different face by property (SIMPLE).

Finally, property (NODE COUNT) has already been established in Lemma 8.16.

□

We list the properties that remain and the lemma that proves them.

(NODE SIZE) (Lemma 8.28, second part), (WEIGHTS BOUND A) (Lemma 8.30),
(NODE TYPES) (Lemma 8.30), (WEIGHTS BOUND B) (Lemma 8.29),
(FACE SIZE) (Lemma 8.31), (WEIGHTS BOUND D) (Lemma 8.33).

8.3.2 properties of nodes

Lemma 8.27 *Let H be a contravening hypermap. For every dart x in a triangular face of H,*

$$0.852 \le \text{azim}(x) \le 1.9.$$

For every dart x in a nontriangular face of H,

$$1.15 \le \text{azim}(x) < \pi.$$

Consequently, if a node $\mathbf{v}$ *has type* (p, q, r)*, then* $(p, q + r)$ *must be one of the following pairs:*

$$(0,3),\ (0,4),\ (0,5),\ (1,2),\ (1,3),\ (1,4),\ (2,1),\ (2,2),\ (2,3),$$
$$(3,1),\ (3,2),\ (3,3),\ (4,0),\ (4,1),\ (4,2),\ (5,0),\ (5,1),\ (6,0),\ (6,1),\ (7,0).$$

Proof The angle bounds are a *computer calculation*[2] [21]. The sum of the azimuth angles around a node satisfies:

$$p(0.852) + (q + r)(1.15) \le 2\pi < p(1.9) + (q + r)\pi,$$

and the pairs satisfying these constraints are listed. □

Lemma 8.28 *Contravening hypermaps satisfy the second part of property* (NODE SIZE) *of tameness. That is, the cardinality of every node is at most seven.*

Proof For every pair $(p, q + r)$ in the list of Lemma 8.27, $p + q + r \le 7$. □

Lemma 8.29 *Let* $\mathbf{v}$ *be a node of type* $(p, q, 0)$ *in a contravening hypermap. Then the property* (BOUND B) *of a admissible weight assignment holds:*

$$\sum_{F \in A} \tau(V, E, F) \ge b(p, q),$$

where the sum runs over the set A of faces that meet the node $\mathbf{v}$.

Proof A *computer calculation*[3] [21] gives a list of nonlinear inequalities for $\tau(V, E, F)$ when F is a triangle or quadrilateral. Each nonlinear inequality has the form

$$\tau(V, E, F) \ge a\,\text{azim}(x) + b$$

for some $a, b \in \mathbb{R}$, where x is the uniquely determined dart at the node $\mathbf{v}$ in the face F. These nonlinear inequalities admit a linear relaxation as follows. For each a, b, there is a corresponding linear inequality

$$t(F) \ge a\, z(F) + b,$$

where $t(F)$ and $z(F)$ are variables indexed by $F \in A$. The linear relaxation asks for the minimum of

$$\sum_{F \in A} t(F)$$

[2] [KCBLRQC]
[3] [KCBLRQC]

subject to these linear inequalities and the constraint

$$2\pi = \sum_{F\in A} z(F).$$

We have executed the linear program as a *computer calculation*[4] [21] for each of the types $(p, q, 0)$ of Lemma 8.27. The given constants are obtained from the (downward rounded) solutions to these linear programs. □

Lemma 8.30 *Every contravening hypermap satisfies properties* (NODE TYPES) *and* (WEIGHT BOUND A) *of tameness: If a node has type* (p, q, r) *with* $p+q+r \geq 6$ *and* $r \geq 1$*, then* $(p, q, r) = (5, 0, 1)$*. Furthermore, assume the type is* $(5, 0, 1)$ *and let* A *be the set of five triangles at the node* $\mathbf{v}$*. Then*

$$\sum_{F\in A} \tau(V, E, F) > a,$$

where $a = 0.63$.

Proof We have also checked these conclusions by computer. The same set of nonlinear inequalities is used as in Lemma 8.29, and the linear relaxation is constructed in the same way as the proof of that lemma. The linear programming bounds exceed the constant tgt in the cases excluded in the conclusion of the lemma. The constant a is the downward rounded solution to the linear program for $(5, 0, 1)$. □

8.3.3 faces

Lemma 8.31 *Property* (FACE SIZE) *holds. That is, every face of a contravening hypermap* $\mathrm{hyp}(V, E_{std})$ *has cardinality at least three and at most six.*

Proof The lower bound holds because the hypermap has no loops or double joins. (See Lemma 4.55.)

Let F be a face of the hypermap of cardinality k. The inequalities $\mathcal{L}(V) \geq 12$,

[4] [KCBLRQC]

$L(h_i) \le 1$, and $\mathrm{card}(V) \le 15$ imply that

$$\begin{aligned} 12 &\le \mathcal{L}(V) \\ &= \sum_{i=1}^{k} L(h_i) + \sum_{i=k+1}^{\mathrm{card}(V)} L(h_i) \\ &\le \sum_{i=1}^{k} L(h_i) + (\mathrm{card}(V) - k) \\ &\le \sum_{i=1}^{k} \frac{h_0 - h_i}{h_0 - 1} + (15 - k), \end{aligned}$$

or

$$\sum_{i=1}^{k} h_i \le -(h_0 - 1)(k - 3) + k h_0.$$

By Lemma 7.38 and Lemma 6.107, the perimeter P of F satisfies

$$2\pi \ge P \ge \sum_{i=1}^{k} \mathrm{arc}(2h_i, 2h_{i+1}, 2) \ge 2 \sum_{i=1}^{k} g(h_i)$$

where $g(h) = \arccos(h/2) - \pi/6$ and where $h_i \in [1, h_0]$.

A calculation of the second derivative shows that the function g is concave. Let $g_1(h) = ah + b \approx -0.61h + 1.13$ be the linear secant approximation to g_1 on $[1, h_0]$:

$$g_1(1) = g(1), \quad g_1(h_0) = g(h_0).$$

Then $a < 0$ and

$$\begin{aligned} 2\pi &\ge 2 \sum_{i=1}^{k} g_1(h_i) \\ &\ge 2a \sum_{i=1}^{k} h_i + 2kb \\ &\ge 2a\,[-(h_0 - 1)(k - 3) + k h_0] + 2kb. \end{aligned}$$

Solving for k, we get $k < 7$. □

8.4 Admissibility

The main result (Lemma 8.33) of this section is a proof that for every contravening hypermap $H = \mathrm{hyp}(V, E)$, the function $\tau(V, E, *)$ is an admissible weight assignment on H.

Lemma 8.32 *Let V be a contravening packing, and let F be any face of* hyp(V, E_{std}). *Then*

$$\tau(V, E_{std}, F) = \tau(V', E', F),$$

where (V', E', F) is the localization of (V, E_{std}) along the face F.

Proof The value $\tau(V, E, F)$ is expressed entirely in terms of $\|\mathbf{v}\|$ for $\mathbf{v} \in V' \subset V$ and in terms of azim(x) for $x \in F$. By Lemma 7.8, the terms azim(x) are the same, whether calculated in terms of the hypermap of (V, E_{std}) or in terms of that of (V', E'). □

Lemma 8.33 *Let V_0 be a contravening packing, and let F be any face of* hyp(V_0, E_{std}) *of cardinality k. Then property* (WEIGHT BOUND D) *holds. That is, $\tau(V_0, E_{std}, F) \geq d(k)$.*

Proof Let V_0 be a contravening packing. By Lemma 8.32, $\tau(V_0, E_{std}, F) = \tau(V, E, F)$, where (V, E, F) is the localization of (V_0, E_{std}) along F.

The tuple (V, E, F) satisfies the conditions of Theorem 7.43. Indeed, every property can be verified in turn. The properties (PACKING) and (ANNULUS) result from the assumption that V is contravening packing.

(NONREFLEXIVE LOCAL FAN) The localization (V, E, F) is a nonreflexive local fan by Lemma 7.9. (DIAGONAL) If $\|\mathbf{v} - \mathbf{w}\| < 2h_0$ and $\mathbf{v} \neq \mathbf{w}$, then $\{\mathbf{v}, \mathbf{w}\} \in E_{std}$. (CARD) The bound $3 \leq \text{card}(F) \leq 6$ follows from Lemma 8.31.

By Lemma 7.44 and Theorem 7.43,

$$\tau(V_0, E_{std}, F) = \tau(V, E, F) \geq d(k).$$

This completes the proof. □

Remark 8.34 It is helpful to keep in mind the origin of the constants $d(k)$. Although the proof of Lemma 8.33 does not produce sharp lower bounds on $\tau(V, E, F)$, the statement of the lemma is motivated by the configurations that follow. Consider a nonplanar polygon whose vertices lie on a sphere of radius 2, with k sides all of length $y_{i,i+1} = 2$, heights $y_i = 2$, and $k - 3$ diagonals of length $2h_0$: $y_{0,j} = 2h_0$ for $j = 2, \ldots, k - 2$ (Figure 8.2). Let V be the set of vertices of the polygon, let (V, E_{ctc}) be its contact fan, and let F be the face of hyp(V, E_{ctc}) representing the "interior" of the polygon. Evaluating τ on these rigid configurations gives

$$\tau(V, E_{ctc}, F) \approx \begin{cases} 0.20612 & k = 4 \\ 0.48356 & k = 5 \\ 0.760993 & k = 6. \end{cases}$$

These calculations suggested the values of constants $d(k)$. The constants $d(k)$ are slightly smaller than these calculated values.[5]

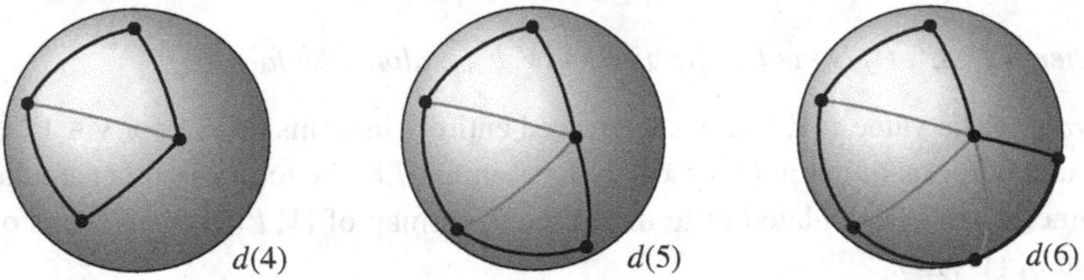

Figure 8.2 The fans associated with these spherical polygons attain values of τ close to the lower bounds $d(k)$. The choice of constants $d(k)$ was based on these examples. The points lie on a sphere of radius 2. The polygon is triangulated with diagonals of Euclidean length $2h_0$.

Definition 8.35 (opposite) The *opposite* of a hypermap (D, e, n, f) is the hypermap (D, fn, n^{-1}, f^{-1}).

Lemma 8.36 *A hypermap is tame if and only if its opposite hypermap is tame as well.*

Lemma 8.37 *Every tame hypermap is a restricted hypermap.*

Proof By definition, a tame hypermap is nonempty, connected, plain, planar, and simple. The edge and node maps have no fixed points. The cardinality of every face is at least three. These are also precisely the defining properties of a restricted hypermap. □

8.5 Linear Programs

This is a short section, but it represents a major part of the proof of the Kepler conjecture. It is short only because the calculations are better expressed as computer code than as published text. The code appears at the project website [21].

The classification result in this section is one of the main results of this book. All of the work to prove the classification algorithm has been completed in the chapter on hypermaps. A list of hypermaps appears at [21]. The following theorem has been established by executing a computer program that generates tame hypermaps. The program generates restricted hypermaps and filters out

[5] Note $\tau(2.1028, 2, 2, 2, 2.52, 2.52) \approx 0.275951 < 0.277433 \approx \tau(2, 2, 2, 2, 2.52, 2.52)$.

those that are not tame. About 25,000 tame hypermaps occur in the output of the program. Several examples appear in Figure 8.1.

Theorem 8.38 *Every tame hypermap is isomorphic to a hypermap in the list [21] or is isomorphic to the opposite of a hypermap in the list.*

Because of this classification, we may attach an explicit linear program to each tame hypermap. For each tame hypermap H there is a configuration space $\mathcal{V}_H$ of all contravening packings $V \subset \mathcal{B}$, whose standard fan is isomorphic to H or to the opposite of H.

A nonlinear optimization problem asks for the maximum of

$$\mathcal{L}(V) = \sum_{\mathbf{v} \in V} L(\|\mathbf{v}\|/2) \tag{8.39}$$

over all $V \in \mathcal{V}_H$.

Theorem 8.40 *Let H be a tame hypermap that appears in the explicit list [21]. Let $V \in \mathcal{V}_H$. Then*

$$\mathcal{L}(V) < 12.$$

Proof The linear program comes as a linear relaxation of this nonlinear optimization problem on $\mathcal{V}_H$. That is, the optimal solution of the linear program has value at least as great as the corresponding nonlinear problem. When a single linear program is not sufficient, branch and bound methods replace the single linear program with a finite sequence of linear progams. About 50,000 linear programs arise. By showing that the value of each linear program is less than 12, we conclude that the maximum of (8.39) is less than 12. □

Theorem 8.41 *The Kepler conjecture holds: no packing of congruent balls in three-dimensional Euclidean space has density greater than the face-centered cubic packing (of density $\pi/\sqrt{18}$).*

Proof We take the precise meaning of the Kepler conjecture to be that given in Remark 6.16. We argue by contradiction. If the Kepler conjecture is false, then by Lemma 6.100, there exists a finite packing $V \subset \mathcal{B}$ such that

$$\mathcal{L}(V) > 12. \tag{8.42}$$

By Lemma 8.16 and Definition 8.17, we may assume that V is a contravening packing. Let (V, E_{std}) be the standard fan of V, and let H be the fan's hypermap. By Theorem 8.25, the hypermap H is tame. By Theorem 8.38, either H or its opposite is isomorphic to a tame hypermap in the classification list. By the definition of $\mathcal{V}_H$, we have $V \in \mathcal{V}_H$. By Theorem 8.40,

$$\mathcal{L}(V) < 12, \quad (8.43)$$

for all $V \in \mathcal{V}_H$. This contradicts Inequality 8.42. □

8.6 Strong Dodecahedral Theorem

The same methods that have been used to prove the Kepler conjecture can be used to prove some other longstanding conjectures in discrete geometry. This section gives a proof of the strong dodecahedral conjecture.

Earlier sections are written in a formal blueprint style. Complete proofs are provided, even for statements that might be viewed as geometrically obvious. In this section, we relax our standards of proof just a bit. What we write is still a proof by traditional mathematical standards, but not as detailed as earlier chapters.

Bezdek has conjectured that the Voronoi cell of smallest surface is the regular dodecahedron with unit inradius. This is the *strong dodecahedral conjecture* [4], [5].

Theorem 8.44 (strong dodecahedral conjecture) *The surface area of a Voronoi cell in a packing is at least the surface area of the regular dodecahedron with unit inradius.*

This section gives a proof of this theorem. We begin with some simple observations.

Remark 8.45 If a packing is saturated, then the surface area of a Voronoi cell is finite. *We may assume without loss of generality that the packing is saturated.* Indeed, consider a new facet F that is created on a Voronoi cell by the addition of a new point to the packing V. Let X be the polygonal boundary of the new facet. The area minimizing surface that has X as a boundary is the facet F. Thus the new facet replaces a surface of larger area with a surface of smaller area. That is, by saturating a packing, the surface area of a Voronoi cell can only decrease.

Remark 8.46 As in the proof of the Kepler conjecture, the truncation of a Voronoi cell is easier to study than the Voronoi cell itself. In order to obtain sharp bounds, the truncation must have no effect on the optimal Voronoi cell. This constraint forces the truncation parameter to be at least the circumradius $\sqrt{3}\tan\pi/5 \approx 1.258$ of the regular dodecahedron. The truncation parameter $\sqrt{2}$

that we use in the proof of the Kepler conjecture satisfies this constraint and is therefore well-suited for the strong dodecahedral conjecture.

Lemma 8.47 *Let $r_0 \geq 0$. The surface area of the Voronoi cell $\Omega(V, \mathbf{u}_0)$ is at least that of $\Omega(V, \mathbf{u}_0) \cap B(r_0)$, where $B(r_0)$ is the ball of radius r_0 centered at $\mathbf{u}_0$.*

Proof The surface element for a parameterized surface $r(\theta, \phi)$ in spherical coordinates is

$$r\sqrt{r_\theta^2 + (r^2 + r_\phi^2)\sin^2\phi}\; d\theta\, d\phi,$$

which is at least the surface element $r_0^2 \sin\phi\, d\theta\, d\phi$ of a sphere of radius r_0, provided $r(\theta, \phi) \geq r_0$. Hence, projection of a surface outside sphere onto the sphere is area decreasing. □

Fejes Tóth's classical dodecahedral conjecture is the corresponding conjecture about volumes rather than surface areas, asserting that the Voronoi cell of smallest volume is the regular dodecahedron of unit inradius. The strong dodecahedral conjecture yields the dodecahedral conjecture as a corollary.

Lemma 8.48 *If the surface area of a Voronoi cell is at least the surface area of a regular dodecahedron with unit inradius, then its volume is also at least that of a regular dodecahedron.*

Proof Let $A_1, \ldots, A_n$ be the areas of the facets of a Voronoi cell. Let $h_1, \ldots, h_n$ be the distances from the affine hulls of the facets to the center of the Voronoi cell. Then $h_i \geq 1$. Assume that $\sum A_i \geq A_D$, where A_D is the surface area of a regular dodecahedron. Then its volume is

$$\text{vol} = \sum A_i h_i/3 \geq \sum A_i/3 \geq A_D/3 = \text{vol}_D,$$

where vol_D is the volume of the regular dodecahedron. □

8.6.1 *D*-cells

The notation follows Section 6.2. Let V be a saturated packing. Let $\Omega(V, \mathbf{u}_0)$ be a Voronoi cell with Rogers's partition

$$\Omega(V, \mathbf{u}_0) = \bigcup \{R(\underline{\mathbf{u}}) \; : \; \underline{\mathbf{u}} \in \underline{V}(3), \quad d_0\underline{\mathbf{u}} = [\mathbf{u}_0]\}.$$

Definition 8.49 (B, D_k-cell) Let B be the ball of radius $\sqrt{2}$ centered at $\mathbf{u}_0$. We define D_k-cells for $k = 1, 2, 3, 4$ for each $\underline{\mathbf{u}} = [\mathbf{u}_0; \ldots; \mathbf{u}_3] \in \underline{V}(3)$ by

$$D_k(\underline{\mathbf{u}}) = \Omega(V, \mathbf{u}_0) \cap \mathrm{cell}(\underline{\mathbf{u}}, k),$$

where $\mathrm{cell}(\underline{\mathbf{u}}, k)$ is the Marchal k-cell of $\underline{\mathbf{u}}$.

A D_k-cell, which is a subset of $\Omega(V, \mathbf{u}_0) \cap B$, is the adaptation of a k-cell to the geometry of the strong dodecahedral conjecture.

Lemma 8.50 *Let V be a saturated packing and let $\mathbf{u}_0 \in V$. If the intersection of a D_i-cell with a D_j-cell is not a null set, then $i = j$ and the two cells are equal. The union of all the D_k-cells at $\mathbf{u}_0$ is $\Omega(V, \mathbf{u}_0) \cap B$.*

Proof This follows from the corresponding facts for cells in Lemma 6.61. Each null set is in fact a subset of a plane. □

8.6.2 surface area and dihedral angle

Every cell $D_k(\underline{\mathbf{u}})$ is eventually radial at $\mathbf{u}_0$ and has a solid angle $\mathrm{sol}(\mathbf{u}_0, D_k(\underline{\mathbf{u}}))$. Every cell $D_k(\underline{\mathbf{u}})$ has an *exposed* surface area $\mathrm{surf}(D_k(\mathbf{u}))$, the area of the intersection of $D_k(\underline{\mathbf{u}})$ and the boundary of $\Omega(V, \mathbf{u}_0) \cap B$. It consists of the sum of the areas of the analytic facets (linear or spherical surfaces) that do not meet the point $\mathbf{u}_0$. The total surface area of $\Omega(V, \mathbf{u}_0) \cap B$ is the sum of the exposed surface areas $\mathrm{surf}(D_k(\mathbf{u}))$.

We use the functions dih_i in (7.41) to introduce a function of six variables $y = (y_1, y_2, \ldots, y_6)$:

$$\mathrm{soly}(y) = \mathrm{dih}_1(y) + \mathrm{dih}_2(y) + \mathrm{dih}_3(y) - \pi.$$

By Girard's formula for the solid angle of a simplex (Lemma 3.23),

$$\mathrm{soly}(y_1, y_2, y_3, y_4, y_5, y_6) = \mathrm{sol}(\mathbf{u}_0, \mathrm{conv}\{\mathbf{u}_0, \mathbf{u}_1, \mathbf{u}_2, \mathbf{u}_3\})$$

when

$$y_i = \begin{cases} \|\mathbf{u}_0 - \mathbf{u}_i\|, & i \in \{1, 2, 3\}, \\ \|\mathbf{u}_j - \mathbf{u}_k\|, & i \in \{4, 5, 6\} \text{ and } \{i - 3, j, k\} = \{1, 2, 3\}. \end{cases}$$

Every cell $D_k(\underline{\mathbf{u}})$ has a set $E(k, \underline{\mathbf{u}})$ of distinguished edges (that is, the edges of the cell that extend from $\mathbf{u}_0$ to a midpoint $(\mathbf{u}_0 + \mathbf{u}_i)/2$) and a dihedral angle $\mathrm{dih}(e)$ for $e \in E(k, \underline{\mathbf{u}})$. Each edge has a length $h(e) \in [1, \sqrt{2}]$.

8.6.3 local inequality

We reduce the strong dodecahedral conjecture to an estimate from an earlier chapter (6.95) and a local inequality.

Definition 8.51 (a_D, b_D, y_D, v_D, f) Define constants a_D, b_D, y_D, and functions v_D, f as follows. Let $y_D \approx 2.1029$ be defined by the condition

$$\mathrm{soly}(2, 2, 2, y_D, y_D, y_D) = \pi/5.$$

For any $\mathbf{u}_i \in \mathbb{R}^3$, let $g(\mathbf{u}_0, \mathbf{u}_1, \mathbf{u}_2, \mathbf{u}_3)$ be the volume of the intersection of the convex hull of $S = \{\mathbf{u}_0, \ldots, \mathbf{u}_3\}$ with set of points closer to $\mathbf{u}_0$ than to any other point in S. When

$$\|\mathbf{u}_0 - \mathbf{u}_i\| = 2 \quad \text{and} \quad \|\mathbf{u}_i - \mathbf{u}_j\| = y \text{ for } i, j \geq 1, \tag{8.52}$$

this volume depends only on y. Write $v(y) = g(\mathbf{u}_0, \ldots, \mathbf{u}_3)$. Set

$$f(y; a, b) = v(y) + a\,\mathrm{soly}(2, 2, 2, y, y, y) + 3b\,\mathrm{dih}(2, 2, 2, y, y, y).$$

The linear system

$$f(y_D; a, b) = 0, \quad \frac{\partial f}{\partial y}(y_D; a, b) = 0 \tag{8.53}$$

has a unique solution in a, b with values $a = a_D \approx -0.581$, $b = b_D \approx 0.0232$.

Note that the regular dodecahedron has volume $20v(y_D)$ and surface area $60v(y_D)$. Also,

$$2\,\mathrm{soly}(2, 2, 2, y_D, y_D, y_D) = \mathrm{dih}(2, 2, 2, y_D, y_D, y_D) = 2\pi/5. \tag{8.54}$$

Lemma 8.55 (local inequality) *For any cell $D_k(\underline{\mathbf{u}})$*

$$\mathrm{surf}(D_k(\underline{\mathbf{u}})) + 3a_D\,\mathrm{sol}(D_k(\underline{\mathbf{u}})) + 3b_D \sum_{e \in E(k,\underline{\mathbf{u}})} L(h(e))\,\mathrm{dih}(e) \geq 0,$$

where L is the function of Definition 6.88. Equality holds precisely when the cell is a null set or a 4-cell with edges $(2, 2, 2, y_D, y_D, y_D)$.

For a cell $D_4(\mathbf{u})$ with parameters of the form (8.52), the local inequality reduces to the inequality $f(y; a_D, b_D) \geq 0$. The constants a_D and b_D are chosen so that $y = y_D$ is a critical point of f with value $f(y_D; a_D, b_D) = 0$. In particular, the local inequality asserts that f has a local minimum at $y = y_D$.

Proof The dimension of the set of all D_k-cells up to rigid motion is $\binom{k}{2}$, which is at most six. This cell inequality is a nonlinear inequality in a small number of variables and is verified by a *computer calculation*[6] [21]. □

Lemma 8.56 *The local inequality and the estimate* (6.95)

$$\sum L(h) \leq 12$$

imply the strong dodecahedral conjecture.

Proof Sum the local inequality over all the D_k-cells in a Voronoi cell. The solid angles sum to 4π and the dihedral angles around each edge sum to 2π:

$$\begin{aligned}
\text{surf}(\Omega) &\geq \text{surf}(\Omega \cap B)\\
&= \sum_{k, D_k\text{-cell}} \text{surf}(D_k(\underline{\mathbf{u}}))\\
&\geq -12\pi a_D - 6\pi\, b_D \sum L(h)\\
&\geq -12\pi a_D - 72\pi\, b_D\\
&= -60\text{soly}(2,2,2,y_D,y_D,y_D)a_D - 180\,\text{dih}(2,2,2,y_D,y_D,y_D)b_D\\
&= 60(\nu(y_D) - f(y_D; a_D, b_D))\\
&= 60\nu(y_D).
\end{aligned}$$

The final term is the surface area of a regular dodecahedron. □

The case of equality occurs only for the regular dodecahedron. We note that the strong dodecahedral conjecture follows from the same estimate (6.96) that is used to prove the Kepler conjecture.

Exercise 8.57 Recently, Musin and Tarasov solved the Tammes problem for $k = 13$ points on a sphere in dimension $n = 2$ [32]. In parting, we leave it as a challenging problem to adapt their solution to the framework of this book to obtain an independent solution to the Tammes problem.

[6] `[TNVWUGK]`

Appendix A

Credits

This book is just a blueprint, which gives instructions about how to construct the formal proof. *Flyspeck* is the name of an ongoing project to construct a formal proof of the Kepler conjecture in the HOL Light proof assistant, along the lines described in this book. The eventual aim of the project is to give a formal verification of the computer portions of the proof as well as the standard text portions of the proof. The project is about 80% complete as of May 2012. The source code for the project and information about the current project status are available at [21].

Here is the fine print about the current project status. (I hope that this status report is out of date by the time this book is printed.) There are four components to the formalization project (text, hypermaps, linear programs, and nonlinear inequalities), at various stages of completion.

1. The English mathematical text in this book has been fully formalized, with the following explicit omissions.
   a. The formal proof does not include the parts of the text (such as remarks, introductory passages, and auxiliary results in Section 8.6) that are not strictly a part of the proof of the Kepler conjecture.
   b. There are three specific sections of this book that describe the relationship between the text and the computer portions of the proof: Section 4.7.4 (Hypermap Algorithm), Section 7.4 (Main Estimate), and Section 8.5 (Linear Programs). They have not yet been formalized.
   c. There are still a few lemmas scattered throughout in the book that remain to be formalized. These lemmas are listed at [21]. Combined, these scattered lemmas amount to fewer than 300 lines of English proof text.
2. Formal verification of the hypermap generation program is complete [33].
3. Formal verification of linear programs is near completion [34] and [41].

The technology to verify linear programs in HOL Light has been developed and an efficient implementation has been made, but the actual formal verifications of the long list of linear programs remains to be made.

4. The technology to make formal verifications of nonlinear inequalities over the field of real numbers has been designed and implemented. These are slow verifications, and the software must be further optimized, before the verification of large collections of inequalities can be considered practical. This research will appear in a forthcoming doctoral thesis by A. Solovyev at the University of Pittsburgh. Earlier work on formal proofs of nonlinear inequalities is mentioned in [24].

The Flyspeck project has been a large team effort over a period of years, and my contributions have been just a fraction of the whole. This appendix cites their contributions. The principal formalizer of each chapter is indicated with an asterisk.

Text Chapter	**Author of formalization work**
Trigonometry	Nguyen Quang Truong*; Rute, Jason; Harrison, John; Vu Khac Ky
Volume	Harrison, John*; Nguyen Tat Thang
Hypermap	Tran Nam Trung*
Fan	Hoang Le Truong*; Harrison, John
Packing	Solovyev, Alexey*; Vu Khac Ky*; Nguyen Tat Thang; Hales, Thomas
Local Fan	Nguyen Quang Truong*; Hoang Le Truong*
Tame Hypermap	Solovyev, Alexey*; Dat Tat Dang; Trieu Thi Diep; Vu Quang Thanh; Vuong Anh Quyen

Code-Verification	**Author of formalization work**
Hypermap Generation	Nipkow, Tobias*; Bauer, Gertrud*
Linear Programs	Obua, Stephen*; Solovyev, Alexey*
Nonlinear Inequalities	Solovyev, Alexey*

Acknowledgments

I am grateful to Sam Ferguson for the years that he spent working on this problem with me. I also thank the early editors Robert MacPherson, Gabor Fejes Tóth, and Jeff Lagarias for the many improvements they brought to the original proof.

Many have worked tirelessly to make the Flyspeck formalization project a reality. I wish to thank Dang Tat Dat, Hoang Le Truong, Nguyen Duc Thinh, Nguyen Duc Tam, Nguyen Tat Thang, Nguyen Quang Truong, Ta Thi Hoai An, Tran Nam Trung, Trieu Thi Diep, Vu Khac Ky, Vu Quang Thanh, Vuong Anh Quyen, Mark Adams, Catalin Anghel, Jeremy Avigad, Gertrud Bauer, John Harrison, Mary Johnston, Laurel Martin, Sean McLaughlin, Tobias Nipkow, Steven Obua, Joe Pleso, Jason Rute, Alexey Solovyev, Erin Susick, Nicholas Volker, Matthew Wampler-Doty, Freek Wiedijk, and Roland Zumkeller.

I would like to give particular thanks to John Harrison for designing and implementing the most amazing piece of software I have ever encountered. He has contributed large bodies of formal mathematics to the libraries of the HOL Light proof assistant in order to make the Flyspeck project possible. Tobias Nipkow became involved in the project at an early stage and directed the graduate work of two students, Bauer and Obua, on the formal verification of computer code for the Flyspeck project. Mark Adams has taught me about how large-scale project should be organized and has skillfully managed Flyspeck group in Hanoi.

When I learned that my NSF proposal[1] had been funded, I considered turning down the funding because I worried that the deliverables I had put into the proposal were beyond my capacity. It was one thing to grandstand and quite another to implement the largest formalization project that had ever been attempted. In retrospect, I am indebted to NSF for its generous support of this project. It has supported this project in all of its aspects, including a two-month workshop on formal proof and the Flyspeck project at the Math Institute in Hanoi during the summer of 2009. This workshop and all that has grown out of it have been crucial for the success of this project.

I am also indebted to the Benter Foundation for a generous grant to rework the original proof along lines proposed by Marchal.

Much of the material from this book was covered in a course on discrete geometry and computers at the University of Pittsburgh and then later in Hanoi. I would like to thank the members of these groups for assisting in the preparation of the book.

[1] This research has been supported by the National Science Foundation under Grants 0503447 and 0804189 as well as a grant from the Benter Foundation.

A draft of this book was written during a sabbatical leave from the University of Pittsburgh, 2007–2008. I wish to thank the many institutions that supported me during this period: the Max Planck Institute in Bonn, the École Normale Supérieure in Paris, the Institute of Math in Hanoi, Radboud University in Nijmegen, and the University of Strasbourg. These visits would not have been possible without the assistance of Grigori Mints, François Loeser, Florence Lecomte, Henk Barendregt, Ha Huy Khoai and Ngo Viet Trung.

This book is dedicated to the memory of my grandfather Wayne B. Hales.

References

[1] Aristotle. *On the heaven.* translated by J.L. Stocks, `http://classics.mit.edu/Aristotle/heavens.html`, 350BC.

[2] A. Barvinok. *A Course in Convexity*, volume 54 of *Graduate Studies in Mathematics*. American Mathematical Society, 2002.

[3] J. Beery and J. Stedall, editors. *Thomas Harriot's Doctrine of Triangular Numbers: the 'Magisteria Magna'*. European Math. Soc., 2008.

[4] K. Bezdek. On a stronger form of Rogers' lemma and the minimum surface area of Voronoi cells in unit ball packings. *J. Reine Angew. Math.*, 518:131–143, 2000.

[5] K. Bezdek and E. Daróczy-Kiss. Finding the best face on a Voronoi polyhedron – the strong dodecahedral conjecture revisited. *Monatshefte für Mathematik*, 145:191–2006, 2005.

[6] M. Bhubaneswar. *Computational Real Algebraic Geometry*. CRC Press, 1997.

[7] N. Bourbaki. *Elements of Sets*. Addison-Wesley, Boston MA, 1968.

[8] W. Casselman. Packing pennies in the plane, an illustrated proof of Kepler's conjecture in 2d. AMS Feature Column Archive, `http://www.ams.org/featurecolumn/archive/cass1.html`, December 2000.

[9] J. H. Conway and N. J. A. Sloane. What are all the best sphere packings in low dimensions? *DCG*, 13:383–403, 1995.

[10] A. Doxiadis and C. H. Papadimitriou. *Logicomix An Epic Search for Truth.* Worzalla Publishing, 2009.

[11] L. Euler. Variae speculationes super area triangulorum sphaericorum. *N. Acta Ac. Petrop.*, pages 47–62, 1797. E698 (Eneström index).

[12] L. Fejes Tóth. *Lagerungen in der Ebene auf der Kugel und im Raum.* Springer-Verlag, Berlin-New York, first edition, 1953.

[13] W. Fulton. *Introduction to Toric Varieties*. Princeton University Press, Princeton NJ, 1993.

[14] C. F. Gauss. Untersuchungen über die Eigenscahften der positiven ternären quadratischen Formen von Ludwig August Seber. *Göttingische gelehrte Anzeigen*, July 1831. also published in *J. Reine Angew. Math.* 20 (1840), 312–320, and *Werke*, vol. 2, Königliche Gesellschaft der Wissenschaften, Göttingen, 1876, 188–196.

[15] G. Gonthier. A computer-checked proof of the four colour theorem. Unpublished manuscript, 2005.

[16] G. Gonthier. Formal proof – the four colour theorem. *Notices of the AMS*, 55 (11):1382–1393, December 2008.

[17] B. Gracián. *The art of worldly wisdom.* Frederick Ungar Publishing, New York NY, 1967.

[18] T. C. Hales. The sphere packing problem. In *Journal of Computational and Applied Math*, volume 44, pages 41–76, 1992.

[19] T. C. Hales. Cannonballs and honeycombs. *Notices of the AMS*, 47(4):440–449, 2000.

[20] T. C. Hales. An overview of the Kepler conjecture. *Discrete and Computational Geometry*, 36(1):5–20, 2006.

[21] T. C. Hales. The Flyspeck Project, 2012. `http://code.google.com/p/flyspeck`.

[22] T. C. Hales and S. P. Ferguson. The Kepler conjecture. *Discrete and Computational Geometry*, 36(1):1–269, 2006.

[23] T. C. Hales and S. McLaughlin. A proof of the dodecahedral conjecture. *Journal of the AMS*, 23:299–344, 2010. `http://arxiv.org/abs/math/9811079`.

[24] T. C. Hales, J. Harrison, S. McLaughlin, T. Nipkow, S. Obua, and R. Zumkeller. A revision of the proof of the Kepler Conjecture. *DCG*, 2009.

[25] J. Harrison. Formalizing an analytic proof of the prime number theorem. *Journal of Automated Reasoning*, 43:243–261, 2009.

[26] J. Harrison. The HOL Light theorem prover, 2010. http://www.cl.cam.ac.uk/˜ jrh13/hol-light/index.html.

[27] R. Kargon. *Atomism in England from Hariot to Newton.* Clarendon Press, Oxford, 1966.

[28] J. Kepler. *The Six-cornered snowflake.* Oxford Clarendon Press, 1966. forward by L. L. Whyte.

[29] J. Lanier. *You are not a gadget.* Alfred A. Knopf, New York, 2010.

[30] J. Leech. The problem of the thirteen spheres. In *Mathematical Gazette*, pages 22–23, February 1956.

[31] C. Marchal. Study of the Kepler's conjecture: the problem of the closest packing. *Mathematische Zeitschrift*, December 2009.

[32] O. R. Musin and A. S. Tarasov. The strong thirteen spheres problem. preprint `http://arxiv.org/abs/1002.1439`, February 2010.

[33] T. Nipkow, G. Bauer, and P. Schultz. Flyspeck I: Tame Graphs. In Ulrich Furbach and Natarajan Shankar, editors, *International Joint Conference on Automated Reasoning*, volume 4130 of *Lect. Notes in Comp. Sci.*, pages 21–35. Springer-Verlag, 2006.

[34] S. Obua. Proving bounds for real linear programs in Isabelle/HOL. In J. Hurd and T. F. Melham, editors, *Theorem Proving in Higher Order Logics*, volume 3603 of *Lect. Notes in Comp. Sci.*, pages 227–244. Springer-Verlag, 2005.

[35] K. Plofker, January 2000. private communication.

[36] C. A. Rogers. The packing of equal spheres. *Journal of the London Mathematical Society*, 3/8:609–620, 1958.

[37] K. Schütte and B. L. van der Waerden. Auf welcher Kugel haben 5, 6, 7, oder 9 Punkte mit Mindestabstand Eins Platz. *Math. Annalen*, 123:96–124, 1951.

[38] K. Schütte and B. L. van der Waerden. Das Problem der dreizehn Kugeln. *Math. Annalen*, 125:325–334, 1953.

[39] W. Shirley. *Thomas Harriot: a biography*. Oxford, 1983.

[40] K. S. Shukla. *The Āryabhaṭīya of Āryabhaṭa with the Commentary of Bhāskara I and Someśvara*. New Delhi: Indian National Science Academy, 1976.

[41] A. Solovyev and T. C. Hales. *Efficient formal verification of bounds of linear programs*, volume 6824 of *LNCS*, pages 123–132. Springer-Verlag, 2011.

[42] J. Spolsky. *Joel on Software*. Apress, 2004.

[43] G. G. Szpiro. *Kepler's Conjecture: How Some of the Greatest Minds in History Helped to Solve one of the Oldest Math Problems in the Worlds*. John Wiley and Sons, New York NY, 2003.

[44] A. Tarski. *A decision method for elementary algebra and geometry*. University of California Press, Berkeley and Los Angeles, Calif., 1951. 2nd ed.

[45] A. Thue. Om nogle geometrisk taltheoretiske theoremer. *Forandlingerneved de Skandinaviske Naturforskeres*, 14:352–353, 1892.

[46] A. Thue. über die dichteste Zusammenstellung von kongruenten Kreisen in der Ebene. *Christinia Vid. Selsk. Skr.*, 1:1–9, 1910.

[47] W. T. Tutte. *Graph Theory*. Encyclopedia of Mathematics and Its Applications. Addison-Wesley Publishing, 1984.

[48] R. J. Webster. *Convexity*. Oxford University Press, 1994.

[49] F. Wiedijk. Jordan curve theorem. `http://www.cs.ru.nl/~freek/jordan/index.html`, referenced 2010.

[50] F. Wiedijk. Formalizing 100 theorems. `http://www.cs.ru.nl/~freek/100/`, referenced 2012.

General index

adapted, 51
adjacent, 216
adjusted, 149
admissible, 237
affine, 35, 36, 133
  dimension, 133
  hull, 35, 133
  independence, 133, 157
algorithmically planar, 108
angle, 39, 55, 125
  azimuth, 51, 52, 118, 241
  dihedral, 43, 48, 52, 65, 176, 254
  total solid, 175
  zenith, 54
annulus, 184
apex, 66
arc, 39, 40
  geodesic, 40
arccosine, 32
arclength, 39, 40
arctangent, 31
  $\arctan_2$, 31
  derivative, 31
  near 0, 32
Aristotle, 150
azimuth, 51, 52, 57, 58, 118, 241
azimuth cycle, 51, 57, 58, 114

ball, 67
ball, open and closed, 63
Bezdek, K., 252
bisector, 147
blade, 37, 113
boundary
  relative, 133

canonical function, 97
cardinality, 237, 241
  finite, 70
carrier, 105
Cauchy–Schwarz inequality, 34, 39, 41
Cayley–Menger
  determinant, 37, 68
cell, 167, 169
cell cluster, 181, 182
circular cone, 62
circular fan, 197, 198
circumcenter, 157
circumradius, 157
claim (italic format of small claims), xii
closed, 135
closed ball, 63
closure, 133
clusters, 181
codimension, 152
collinear, 37
combinatorial component, 77, 117
complement, 88
component
  combinatorial, 77, 87, 127
  topological, 126, 131, 138
computer calculation, 251
  5202826650 a, 234
  BIEFJHU, 191
  KCBLRQC, 246, 247
  OXLZLEZ, 183
  TNVWUGK, 256
  TSKAJXY, 182
  TVAWGDR, 240
  UKBRPFE, 192
  UPONLFY, 227
  WAZLDCD, 192
  notation, xii

cone
  right-circular, 65
conforming, 126, 128, 243
conic cap, 66, 67
connected, 77, 117, 120, 127, 138
  combinatorial component, 117
  topological component, 117, 118, 120, 123, 124
constraint system, 216
contact fan, 238
contour
  loop, 86
  Moebius, 88
  path, 86
contravening, 17, 20, 243
convex, 36, 126, 133, 188
convex hull, 36, 162, 255
coordinate systems, 52, 54, 55
  cylindrical coordinates, 52
  polar coordinates, 51
  spherical coordinates, 54
coplanar, 37, 65
corrected volume, 149
correction term, xiii
cosine, 27
  derivative, 28
  law of cosines, 40
  roots, 29
  series definition, 27
  spherical law of cosines, 45, 47
cover, 222
critical edge, 181
cross product, 42
cyclic
  list, 86
  permutation, 55, 57, 74, 196
  set, 57, 58
cylindrical coordinates, 52

D-cell, 254
dart, 74, 116, 118, 119, 123, 138
  degenerate, 77
  isolated, 116
  nondegenerate, 77
  set, 76
decomposition
  Delaunay, 15, 167
  Marchal, 18, 167
  Rogers, 150, 253
  Voronoi, 9, 146, 167, 252
deformation, 203, 224
degenerate, 77, 79
Delaunay, *see* decomposition
determinant, 38
  Cayley–Menger, 37, 68
Dhammapada, 77
diagonal, 213
diagonal cover, 222, 224
dihedral, 94
dimension, 133
disjoint sum decomposition, 119
dodecahedral conjecture, 252
dot product, 34
double join, 93, 238

ear, 217
edge, 77, 133, 134
  fan, 114
  graph, 114
  length, 38, 43, 254
  map, 74
  walkup, 79
equivalence relation, 76
Euclidean space, 33
Euler characteristic relation, 84
Euler's formula for solid angle, 68
eventually, *see* radial
exposed, 254
extension, 98
extremal edges, 175
extreme point, 133, 134, 175

face, 77, 133–135, 238
  attribute, 128
  map, 74
  walkup, 79
face-centered cubic, *see* FCC
facet, 133, 135
fan, 112, 113, 118, 124, 137
  contact, 238, 241
  local, 195, 198, 203, 206, 207, 211
  standard, 238
fan, 113
FCC, 7, 10, 148
  compatible, 149, 179
  pattern, 10
Fejes Tóth, L., 253
final, 108
fixed point, 77
flag, 96, 97
Flyspeck, 257
frame, 51, 57
free, 225
frustum, 65, 67
fully surrounded, 121, 243

generation, 95
generic fan, 197, 198
Girard's formula, 48, 68, 125, 188
Google Code project hosting, xii
graph record, 107
great circle, 46
group, 164

half-plane, 37, 191
half-space, 37, 65, 125, 147, 191
  open, 37
Harriot, T., 48
HCP, 10, 148
  pattern, 10
Heron's formula, 41
hexagonal layer, 11
hexagonal-close packing, *see* HCP
HOL Light, xiii
hull
  affine, 35, 133
hypermap, 72, 74, 112, 114, 118, 119, 123, 124, 127, 138
  algorithm, 96
  connected, 127
  contravening, 241
  dihedral, 94
  opposite, 250
  plain, 83
  planar, 89, 192
  planar index, 84
  restricted, 96
  simple, 127
  subquotient, 92
  tame, 235, 236
hyperplane , 133

I (use of personal pronoun), xii
interior, 196
  point, 138
  relative, 133
interior point, 133
Isabelle/HOL, 95
isolated, 116, 241
isomorphism
  hypermap, 91
  proper, 105
  tame hypermap classification, 251
  torsor, 216

Java, 95
Jordan curve theorem, 112

Kepler conjecture, 148, 150, 251

latitude, 54
lattices, x
law of cosines, 14, 40
law of sines, 42
lead into, 120
length, 152
level, 223
line, 37
linear stretch, 63
linearly independent, 158
list, 75
listing, 105
local fan, 195
local inequality, 255
localization, 194, 196
longitude, 26, 52
loop, 86, 238
lunar fan, 197, 198
lune, 64, 65

map, 105
Marchal cell, 18, 167
Marchal, C., 167
marked hypermap, 98
Mathematica, 95
measurable, 62
measure, 61, 62, 146
  Lebesgue, 62
merge, 79, 81
metric space, 117
minimal counterexample, 223
ML, 95

named property
  angle, 195
  bijection, 126
  bound a, 237
  bound b, 237
  bound d, 237
  cardinality, 113
  circular fan, 197
  diagonal, 126
  dihedral, 195
  face, 195
  face count, 238
  face size, 238
  fan, 195
  generic fan, 197
  half-space, 126
  intersection, 113
  linear stretch, 63
  lunar fan, 197
  no double joins, 238
  no loops, 238

node count, 238
node size, 238
node types, 238
nondegenerate, 238
nonnegative, 62
nonparallel, 113
null difference, 62
null set, 62
origin, 113
planar, 238
primitives, 63
simple, 238
solid angle, 126
surroundedness, 126
translation, 63
union, 62
wedge, 195
weights, 238
natural numbers, 26, 27
negligible, 148, 149, 179
node, 77, 134, 237
map, 74
properties, 245
type, 236
walkup, 79
nondegenerate, 77, 238
nonreflexive, 195
local fan, 198
norm, 34
normal family, 91, 92
null set, 62, 254

open ball, 63
open half-space, 37
opposite, 250
orbit, 77, 80
order, 74
lexicographic, 55
total, 55
orthogonal frame, 54
orthogonality, 43
overloaded, 151

packing, 112, 145, 146
finite, 241
parallel, 37
parallelepiped, 37
partial perimeter, 211
partition, *see* decomposition
path, 75, 77, 86
injective, 76
maximal, 92
path connected, 118
pencil and pen heuristic, 96
perimeter, 211
nonreflexive local fan, 211
periodicity, 29
permutation, 73, 164
plain, 77
planar, 21, 83, 112, 114, 127, 238
algorithmic, 108
graph, 21, 72
index, 84
map, 72
plane, 37, 62, 77, 121
graph, 21, 72, 74
polar
coordinate, 120
coordinates, 51
cycle, 55, 57
fan, 209
triangle, 46
polygon, 185, 188
polyhedron, 134, 135, 137, 138, 189, 192
polysemes (face, edge, node), 134
positive, 51
primitive region, 62, 66
proper, 133
protracted, 213

quadrilateral, 236
quotient dart, 92

radial, 64
eventually, 63, 70
set, 61, 63
real analysis, 27
real arithmetic, 27
rearrangement, 164, 173
record, 108
rectangle, 67
reflex angle, 195
regular dodecahedron, 255
surface area, 255
volume, 253, 255
relative boundary, 133, 135
relative interior, 133
restricted hypermap, 96
right-circular cones, 65
Rogers, *see* decomposition
rotation, 55

saturated, 15, 146
set (of a list), 105
simple, 77, 238
sine, 27

law of sines, 42
series definition, 27
slice
fan, 206, 207
torsor, 221
solid angle, 63, 64
solid triangle, 65, 67
sphere, 62, 117
sphere packing problem, 150
spherical
coordinates, 54, 120
law of cosines, 45, 47
triangle, 45
triangle inequality, 59
split, 79
stable, 216
standard, 213
standard fan, 20, 238
standard main estimate, 215
steps, 75
straight fan node, 198, 226
strong dodecahedral conjecture, 252
sublist, 75, 86
subquotient, 91, 92
surface area, 253
exposed, 254
surrounded, 241
symmetric difference, 62

tame, 17, 21, 235, 238
contravention, 243
hypermap, 21, 236
tangent, 30
target, 236
Tarski arithmetic, 27, 66
tetrahedron, 43, 65, 67
topological component, 117
torsor, 216
total solid angle, 175
transfer, 224
transform, 101
translation, 63
triality relation, 75
triangle, 236
Euler, 49
spherical, 45, 48, 59
triangle attributes, 125
triangle inequality, 35, 41
trigonometry, 25
addition formula, 28
arccos, 33
arctan, 33
circle identity, 28
identities, 28
inverse, 32
law of cosines, 40
law of sines, 42
periodicity, 30
spherical, 45
tangent, 30
type, 236

unit list, 76

vector, 33
addition, 34
cross product, 42
dot product, 34
norm, 34
projection, 43, 57
scalar multiplication, 34
subtraction, 34
zero, 33
vector space, 34
vertices, 134
visits, 76
volume, 63
primitive, 63
Voronoi, *see* decomposition

walkup, 78, 79, 83
degenerate, 79
double, 81
we (use of personal pronoun), xii
weakly saturated, 189
wedge, 64, 65, 67, 69, 118
weight, 182, 238
admissible assignment, 237
assignment, 236, 237
total, 237

zenith, 54

Notation index

$\sim$ (equal up to positive scalar), 53, 209
$\sim$ (equivalence relation), 76
$\binom{n}{k}$ (binomial coefficient), 175
$[Y]$ (topological components), 118
$[\mathbf{u}, \mathbf{v}, \mathbf{w}] = (\mathbf{u} \times \mathbf{v}) \cdot \mathbf{w}$, 211
$*[\mathbf{v}, \mathbf{w}]$ (slicing a fan), 206
$\angle$ (dart angle), 196, 206
$\#c$ (number of components), 77
$\#h$ (number of orbits), 77
$\cdot$ (dot product), 34
$*$ (wildcard symbol), xii
$*'$ (polar local fan (V', E', F')), 209
$*^c$ (complement), 88
$\times$ (cross product), 42
$\perp$, 159
$\dashrightarrow_p$ (input-output hypermap-generating relation), 108

A (Voronoi cell face area), 253
A (index set), 246
A (plane), 37
A (set of faces), 246
A (set of triangles), 237, 247
ABC (triangle), 46
$A(\mathbf{u}, \mathbf{v})$ (bisector), 147
$A_+(\mathbf{u}, \mathbf{v})$ (half-space), 147
aff (affine hull), 133
$\mathrm{aff}_\pm$, $\mathrm{aff}^0_\pm$, 36, 65
$a = 0.63$ (tame parameter), 247
a_D (dodecahedral parameter), 255
arc, 40
arccos, 32, 33
arctan, 31, 33
$\arctan_2$, 31
arc_V, 39
azim, 52, 58, 118
$\mathrm{azim}(H, \mathbf{v})$, 206

$B(\mathbf{v}, r)$ (open ball), 17, 63, 146
$\bar{B}(\mathbf{v}, r)$ (closed ball), 63
b (tame parameter), 237
b_D (dodecahedral parameter), 255
$\mathcal{B}$, 184
$\mathcal{B}_s$, 217

C (set), 63
C (wedge), 69
$C(S)$ (cell-like subset of $\mathbb{R}^3$), 170
$C_\pm$, $C^0_\pm$ (blade), 113
c_i (ranking functions on a packing), 242
CAP, 66, 70
$c_{\mathrm{stab}} = 3.01$, 216
CL (cell cluster), 182
cl (closure), 133
cluster, 182
conv, 36
cos, 27

D (dart set), 74, 92, 116
D (determinant), 38
D (spherical disk), 190
D_k (D-cell), 254
d (real parameter), 237, 250
$d(s, \mathbf{v})$, 220
$d(\mathbf{u}, \mathbf{v})$ (metric on $\mathbb{R}^3$), 59
d_j (truncation of lists), 152
dih, 43, 52, 65, 70, 176, 254
dih_i, 213
dih_V, 43
dim aff (affine dimension), 133, 151

E (edge set of a fan), 113
E_P (edge), 137
E_{ctc} (edge set of contact fan), 238
E_{std} (edge set of standard fan), 238

E' (polar edge set), 209
$E(\mathbf{v})$ (set of edges adjacent to $\mathbf{v}$), 114, 241
$E(X)$, 175
EC (critical edges), 182
E (frame), 51
equi (intersection of spheres), 170
e (edge map), 74
$\mathbf{e}_i$ (orthonormal vectors), 52

F (hypermap face), 94, 125, 249
F (polygon), 246
F' (polar face), 209
$\mathbf{F}$ (subquotient bijection), 92
f (function name), 57
f (face map), 74
FR (frustum), 65, 70

G (isomorphism of hypermaps), 91
G (negligible function), 149
g (function name), 183
g (graph record), 108
g (triangle area), 188

(H, τ), 245
H (hypermap), 84, 96, 243, 245, 246, 251
$H/\mathcal{L}$ (subquotient), 92
h (circumradius), 160
h (cylindrical coordinate), 52
h (half-edge length), 176
h (permutation), 77
$h_+ = 1.3254$, 177
$h_- \approx 1.23175$, 181
$h_0 = 1.26$, 181, 213
hyp (hypermap), 116, 238

I (identity map), 73
I (real interval), 203
I (torsor), 216
$I[p, q]$ slice (of a torsor), 221

J, 216

k (cardinality of a face), 128, 249, 250

L (dart path), 76
L (linear function), 181, 213, 244
$\mathcal{L}$ (normal family of loops), 92
ℓ (level function), 223
$\mathcal{L}(V)$ (estimation of a packing), 17, 235, 243, 251
$\ell_H(x)$ (listing of dart x), 105

M (Marchal's quartic), 177
m (face map exponent), 99
$m_1 \approx 1.012$, 177
$m_2 \approx 0.0254$, 177
map(ϕ, ℓ) (map of a function over a list), 105

N (integer invariant of a fan), 128
$\mathbb{N} = \{0, 1, \ldots, \}$, 27
n (integer variable), 58
node : $D \to V$ (node of a dart), 119, 196
n (node map), 74

O (Landau's big O), 180
$O(h, x)$ (orbit of x under h), 80, 105

P (dart path), 76, 90
P (polyhedron), 135
p (Euler solid angle numerator), 49
p (face map exponent), 99
p (trigonometric expression), 47
$\mathbf{p} \in \mathbb{R}^3$, 62
per (perimeter), 211
(p, q, r) (node parameters), 236, 246
$p_{\mathbf{v}}$, 236

R (Rogers simplex), 154
$\mathbb{R}$ (field of real numbers), 27
r (polar, cylindrical, and spherical radius), 32, 51, 52, 54, 65
rcone, rcone^0, 65
$\mathbb{R}^N$, 33
reg (area of regular spherical polygon), 191
ri (relative interior), 133

S (finite subset of $\mathbb{R}^3$), 157
S (flag, set of darts), 97
$S(H, L, x)$ (flag set), 97
$S^2(r)$ (sphere of radius r), 117, 204
s (constraint system), 216
s (real variable), 124
seed$_p$, 108
set (of a list), 105
sin, 27
S_{main} (main estimate constraint systems), 217
$\text{sol}_0 = 3\arccos(1/3) - \pi$, 177, 213, 244
sol (solid angle), 64, 254
soly (solid angle as a function of edges), 254
sol(V, E, F) (formal solid angle), 208
split$_c$, split$_f$, 85
surf (surface area), 254
Sym (symmetric group), 164, 166

T (regular tetrahedron), 150
T (rotation), 56
T (transform), 101
$T_{\mathbf{v}}$ (linear stretch), 63
t (real variable), 246
tan, 30
TET (tetrahedron), 65
tgt = 1.541, 236, 244, 247
TRI (solid triangle), 65, 66

tsol, 175
U_x (topological component), 120
U_F (topological component), 123, 125, 243
(V, E) (fan), 113, 238
V (packing), 145, 149, 238
V (subscript marking vector functions), 40, 43
V' (packing), 241
V' (polar node set), 209
$V(X) = V \cap X$, 175
$V \subset \mathbb{R}^n$, 36
$V(\mathbf{p}, r) = V \cap B(\mathbf{p}, r)$, 146
$\mathcal{V}$ (configuration space), 242
$\mathcal{V}_H$ (configuration space), 251
$\mathbf{v} \in \mathbb{R}^3$, 37, 39, 65, 69, 237
$\mathbf{v}(t)$ (deformation of $\mathbf{v}$), 203, 225
vol, 62
vol_D (volume of dodecahedron), 253
W, W^0 (wedge), 64, 65
W_F (topological component of a facet), 138
W (walkup), 79, 85
W^0_{dart} (wedge), 118, 120, 206
$W^0_{\mathrm{dart}}(F, \mathbf{v})$, $W_{\mathrm{dart}}(F, \mathbf{v})$, 196
W_{dart} (closure of W^0_{dart}), 118
$\mathbf{w} \in \mathbb{R}^3$, 39, 65
wt (weight), 182
X (measurable set), 62
$X(V, E) \subset \mathbb{R}^3$ (union of fan blades), 118
x (dart), 78, 79, 246
(x, y) (Cartesian point), 32
Y (measurable set), 62
$Y(V, E) \subset \mathbb{R}^3$ (fan complement), 118
y (dart), 99
y_D (dodecahedral parameter), 255
z (dart), 99
z (real variable), 246
α (angle), 48, 70, 188
β (angle), 48, 188
β, β_0 (bump), 182
Γ, 182
γ (angle), 40, 53, 188
γ (packing inequality), 178
Δ, 37, 38, 47, 51
$\delta(V, \mathbf{p}, r)$, 149
$\delta \in \mathbb{R}$, 123
$\delta \in \{0, 1\}$, 223
$\epsilon \in \mathbb{R}$, 118, 123
ε (edge of a fan), 113
η (circumradius), 241
θ (polar, cylindrical, and spherical angle), 32, 51–54, 56, 69
ι (planar index), 84
ν_D (dodecahedral function), 255
ξ (cell parameter), 168
ρ (permutation on nodes of a local fan), 196
ρ (permutation), 164
ρ (rotation of lists), 106
ρ_0 (real-valued function), 213
σ (azimuth and polar cycle), 55–58
$\sigma = \pm 1$, 220
τ (weight assignment), 213, 237, 244–246, 250
τ^*, 220
τ_{tri}, 213, 227, 228, 233
$\tau_0 \approx$ tgt, 177
υ (polynomial), 38, 41, 47
$\check{\varphi}_{can}$ (canonical function), 97
φ (path), 124
ϕ (proper isomorphism between sets of lists), 105
ϕ (zenith), 54
ψ, 52, 56
Ω (Voronoi cell), 146, 253
$\Omega(V, W)$ (intersection of Voronoi cells), 151
ω (extreme points of Rogers simplex), 154

Printed in the United States
by Baker & Taylor Publisher Services